SPRINGER HANDBOOK OF AUDITORY RESEARCH

Series Editors: Richard R. Fay and Arthur N. Popper

T0180611

SPRINGER HANDBOOK OF AUDITORY RESEARCH

Continued after index

Richard J. Salvi
Arthur N. Popper
Richard R. Fay

Editors

Hair Cell Regeneration, Repair, and Protection

 Springer

Richard J. Salvi
Center for Hearing and Deafness
University of Buffalo
Buffalo, NY 14214
USA
salvi@buffalo.edu

Arthur N. Popper
Department of Biology
University of Maryland
College Park, MD 20742
USA
apopper@umd.edu

Richard R. Fay
Parmly Hearing Institute
6525 North Sheridan Road
Loyola University Chicago
Chicago, IL 60626
USA
rfay@luc.edu

Series Editors:
Richard R. Fay
Parmly Hearing Institute
6525 North Sheridan Road
Loyola University Chicago
Chicago, IL 60626
USA

Arthur N. Popper
Department of Biology
University of Maryland
College Park, MD 20742
USA

ISBN: 978-1-4419-2519-0   e-ISBN: 978-0-387-73364-7

Cover illustration: The image includes parts of Figures 5.2 and 2.5 appearing in the text.

Printed on acid-free paper

9 8 7 6 5 4 3 2 1

springer.com

Contents

Contributors

JEFFREY T. CORWIN
Department of Neuroscience, University of Virginia School of Medicine, Charlottesville, VA 22908, USA, Email: jcorwin@virginia.edu

MICHEAL L. DENT
Department of Psychology, University at Buffalo–SUNY, Buffalo, NY 14260, USA, Email: mdent@buffalo.edu

ROBERT J. DOOLING
Department of Psychology and Center for the Comparative Evolutionary Biology of Hearing, University of Maryland, College Park, MD 20742, USA, Email: dooling@psyc.umd.edu

ANDREW FORGE
Centre for Auditory Research, UCL Ear Institute, University College London, London WC1X 8EE, UK, Email: a.forge@ucl.ac.uk

MATTHEW C. HOLLEY
Department of Biomedical Science, University of Sheffield, Sheffield, S10 2TN, UK, Email: m.c.holley@sheffield.ac.uk

AMANDA M. LAUER
Department of Psychology and Center for the Comparative Evolutionary Biology of Hearing, University of Maryland, College Park, MD 20742, USA, Email: alauer1@umd.edu

JASON R. MEYERS
Department of Molecular, Cellular and Developmental Biology, University of Michigan, Ann Arbor MI 48109, USA, Email: jrmeyers@umich.edu

ELIZABETH C. OESTERLE
Virginia Merrill Bloedel Hearing Research Center, Department of Otolaryngology, University of Washington, Seattle, WA 98195-7923, USA, Email: oesterle@u.washington.edu

MARCELO N. RIVOLTA
Centre for Stem Cell Biology, University of Sheffield, Sheffield, S10 2TN, UK,
Email: m.n.rivolta@sheffield.ac.uk

BRENDA M. RYALS
Department of Communication Sciences and Disorders, James Madison
University, Harrisonburg, VA 22807, USA, Email: ryalsbm@cisat.jmu.edu

RICHARD J. SALVI
University of Buffalo, Center for Hearing and Deafness, Buffalo, NY 14214,
USA, Email: salvi@buffalo.edu

JAMES C. SAUNDERS
University of Pennsylvania, Philadelphia, PA 19104, USA, Email:
saunderj@mail.med.upenn.edu

JENNIFER S. STONE
Virginia Merrill Bloedel Hearing Research Center, Department of Otolaryngology, Seattle, WA 98195 7923, USA, Email: stoner@u.washington.edu

THOMAS R. VAN DE WATER
Department of Otolaryngology, Cochlear Implant Research Program, University
of Miami Ear Institute, University of Miami, Miller School of Medicine, Miami,
FL 33136-1015, USA, Email: tvandewater@med.miami.edu

Series Preface

The Springer Handbook of Auditory Research presents a series of comprehensive and synthetic reviews of the fundamental topics in modern auditory research. The volumes are aimed at all individuals with interests in hearing research, including advanced graduate students, postdoctoral researchers, and clinical investigators. The volumes are intended to introduce new investigators to important aspects of hearing science and to help established investigators to better understand the fundamental theories and data in fields of hearing that they may not normally follow closely.

Each volume presents a particular topic comprehensively, and each serves as a synthetic overview and guide to the literature. As such, the chapters present neither exhaustive data reviews nor original research that has not yet appeared in peer-reviewed journals. The volumes focus on topics that have developed a solid data and conceptual foundation rather than on those for which a literature is only beginning to develop. New research areas will be covered on a timely basis in the series as they begin to mature.

Each volume in the series consists of a few substantial chapters on a particular topic. In some cases, the topics will be ones of traditional interest for which there is a substantial body of data and theory, such as auditory neuroanatomy (Vol. 1) and neurophysiology (Vol. 2). Other volumes in the series deal with topics that have begun to mature more recently, such as development, plasticity, and computational models of neural processing. In many cases, the series editors are joined by a coeditor having special expertise in the topic of the volume.

RICHARD R. FAY, Chicago, IL
ARTHUR N. POPPER, College Park, MD

Volume Preface

The human brain's ability to sense and interpret acoustic events taking place in remote or nearby locations in the external environment is mediated by highly specialized and extremely sensitive sensory hair cells located in the inner ear. These cells transduce acoustic information from the environment into a pattern of neural activity that can be interpreted by sophisticated neural networks located at multiple levels of the central nervous system. It has long been known that hair cell loss in mammals due to aging, ototoxic drugs, acoustic trauma, infection, or genetic factors results in permanent hearing loss or balance problems. Over the past 50 years, efforts to find a cure for deafness have focused on hardware and engineering solutions. While much effort has been made to use electronic means to improve hearing, the next giant step toward restoring hearing to the profoundly deaf will involve regenerating the damaged biological structures in the inner ear, in particular the hair cells and spiral ganglion neurons. The major clinical advances in hearing and balance that will occur in the 21st century will involve biologically based medical innovations that were set into motion during the past few decades by the discovery of hair cell regeneration and by the recognition that stem cells exist in many regions of the nervous system, including the inner ear.

These discoveries, and the potential for helping people with hearing loss, are the focus of this volume. In Chapter 1, Salvi reviews the history of studies on hair cell regeneration and provides an overview of current knowledge as well as new technologies to promote regeneration and repair. The recognition that hair cell regeneration can occur in nonmammals gave way to ground breaking studies using gene therapy to simulate hair cell regeneration in mammals. The history of the field, as well as what is known about the morphology associated with regeneration and repair of sensory hair cells, are the focus of Chapter 2 by Meyers and Corwin. One of the fundamental issues examined is whether regenerated hairs arise from repair of damaged cells, conversion of support cells to hair cells, or proliferation of support cells that differentiate to either hair cells or replacement support cells.

One of the most important areas that stimulated studies of damage, regeneration, and repair has been in the avian auditory system. In Chapter 3, Salvi and Saunders describe the remarkable recovery of function of the avian auditory system following acoustic trauma and ototoxic insult. Physiological studies show significant recovery, with only minor deficits except for cases in which the supporting cells are destroyed. In Chapter 4, Dooling, Dent, Lauer, and Ryals go into considerable detail about actual recovery of hearing function following loss of hair cells. Behavioral measures of hearing, the gold standard, show almost complete recovery of function on simple measures such as threshold as well as highly sophisticated measures that involve discrimination of complex vocalizations.

The mechanisms involved in proliferation, differentiation, and regeneration are discussed in detail by Oesterle and Stone in Chapter 5. The roles that growth factor, intercellular signaling, intracellular signaling and differentiation factors play in proliferation, conversion, and repair are carefully considered. In Chapter 6, Forge and Van De Water consider ways to protect sensory hair cells from damage so that regeneration is not needed. The modes of cell death are reviewed and various strategies for blocking cell death such as antioxidants, inhibition of apoptosis, and small molecules that block genes or enzymes in the cell death pathway are considered. Finally, in Chapter 7, Rivolta and Holley discuss new experimental approaches that may aid in understanding cell death, cell repair, proliferation, and differentiation. The use of gene array technologies and inner ear cell lines may provide more efficient and comprehensive methods for understanding apoptosis, repair, and regeneration.

As is often the case, new volumes in the Springer Handbook of Auditory Research amplify and extend materials discussed in earlier volumes in the series. While the current volume concerns regeneration and repair, engineering methods have been quite successful in dealing with deafness. In particular, cochlear implants have been a widely used approach and this was covered in depth in Vol. 20 of the series, *Cochlear Implants* (edited by Zeng, Popper, and Fay). The genetics of the ear and of hearing loss was discussed in detail in Vol. 14, *Genetics and Auditory Disorders* (edited by Keats, Popper, and Fay). While the current volume focuses on hair cells, Vol. 23, *Plasticity of the Auditory System* (edited by Parks, Rubel, Fay, and Popper), includes chapters that consider overall plasticity at many levels of the auditory system. Mechanisms of damage to the auditory system is considered at length in in Vol. 31, *Auditory Trauma, Protection, and Repair* (edited by Schacht, Popper, and Fay). Finally, the physiology and function of sensory hair cells is discussed in many chapters of Vol. 27, *Vertebrate Hair Cells* (edited by Eatock, Fay, and Popper).

RICHARD J. SALVI, BUFFALO, NY
ARTHUR N. POPPER, College Park, MD
RICHARD R. FAY, Chicago, IL

1
Overview: Regeneration and Repair

RICHARD J. SALVI

1. Introduction

1.1 Hair Cells and the Acoustic World

The human brain's ability to sense and interpret acoustic events taking place in remote or nearby locations in the external environment is mediated by highly specialized and extremely sensitive sensory hair cells located in the inner ear. Although the external ear and middle ear play important roles in collecting, amplifying, and relaying acoustic information from the environment to the inner ear, the resulting mechanical vibrations of the basilar membrane are of little value unless they can be transduced by the sensory hair cells into a pattern of neural activity that can be interpreted by sophisticated neural networks located at multiple levels of the central nervous system. The sensory hair cells in the inner ear represent the obligatory entry point for gaining access to the central auditory or vestibular systems. It has long been known that hair cell loss in mammals due to aging, ototoxic drugs, acoustic trauma, infection, or genetic factors results in permanent hearing loss or balance problems. Humans suffering from profound hearing loss due to massive loss of cochlear hair cells are shut off from the world of music and oral communication. Profoundly deaf individuals who are unable to communicate orally can experience a sense of social isolation when trying to interact with the hearing world.

Over the past 50 years, efforts to find a cure for deafness have focused on hardware and engineering solutions. The crowning achievement of this effort has been the multichannel cochlear implant. A microphone at the front end of the cochlear implant converts sound into an electrical signal; this mechanical to electric transduction process is reminiscent of the one that takes place in hair cells to initiate hearing. The electrical output of the microphone is fed to a speech processor that segregates the electrical signal into approximately 16 frequency channels; the electrical output of each channel is relayed to an electrode located near the low-, mid-, or high-frequency region along the length of the cochlea, much like the keys on a piano. While modern cochlear implants have done a remarkable job in enhancing speech comprehension in the profoundly deaf, the perceptual qualities of the "electrically evoked sound" is inferior to the natural sound of speech and music conveyed to the auditory nerve by the hair cells.

1

The next giant step toward restoring hearing to the profoundly deaf will involve regenerating the damaged biological structures in the inner ear, in particular the hair cells and spiral ganglion neurons. The major clinical advances in hearing and balance that will occur in the 21st century will involve biologically based medical innovations that were set into motion during the past few decades by the discovery of hair cell regeneration and the recognition that stem cells exist in many regions of the nervous system, including the inner ear.

1.2 Zeitgeist, Regeneration, and Repair

The history of science is often impeded by roadblocks, intellectual and technical, that hinder the advancement of our thinking and imagination. Our views of the world are often constrained by existing knowledge that is accepted as fact by the majority of scientists: the so-called Zeitgeist or spirit of the times. The collective knowledge and prevailing views of the majority can have a profound impact on how new scientific findings are viewed and interpreted for years or even centuries. New data and theories are often immediately rejected by the majority if the findings go against the prevailing view. Indeed, individual scientists may reject their own findings and consider them artifacts or experimental errors if the data contradict prevailing beliefs, opinions, or knowledge.

The ancient theory of sensory processing known as the "principle of likeness" postulated that sensory stimuli in the environment evoked activity within the sensory organ of the same kind (Gitter 1990; Sente 2004). On the basis of this principle, Empedocles proposed the theory of implanted air in the 4th century B.C. whereby sound in the environment evoked a similar activity within the ear to induce the sensation of hearing. The doctrine provided an intellectual framework, though a distorted one, for interpreting the gross anatomical and microscopic observations that were made over the next 2000 years. In the 1500s, Coiter rejected the theory of "implanted air" based on careful anatomical observations of the middle ear and Eustachian tube. Nevertheless, the principle of likeness persisted for another 100 years until it was finally put to rest by detailed examination of the inner ear with the compound microscope and advances in neurophysiology.

My first encounter with the Zeitgeist blinded my thinking on hair cell regeneration. In the late 1970s, I had been studying the effects of acoustic trauma on the firing patterns of auditory nerve fibers in mammals. The dogma at the time was that hair cell loss was permanent and irreparable. At the 1978 meeting of the Acoustical Society of America, I attended a special session on comparative studies of hearing in vertebrates in which Robert Capranica reviewed his work on anurans. Capranica ended his presentation with a provocative finding showing that when ototoxic aminoglycosides were applied to the frog's inner ear, they completely abolished neurophysiological activity from the ear (Capranica 1978). Surprisingly, the frog's "hearing" recovered after a few weeks. This was an unexpected finding that defied any conventional explanation. Several thoughts raced through my mind. Were frog hair cells incapable of being destroyed by aminoglycosides? Were aminoglycoside antibiotics capable of causing only transient damage to the hair cells or neurons? If frog hair cells could not be destroyed with aminoglycoside antibiotics, maybe they could be destroyed with high-level acoustic stimulation. To test the later hypothesis, we started collaborating with the Capranica lab. Capranica's group drove the frogs from Ithaca, NY to our labs in Syracuse and we exposed the frogs to high-intensity impulse noise in the range of 155–165 dB peak sound pressure level (SPL). Exposures at these levels had caused massive hair cell loss in mammals, and I expected they would do the same in frogs. Afterwards, the frogs were driven back to Cornell University and 1–2 months later their hearing was tested via electrophysiological methods. Surprisingly, the auditory function of the noise-exposed frogs was completely normal. Did the frogs have a potent, long-lasting acoustic reflex that they were using to thwart our acoustic trauma? We tried more vigorous noise exposures several times, but nothing seemed to work. Because normal hearing always returned in the frogs, we eventually dropped the project, considering it a complete failure. Had we examined the frog's inner ear immediately after treatment with aminoglycoside antibiotics or acoustic overstimulation, we most likely would have seen missing sensory hair cells after the traumatic event and if we had waited a few weeks we would have observed a normal sensory epithelium filled with newborn hair cells (Baird et al. 1993). I never imagined that hair cells could regenerate after aminoglycoside treatment or acoustic trauma and we missed the opportunity to discover hair cell regeneration in nonmammals. The frog trauma data did not conform to our view of the world and therefore was ignored. Less than a decade later, Corwin, Cotanche, Rubel, Ryals, and other showed that hair cells regenerated in the avian ear after acoustic overstimulation and ototoxicity (Corwin and Cotanche 1988; Ryals and Rubel 1988). A new chapter in auditory neuroscience had started. The Zeitgeist has shifted 180 degrees, and the possibility that hair cell regeneration could be stimulated to occur in mammals via gene therapy became a realistic and exciting possibility (Zheng and Gao 2000; Izumikawa et al. 2005). Looking backwards through the rear view mirror of time, it is interesting to note that hair cell regeneration in amphibians had already been discovered by Stone in the 1930s, but it was largely overlooked only to be rediscovered and embraced 50 years later (Stone 1933, 1937).

1.3 Regeneration in Fish and Amphibians

The first reports of hair cell regeneration date back to the 1930s when it was discovered that after tail amputation or reamputation, the amphibian lateral line organs on the body surface would grow back by forming a regenerative placode that migrated into the regenerating tail where it formed a neuromasts with new hair cells that received afferent innervation (Stone 1933, 1937; Speidel 1947; Wright 1947; Jones and Corwin 1996). These findings indicate that amphibians possess stem cells that self-renew and differentiate into hair cells and support cells.

Historically, much of the research in auditory neuroscience has focused on the anatomy, physiology, and development of the mammalian auditory system. Unfortunately, many of the early studies in nonmammals were ignored or overlooked by mainstream auditory neuroscientists. Beginning in the early 1970s, renewed interest in the morphological development of the inner ear of amphibians and fish revealed the proliferation of new hair cells and the expansion of the sensory epithelium between birth and adulthood (Lewis and Li 1973; Li and Lewis 1979; Popper and Hoxter 1984). Studies with tritiated thymidine, an amino acid that is incorporated into the DNA of dividing cells, showed that new hair cells and supporting cells were added at the periphery of the sensory epithelium of elasmobranchs and amphibians (Corwin 1981, 1985). Authors of later studies found evidence of new hair cells and supporting cells within the hair cell epithelium of teleost fish, turtles, and chicken vestibular system (Jargensen and Mathiesen 1988; Lombarte and Popper 1994; Severinsen et al. 2003).

1.4 Regeneration in Birds

All the hair cells in the avian and mammalian cochlea ear arise during embryogenesis (Ruben 1967; Tilney et al. 1986; Katayama and Corwin 1989). Therefore the discovery that damaged hair cells in the chicken cochlea were replaced by newborn hair cells after acoustic trauma or aminoglycoside ototoxicity (Cruz et al. 1987; Girod et al. 1991; Hashino et al. 1992) generated considerable excitement and greatly accelerated research on this topic, as reflected by hundreds of articles on this topic during the last two decades (Corwin and Cotanche 1988; Ryals and Rubel 1988). Unlike those in the cochlea, hair cells in the normal avian vestibular system show a low level of hair cell regeneration that is balanced by ongoing hair cell death (Jargensen and Mathiesen 1988; Kil et al. 1997). In both the avian cochlea and vestibular organs, hair cell loss serves as the trigger that initiates a carefully orchestrated process of supporting cell proliferation and differentiation; however, both processes are confined to the damaged region (Girod et al. 1989; Hashino et al. 1995; Stone and Rubel 2000). These results suggest that newborn hair cells suppress the proliferation of additional hair cells, possibly through cell–cell contacts. The appearance of the pairs of dividing cells suggested that one member of the offspring differentiates into a hair cell while the other becomes a supporting cell, thereby replenishing the pool of progenitor cells. While proliferation and differentiation of supporting cells play a role in regeneration (Hashino and Salvi 1993), new hair cells appear in the newt in the presence of mitotic blockers (Taylor and Forge 2005). During avian hair cell regeneration, fewer than half of support cells and hair cells in the regenerated region are labeled with mitotic markers (Roberson et al. 1996). These results suggest that many new hair cells arise from phenotypic conversion (Taylor and Forge 2005). Conversion may also play an important role in generating new hair cells in birds and amphibians (Adler and Raphael 1996; Steyger et al. 1997).

1.5 Neurons and Other Structures Involved with Avian Hair Cell Regeneration

After acoustic trauma or aminoglycoside damage, newborn hair cells develop and migrate toward the luminal surface of the sensory epithelium and primitive stereocilia bundles emerge from their apical surface. The stereocilia bundles begin to develop, but even after 1 month, the bundle organization and orientation are still immature. It takes approximately 90 days before the stereocilia bundles on the regenerated hair cells regain their normal appearance and orientation in young chicks (Duckert and Rubel 1990, 1993). However, when hair cell regeneration was studied in adult birds, the bundles were still disoriented after a 142-day recovery period (Marean et al. 1993).

Approximately 3 days after the regenerated hair cells appear, afferent and efferent synapses can be seen contacting the basal pole of the cell (Ryals and Westbrook 1994) thereby providing the neural circuitry for delivering information to and from the brain. However, careful anatomical analyses raise some cautionary notes. The number of cochlear ganglion cells shows a mild-to-moderate decline between 30 and 90 days postexposure and possibly longer (Ryals et al. 1989). The factors mediating the loss of cochlear ganglion cells after acoustic trauma are currently unknown. The loss could be due to excitotoxic damage (Sun et al. 2001) or temporary lack of neurotrophic support between the time when the original hair cells were lost and the newborn hair cells had fully matured (Ernfors et al. 1995; Fritzsch et al. 1997). In cases in which severe excitotoxicity has occurred, cochlear ganglion cells do not appear to regenerate (Sun et al. 2001). Acoustic trauma also damages the upper fibrous layer and lower honey comb layer of the tectorial membrane. The lower honeycomb layer regenerates within a week or two; however, the upper fibrous layer does not regenerate (Cotanche 1987; Trautwein et al. 1996). Importantly, in cases of severe acoustic trauma in which there is extensive damage to both hair cells and supporting cells, hair cell regeneration fails to occur due to the destruction of progenitor cells (Muller et al. 1996).

The tegmentum vasculosum in the avian ear plays an important role in generating the endolymphatic potential (EP) and establishing the high potassium concentration in the endolymph. Acoustic trauma damages the dark cells in the tegmentum vasculosum of adult quail, but the tegmentum completely recovers after 4 days (Poje et al. 1995). Some results suggest that the repair of the tegmentum vasculosum plays an important role in the recovery of hearing after acoustic trauma.

1.6 Regeneration in the Mammalian Inner Ear

In the mammalian cochlea, where the supporting cells are highly specialized, there is no convincing evidence of proliferation or hair cell regeneration (Meyers and Corwin, Chapter 2) except under special conditions designed to stimulate proliferation and regeneration (Oesterle and Stone, Chapter 5). In the vestibular system of adult animals, there is evidence for limited hair cell regeneration after aminoglycoside treatment. New hair cells with immature stereocilia bundles reminiscent of regenerating avian hair cells have been observed in guinea pigs and human vestibular epithelia (Forge et al. 1993, 1998; Warchol et al. 1993). Many of the new hair cells presumably differentiated from supporting cells because the numbers of support cells were reduced in regions of regeneration. Cell proliferation most likely plays a minor role in producing new hair cells because there were relatively few [^{3}H]thymidine-labeled cells in the damaged region. Finally, one cannot rule out the possibility that some of the new hair cells with immature stereocilia are actually hair cell bodies that have survived the ototoxic insult and are in the process of rebuilding their stereocilia (Sobkowicz et al. 1995; Zhao et al. 1996; Schneider et al. 2002; Rzadzinska et al. 2004; Forge and Van de Water, Chapter 6).

2. Physiological Function After Avian Hair Cell Regeneration

Although avian hair cells regenerate, it was not clear if there would be complete recovery of physiological function (Saunders and Salvi, Chapter 3) because of the loss of cochlear ganglion neurons and lack of regeneration in the upper fibrous layer of the tectorial membrane after acoustic trauma. Gross potentials, evoked potentials, single-unit recordings, and otoacoustic emissions have all been used to assess different aspects of auditory function.

When hair cells are destroyed with aminoglycoside antibiotics, physiological thresholds generally recover to normal levels at low frequencies whereas at high frequencies thresholds only partially recover, resulting in residual thresholds shifts (Tucci and Rubel 1990; Chen et al. 1993). Recovery of physiological thresholds lags behind the emergence of regenerated hair cells by 3–4 months.

2.1 Physiological Thresholds

In cases of severe acoustic trauma where there is complete destruction of supporting cells and sensory cells on the basilar membrane, there is little or no recovery of the compound action potential (CAP) or the spontaneous or sound-evoked discharge patterns of cochlear ganglion neurons (Muller et al. 1996, 1997). However, in cases of moderate acoustic trauma where there is almost complete hair cell regeneration, physiological thresholds measured with different techniques such as the CAP, auditory evoked response, and single-fiber recordings from cochlear ganglion neurons all show significant threshold elevation (40–60 dB) immediately after the exposure followed by complete or nearly complete recovery 1–4 months postexposure (Henry et al. 1988; Chen et al. 1996; Saunders et al. 1996; Muller et al. 1997). Single fiber spontaneous discharge rates, phase-locking, and frequency tuning are nearly normal after a few weeks or months postexposure. However, discharge rate-intensity functions and two-tone rate suppression continue to exhibit residual deficits after 1 month of recovery (Saunders et al. 1996; Chen et al. 2001; Lifshitz et al. 2004; Furman et al. 2006). These deficits may be related to the lack of regeneration of the upper fibrous layer of the tectorial membrane.

The endolymphatic potential (EP), which is approximately +18 mV in adult birds, provides a positive driving force for moving potassium down its concentration gradient into the hair cells. The effects of acoustic trauma and aminoglycosides on the EP appear to be different in young chicks compared to adults. The EP in young chicks shows a large decline after acoustic trauma and then recovers over time, consistent with other physiological measures (Poje et al. 1995). However, acoustic trauma and aminoglycosides failed to produce a decrease in the positive EP in adult chickens, suggesting the EP was more resistant in adults (Chen et al. 1995; Trautwein et al. 1997).

2.2 Otoacoustic Emissions and Avian Hair Cell Regeneration

Distortion product otoacoustic emissions (DPOAEs), a nonlinear response presumably generated by an active process in the hair cells, show a significant decline immediately postexposure in both young chicks and adult animals. DPOAE amplitudes completely recover in young chicks after acoustic trauma and aminoglycoside damage (Ipakchi et al. 2005). However, DPOAE amplitude only partially recovers in adult chickens; residual deficits are evident in adults chicken even after a 16-week recovery period (Trautwein et al. 1996; Chen et al. 2001). It was originally suggested that the persistent reduction in DPOAE amplitude in adults might be related to residual damage to the upper fibrous layer of the tectorial membrane; however, this explanation may not be valid because the fibrous layer is also missing in young chicks that show complete recovery. Collectively, the DPOAE and EP data suggest that the ears of young birds are more resistant or better able to recover from acoustic trauma than are adult birds.

DPOAEs are thought to arise from an active feedback mechanism that enhances the sensitivity and frequency selectivity of the inner ear. In mammals, the active feedback mechanism, or electromotile response, arises from the motor protein, prestin, located in the lateral wall of outer hair cells (OHCs) (Hofstetter et al. 1997; Zheng et al. 2000a; Liberman et al. 2002). Because birds have robust DPOAEs, but lack OHCs, can the short and tall hair cells in avian ears also generate an electromotile response? Surprisingly, when AC current is applied to the avian ear, a strong, electrically evoked otoacoustic emission emerges from the cochlear into the ear canal (Chen et al. 2001). Selective destruction of avian hair cells with kanamycin greatly reduces otoacoustic emission amplitude. Emission amplitude partially recovers as the hair cells regenerate, suggesting that avian hair cells are the source of the electrically evoked emission. Unlike that of mammals, however, the avian hair cell soma does not elongate or contract in response to electrical stimulation (He et al. 2003). Thus, avian electrically evoked otoacoustic emissions must arise elsewhere, possibly in the hair cell stereocilia bundle (Hudspeth et al. 2000).

2.3 Behavioral Thresholds After Avian Hair Cell Regeneration

Behavioral measures of hearing, though difficult and time consuming to obtain, provide the gold standard for evaluating the recovery of function that occurs when hair cells regenerate. Studies of noise-induced hearing loss in birds were carried out almost a decade before it was known that hair cells could regenerate. Because anatomical data were lacking, it is unclear what role hair cell loss and regeneration played in the temporary or permanent threshold shifts observed (Saunders and Dooling 1974; Hashino et al. 1988). More recent studies in several avian species suggest that when noise-induced temporary threshold shifts are less than 60 dB, pure tone behavioral thresholds fully recover after 2–3 weeks. Complete recovery of pure tone thresholds is associated with hair cell regeneration and normal or near-normal hair cell numbers and morphology. Hearing thresholds return to normal or near-normal levels even when a few hair cells are missing and the upper fibrous region of the tectorial membrane is missing (Niemiec et al. 1994b; Saunders et al. 1995; Ryals et al. 1999). Conversely, when noise-induced temporary thresholds shifts exceed 65–70 dB, behavioral thresholds recover slowly and birds sustain permanent hearing loss. Permanent threshold shifts are generally associated with large numbers of missing or damaged hair cells and loss of the upper fibrous layer of the tectorial membrane.

Aminoglycoside antibiotics only destroy hair cells, but unlike the effects of acoustic trauma, there is no obvious damage to the tectorial membrane. Damage begins in the basal, high-frequency region of the cochlea and gradually spreads toward the apex. Because aminoglycoside damage is largely confined to the hair cells and most hair cells regenerate, hearing thresholds might be expected to recover completely after regeneration. Several aminoglycoside studies have found large thresholds shifts immediately after aminoglycoside treatment; threshold shifts are greater at high versus low frequencies (Marean et al. 1993; Dooling et al. 1997, 2006). Recovery of thresholds began almost immediately after the end of treatment, and several months later low-frequency thresholds returned to near-normal levels, but high-frequency thresholds remain elevated at 20–30 dB. This persistent elevation of high frequency thresholds is consistent with physiological results (Chen et al. 1993). Some results suggest that the residual threshold shifts are associated disoriented stereocilia bundles on the regenerated hair cells (Marean et al. 1993).

2.4 Auditory Discrimination After Avian Hair Cell Regeneration

Hair cell loss leads to significant auditory processing deficits in mammals (Salvi et al. 1983). Similar deficits would be expected to occur in birds when the hair cells are missing or damaged, but afterwards hearing should recover when the hair cells regenerate. Temporal integration refers to the ability of the auditory system to integrate acoustic energy over time. Temporal integration is reflected by the fact that thresholds improve approximately 15 dB as signal duration increases from 10 to 500 ms. Immediately after acoustic trauma, thresholds are elevated and remain constant with stimulus duration indicating a loss of temporal integration. However, as hair cells regenerate and hearing recovers, normal temporal integration is restored (Saunders and Salvi 1993). Amplitude modulation transfer functions have been used to assess auditory temporal resolution before and after aminoglycoside induced hair cell loss (Marean et al. 1998). Amplitude modulation thresholds are temporarily impaired, but eventually recover.

Behavioral tone-on-tone masking patterns can be used to assess the frequency resolving power of the auditory system. Chicken tone-on-tone masking patterns have a narrow, inverted V-shape. Tone-on-tone masking patterns measured several months after acoustic trauma are nearly identical to those measured before trauma, suggesting that the internal auditory filters completely recover after the hair cells regenerate (Saunders and Salvi 1995). Conversely, after aminoglycoside treatment, the starling auditory filter widths measured with notched noise maskers were wider than normal immediately posttreatment, but eventually recovered except for two of four birds that showed a persistent broadening of filter shape at the high frequencies (Marean et al. 1998). These residual deficits may be related to the disoriented stereocilia bundles on the regenerated hair cells (Marean et al. 1993). Frequency difference limens and intensity difference limens also recovered after hair cells regenerated from aminoglycoside treatment (Dooling, Dent, Lauer, and Ryals, Chapter 4). Songbirds temporarily deafened by aminoglycosides initially experience difficulties discriminating among a set of similar vocalizations and have problems recognizing complex species-specific vocalizations after the initial stage of hair cell regeneration. However, after months of recovery they regain their ability to make these discriminations even though they have mild residual hearing loss (Dooling et al. 2006). Aminoglycoside deafening also disrupts the vocalizations of songbirds (Manabe et al. 1998). The impairment is maximal during the period of greatest hearing loss, but rapidly recovers.

Collectively, the behavioral measures of hearing obtained from birds in which most hair cells regenerate suggest that hearing sensitivity returns to normal or near normal levels with only minor residual deficits at high frequencies. The ability to detect or discriminate changes in intensity, frequency, and the temporal features of sounds, and to recognize complex species-specific vocalizations return to normal or near normal levels after the hair cells regenerate.

3. Factors Stimulating Proliferation and Regeneration

Hair cell loss triggers robust cell proliferation and hair cell regeneration in the avian cochlea, but not in mammals. Damage to the mammalian vestibular organs results in limited cell proliferation, but because the number of mitotically labeled

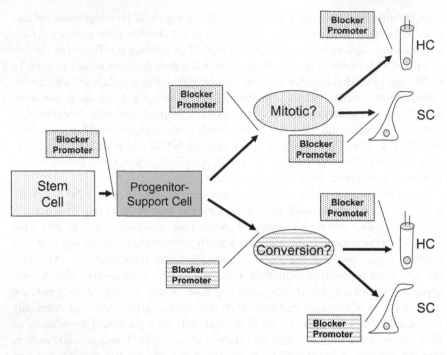

FIGURE 1.1. Schematic illustrating stages through which stem cells and progenitors cells advance to become hair cells (HC) or supporting cells (SC). Blockers and promoters can act at multiple stages to suppress or promote differentiation.

hair cells is far less than the number of new hair cells the production of new hair cells may come from repair or nonmitotic cell conversion (Oesterle and Stone, Chapter 5). What factors are involved in cell proliferation, conversion, and differentiation? Are there factors that block or promote proliferation, conversion, and differentiation, and where might these blockers or promoters reside among the various cell types in the inner ear (Fig. 1.1)?

3.1 Proliferation in Nonmammals

In response to damage, support cells along the basal surface of the epithelium in nonmammals divide; some of the progeny become support cells and others migrate to the luminal surface of the epithelium, where they differentiate into mature hair cells (Balak et al. 1990; Hashino and Salvi 1993; Jones and Corwin 1993; Presson 1994). The elegant work performed in the olfactory system has provided important insights on the type of progenitor cells and some of the factors that regulate proliferation and cell lineage (Huard et al. 1998; Murray and Calof 1999; Schwob 2002). The fact that hair cells can be replenished in the avian inner ear after multiple injuries or continued hair cell loss due to genetic mutation suggests that avian progenitor cells can self-renew (Marean et al. 1993; Gleich et al. 1994; Niemiec et al. 1994a). The asymmetric division of avian support cells into both hair cells and support cells is consistent with the presence of stem cells. In vestibular epithelium of fish there is evidence for two pools of dividing cells, a pool of rapidly dividing progenitors cells and a slowly dividing stem cell pool (Presson et al. 1995). Identification of the specific cell types within these specialized pools has proved difficult because of the lack of cell specific morphologic and immunocytochemical markers.

3.2 Cell Signaling in Proliferation in Nonmammals

The supporting (progenitor) cells in the undamaged avian cochlea express the growth arrest (G_0) marker statin, but lack the proliferation cell marker (gap 1 phase [G_1] and later) proliferating cell nuclear antigen (PCNA). However, when hair cell proliferation is triggered by acoustic trauma or aminoglycosides that damages only the base of the cochlea, PCNA is expressed within and outside the damaged region and statin expression is decreased (Bhave et al. 1995; Sliwinska-Kowalska et al. 2000). These changes signal a movement from growth arrest toward the gap 1 phase (G_1) of the cell cycle throughout the epithelium except in the damaged region where cells are allowed to transit into the DNA synthesis (S) phase. The local signals that promote S-phase entry are not yet known, but are likely to be associated with hair cell death or removal of damaged hair cells (Hashino et al. 1995; Williams and Holder 2000). In birds, signals from hair cells presumably prevent nearby supporting cells from dividing; however, this scenario does not apply to fish that add new hair cells to the periphery and the core of the epithelium throughout life (Popper and Hoxter 1984).

3.3 Growth Factors

Growth factors, which can stimulate proliferation in some stem/progenitor cells, have been tested in the inner ear (Oesterle and Stone, Chapter 5). Insulin and insulin-like growth factor-1 (IGF-1), a secreted hormone, induces proliferation in the mature avian vestibular system, but not in the cochlea (Oesterle et al. 1997). Mixed results have been obtained with transforming growth factor-α (TGF-α), which binds to the epidermal growth factor (EGF) receptor. TGF-α induces proliferation in isolated cultures of the avian vestibular epithelia, but not in organotypic cultures (Oesterle et al. 1997; Warchol 1999). While some growth factors can stimulate proliferation, others, such as fibroblast growth factor-2 (FGF2), can suppress proliferation in cultured vestibular epithelium (Oesterle et al. 2000). Fibroblast growth factor receptor (FGFR3) has been implicated in proliferation in the avian auditory system. FGFR3 is heavily expressed in supporting cells, but is greatly downregulated after hair cells are damaged, and then upregulated again after hair cells regenerate (Bermingham-McDonogh et al. 2001).

3.4 Cell–Cell Contacts and Proliferation

Because cell proliferation is initiated after hair cell death and ceases after hair cells regenerate, cell–cell contacts may play an important role in proliferation in nonmammals (Hashino et al. 1995). So far, there are limited data on the topic. In the chicken utricle, antibody blocking of N-cadherin, which is expressed at cell–cell junctions, suppressed proliferation in organ cultures (Warchol 2002).

3.5 Intracellular Signaling and Proliferation

A number of intracellular signaling pathways have been shown to regulate proliferation (Oesterle and Stone, Chapter 5). One of the earliest to be identified was cAMP. In birds, high levels of forskolin increase intracellular cAMP and increase proliferation, presumably by acting on protein kinase A (PKA; Navaratnam et al. 1996). Brief treatments with forskolin also increase proliferation in mammalian vestibular organ, but it is thought to occur because of increased expression of growth factor receptors on the plasma membrane (Montcouquiol and Corwin 2001b). Several other signaling molecules such as mitogen-activated protein kinase (MAPK), protein kinase C (PKC), phospho-inositide 3-kinase (PI-3K), and target of rapamycin (TOR) slow proliferation, the latter two having effects in both mammals and birds (Montcouquiol and Corwin 2001a; Witte et al. 2001).

4. Differentiation of Supporting Cells and Hair Cells

The progeny of dividing progenitor/stem cells must decide their fate and undertake numerous biochemical and morphological changes in order to differentiate into the appropriate cell type and establish contacts with their neighbors. Avian support cells have features distinct from those of hair cells. Support cells, which constantly secrete tectorins from the apical surface, generate the tectorial membrane; they contain connexons to form intercellular gap junctions to transport molecules between cells, and they contain cytokeratins, a support cell marker (Goodyear et al. 1996; Stone et al. 1996; Forge et al. 2003). A unique morphologic feature of hair cells is the stereocilia bundle, which is rich in actin, fimbrin, and myosins VI and VIIa (Lee and Cotanche 1996; Hasson et al. 1997). Avian hair cells also express TUJ1 β-tubulin and calmodulin.

4.1 Factors Regulating Differentiation

The molecular signals that cause progenitor cells to differentiate into hair cells and supporting cells are complex and not well understood at present (Oesterle and Stone, Chapter 5). Several extrinsic factors have been shown to contribute to supporting cell and hair cell differentiation during normal development or regeneration (Eddison et al. 2000; Kelley 2003). One of the best studied is the extracellular receptor, Notch, that is activated by Delta or Jagged/Serrate ligands expressed on adjacent cells. When ligands bind to Notch it increases the level of HES, a repressor transcription factor that reduces the expression of proneural/sensory genes (Kageyama and Ohtsuka 1999; Kageyama et al. 2005).

Several intrinsic transcription factors such as *Ngn1*, *Atoh1*, *HES1*, and *HES5* are known to control the expression of genes involved in differentiation (Fekete and Wu 2002). *Atoh1* (also *Math1*) is essential for the specification and differentiation of hair cells during development and in nonmitotic hair cell regeneration. *Atoh1* expression emerges before obvious signs of hair cell damage and persists during cell division. During differentiation, *Atoh1* is downregulated in support cells, but retained in hair cells (Cafaro et al. 2007). *HES1* and *HES5*, negative regulators of *Atoh1*, drive dividing cells toward supporting cell lineage (Zheng et al. 2000b; Zine et al. 2001). $HES1^{-/-}$ mice have increased number of inner hair cells (IHCs) whereas $HES5^{-/-}$ have increased numbers of OHCs; mice lacking both *HES1* and *HES5* have increased numbers of vestibular hair cells. *Prox1*, a transcription factor that promotes progenitor cell proliferation and differentiation, is expressed in developing regions of the chick otocyst that are associated with hair cells and neurons.

4.2 Regeneration in Mammals

Because hair cells do not regenerate in the mammalian cochlea and only limited regeneration occurs in the vestibular epithelium, there has been intense interest in identifying methods to stimulate regeneration. Proliferation is actively suppressed in the mammalian inner ear by the cyclin-dependent kinase inhibitors *p19Ink4d* (*Ink4d*) and *p27Kip1* (*Kip1*) (Chen and Segil 1999; Chen et al. 2003). Supporing cells normally express high levels of *p27Kip1*, but when this gene is knocked out, postanal proliferation and supernumerary hair cells are observed (Chen and Segil 1999). When *p19Ink4d* is knocked out, hair cells reenter the cell cycle but later die (Chen et al. 2003). In addition, deletion of *Rb1* coding for the retinoblastoma protein causes differentiated hair cells to divide, differentiate, and become functional (Sage et al. 2005).

Genes that positively regulate the generation of hair cells have been used in gene therapy studies. Rat neonatal organ cultures transfected with a *Math1–EGFP* plasmid stimulated the conversion of supporting cells in the cochlea (inner sulcus) and utricle into hair cells (Zheng and Gao 2000). Transfection of adult rat utricular cultures with an adenovirus expressing *Hath1* (human homolog of *Atoh1*) resulted in the conversion of many support cells to hair cells (Shou et al. 2003). The proof of concept studies carried out in vitro set the stage for hair cell regeneration and hearing restoration in vivo. When guinea pigs were deafened with an ototoxic drug cocktail that destroyed all hair cells and the cochlea subsequently transfected with an adenovirus expressing *Atoh1* (*Math1*), support cells near the original sensory cells converted into hair cells (Izumikawa et al. 2005). Moreover, physiological hearing thresholds were better in the adenovirus-treated ears than in untreated ears. Another promising approach to stimulating hair cell regeneration involves the use of cell-permeable small molecules that can negatively or positively modulate genes involved in hair cell regenerations. Work on Alzheimer's disease has helped to identify several small molecules that inhibit Notch signaling. Gamma secretase, a small molecule that suppresses Notch, reduces the expression of *Hes5* and leads to the production of ectopic hair cells in murine neonatal cultures (Tang et al. 2006; Yamamoto et al. 2006). While gene therapy and the use of small molecules that promote the conversion of support cells to hair cells are promising approaches, a major obstacle that remains to be solved is the need to replenish the support cells that have undergone conversion. Supporting cells, especially those in the highly differentiated and organized cochlea, perform important functions. Therefore, future efforts need to employ a more comprehensive approach that enhances cell proliferation in damaged regions coupled with signaling that converts some of the progeny to hair cells and others to support cells.

5. Repair and Protection

Considerable evidence has emerged indicating that damaged hair cells undergo repair after a nonlethal insult (Sobkowicz et al. 1996; Baird et al. 2000; Watson et al. 2007). Over the past decade, there has been a growing interest in the biological basis of cell death and why some cells die while others survive. This has led to rapid growth in the number of studies dealing with cell death and cell survival signaling and how these pathways can be influenced pharmacologically and by virally mediated gene transfer (Forge and Van de Water, Chapter 6). Sufficient progress has been made that this knowledge is being exploited clinically to protect hair cells from noise, ototoxic, and age-related damage in order to prevent hearing loss.

5.1 Cell Death

Three major cell death pathways are now recognized: necrosis, apoptosis, and autophagy (Debnath et al. 2005); the first two are known to occur in the inner ear (Hu et al. 2006). Necrosis is characterized morphologically by swelling of the cell body and organelles, rupture of the cell membrane, and extrusion of the cell's contents into the extracellular space that leads to an inflammatory response. Apoptosis, a major form of programmed cell death, is characterized by a series of internal biochemical processes that lead to nuclear condensation and fragmentation, plasma membrane blebbing, and breakdown of the cell into apoptotic bodies that are engulfed by nearby cells or macrophages (Savill et al. 1993; Monks et al. 2005). Autophagy, recently recognized as a form of programmed cell death, involves transporting structures targeted for destruction into autophagosomes and moving them into lysosomes or vacuoles where they are hydrolyzed (Reggiori and Klionsky 2005).

5.2 Apoptosis

Caspases, a family of cysteine proteases, play a major role in apoptosis; they are present in cells in their inactive pro-caspase form. Pro-caspases are activated (cleaved) by one or more initiator caspase (e.g., caspase-8 or caspase-9), which in turn activate effector caspases (e.g., caspase-3 or caspase-7) that destroy DNA and other cellular components. Activation of initiator caspase 9, part of the intrinsic cell death pathway, requires the release of cytochrome c from the mitochondria and binding to apoptotic protease activating factor 1 (APAF-1), leading to activation of executioner caspase 3 (Katoh et al. 2004). Initiator caspase-8, part of the extrinsic cell death pathway that is activated by death receptors on the cell's membrane, activates executioner caspase-3. Activated caspase-8 can also cross over into the intrinsic cell death pathway and activate caspase-9. Several caspase-independent cell death pathways have been identified, including calpains, calcium-activated proteases known to be involved in some forms of cell death in the inner ear (Bartus et al. 1995; Wang et al. 1999; Ding et al. 2001; Danial and Korsmeyer 2004).

5.3 Hair Cells and Antioxidant Defenses

Damage to cochlear hair cells whether by noise, ototoxic drugs, or aging follows a stereotypic pattern, with damaging progressing from base-to-apex and from OHCs to IHCs. These vulnerability gradients persist in vitro and apparently are due to the intrinsic differences in susceptibility to traumatic stress (Richardson and Russell 1991; Zheng and Gao 1996). The intrinsic process of cellular respiration results in the production of reactive oxygen species such as the superoxide and hydroxyl radicals that oxidize and damage other molecules within cells (Halliwell et al. 1992). To eliminate these highly toxic reactive oxygen species and promote cell survival, cells deploy a system of antioxidant enzymes such as superoxide dismutase and glutathione to scavenge and inactivate reactive oxygen species (Fuchs et al. 1989). Traumatic agents such as noise or ototoxic drugs generate reactive oxygen species that can overwhelm the cell's antioxidant defense system (Clerici et al. 1996; Jacono et al. 1998; Ohlemiller et al. 1999a). The antioxidant defense systems of hair cells in the base of the cochlear are weaker than those in the apex; this presumably explains why hair cells in the base of the cochlea are more susceptible to traumatic insult and aging (Sha et al. 2001b).

5.4 Spiral Ganglion Neuron Survival Factors

The spiral ganglion neurons provide the only pathway for relaying information from hair cells to the central auditory system. The long-term survival of spiral ganglion neurons is critical for the restoration of hearing through hair cell regeneration or a cochlear implant. Most spiral ganglion neurons survive if damage is confined to the OHCs. However, spiral ganglion neurons slowly degenerate over months or years after the IHCs are destroyed, suggesting that neuron survival is dependent on trophic support from the hair cells (Xu et al. 1993; McFadden et al. 2004). Studies from knockout mice indicate that neurotrophin-3 (NT-3) is the major survival promoting factor for type I neurons in the cochlea while brain-derived neurotrophic factor (BDNF) promotes the survival of type II neurons (Ernfors et al. 1995; Fritzsch et al. 2004). Deletion of BDNF plus NT-3 resulted in complete loss of spiral ganglion neurons.

5.5 Genetic Factors Related to Susceptibility and Protection

Strategies aimed at preventing hearing loss would undoubtedly benefit from an understanding of genetic factors that make individuals susceptible or resistant to ototraumatic insults or age-related hearing loss. Mice carrying age-related hearing loss (*Ahl*) alleles develop early-onset hearing loss. Many of the mice with *Ahl* alleles are also susceptible to acoustic trauma and ototoxic insult (Johnson et al. 1997; Davis et al. 2001; Wu et al. 2001; Keithley et al. 2004). Mice lacking the gene for copper/zinc superoxide dismutase are more suscep-tible to hearing loss from aging and noise exposure (McFadden et al. 1999; Ohlemiller et al. 1999b) whereas overexpression of this gene confers protection against aminoglycoside ototoxicity, but not against aging (Sha et al. 2001a; Coling et al. 2003). A number of other genes are associated with early or late-onset hearing loss and these mutations could modify other forms of ototrau-matic injury. Humans with maternally inherited mutations in certain mitochon-drial genes are hypersensitive to aminoglycosides (Prezant et al. 1993; Guan et al. 2000). The transcription factors Brn 3c (Pou 4f3) and Barhl 1 expressed in hair cells have been implicated in progressive hearing loss (Vahava et al. 1998; Li et al. 2002). Disruption of the *Kcc4* gene for the K–Cl cotransporter, expressed in supporting cells, is associated with progressive early-onset hair cell loss (Boettger et al. 2002). Genes for connexin (cx) 26, 30, and 31 are associated with congenital or late onset hearing loss (Xia et al. 1998; Grifa et al. 1999; Bitner-Glindzicz 2002).

5.6 Protection by Sound Conditioning and Temperature Stress

High-level noise exposures damage the inner ear, so it would seem illogical to suggest that prior sound exposure (conditioning exposure) could reduce the risk of hearing loss from later noise exposures. The first study to show a protective sound conditioning effect was published by Miller and colleagues in 1963 (Miller et al. 1963). They noise-exposed cats at 115 dB for 7.5 min per day for 17 days and used behavioral techniques to measure the hearing loss just after the animals were removed from the noise. The hearing loss after the first day of the exposure was on the order of 35 dB, but gradually declined to

15 dB by day 17. At the time, the authors considered the reduction of hearing loss during the series of daily noise exposures as a curiosity or behavioral artifact. However, more than a decade later, investigators performed similar experiments and reported that the auditory system became more "resistant" to hearing loss over the course of prolonged interrupted noise exposure (Clark et al. 1987; Boettcher et al. 1992). Other have employed moderate intensity noise, either continuous or interrupted, to "condition" the ear. Afterwards, the sound-conditioned group and a control group were exposed to high-intensity noise known to cause significant hearing loss and hair cell loss. Animals that received sound conditioning sustained less hearing loss and hair cell loss than the control group (Canlon et al. 1988; Subramaniam et al. 1991; Canlon 1997; Canlon and Fransson 1998). Unilateral conditioning exposures protect only the exposed ear, suggesting that the effects are local and not systemic (Yamasoba et al. 1999). Sound conditioning persists over many weeks, suggesting that the conditioning exposure induced a long-lasting biological change to the inner ear (McFadden et al. 1997). Conditioning exposures upregulate glutathione, γ-glutamyl cysteine synthetase, and catalase, suggesting that the antioxidant defense system plays a role in sound conditioning (Jacono et al. 1998). The stress protein NFκB, glucocorticoids, and glucocorticoid receptors have also been implicated in resistance and sound conditioning (Tahera et al. 2006; Canlon et al. 2007).

Heat shock proteins, which are expressed in cells during heat, cold, and other forms of stress including exposure to noise, cisplatin, and aminoglycosides, were first discovered in *Drosophila melanogaster* in the early 1960s (Ritossa 1996; Ni et al. 2005; Coling et al. 2007). Heat shock proteins assist cells to recover from stress by repairing damaged proteins (protein refolding) or eliminating proteins that are damaged, thereby enhancing the ability of damaged cells to survive (Michils et al. 2001; Winter and Jakob 2004). Heat shock proteins are upregulated in the cochlea in response to hyperthermia (Lim et al. 1993; Yoshida et al. 1999) (Forge and Van de Water, Chapter 6). The upregulation of heat shock proteins by hyperthermia protects against acoustic trauma in vivo, and their subsequent downregulation is correlated with greater susceptibility to noise-induced hearing loss. Hyperthermia-induced upregulation of heat shock proteins also protects against cisplatin- and aminoglycoside-induced ototoxicity in vitro (Cunningham and Brandon 2006). Hyperthermia upregulates the stress-activated transcription factor HSF-1 that regulates expression of heat shock factor genes. Not surprisingly, HSF-1 knockout mice are more susceptible to noise-induced hearing loss (Fairfield et al. 2005). Nonsteroidal anti-inflammatory drugs such as sodium salicylate, an antioxidant (discussed later), activates HSF-1 (Housby et al. 1999). Whole-body hypothermia also induces HSF-1 and heat shock proteins in many organs including the brain (Cullen and Sarge 1997) and reduces hearing loss and/or hair cell loss from ischemic reperfusion injury, electrode insertion trauma, and acoustic trauma (Drescher 1974; Henry and Chole 1984; Watanabe et al. 2001; Balkany et al. 2005).

5.7 Antioxidants and Protection

Because reactive oxygen and reactive nitrogen molecules are generated in the inner ear in response to ototoxicity and acoustic trauma, exogenous antioxidants or compounds that upregulate the bodies antioxidant defenses have been evaluated as otoprotectants. The list of compounds used to scavenge reactive oxygen species (e.g., $\bullet OH$, H_2O_2, $O_2\bullet$) and reactive nitrogen species ($ONOO^-$, NO) and protect against inner ear damage (see Table 6.1, Forge and Van De Water, Chapter 6) include salicylate (aspirin), acetyl-L-carnitine (ALCAR), α-tocopherol (vitamin E), L-N-acetylcysteine (L-NAC), ebeselen and allopurinol, D-methionine, M40403, and R-PIA (Campbell et al. 1996; Hu et al. 1997; Sha and Schacht 1999; Kopke et al. 2000, 2002; McFadden et al. 2003; Fetoni et al. 2004; Lynch et al. 2004, 2005). Some of these compounds provide protection only against one or two ototraumatic agents and some are effective only when administered in combination. In addition to scavenging reactive oxygen and reactive nitrogen molecules, some compounds have antiapoptotic properties. Because antioxidants intervene during the early stages of ototrauma they may be effective in limiting the amount of damage to the inner ear.

5.8 Caspase Inhibition Confers Protection

Once the inner ear has been damaged, programmed cell death moves forward by activating initiator and effector caspases. Sublethal cell damage can be prevented from moving toward the lethal stage by inhibiting the action of caspases, either individually or collectively (Forge and Van De Water, Chapter 6). Caspase inhibitors administered in vitro suppress aminoglycoside- and cisplatin-induced damage to hair cell death or spiral ganglion neurons (Liu et al. 1998; Forge and Li 2000; Cunningham et al. 2002; Matsui et al. 2002). Intracochlear or systemic administration of caspase-9 and caspase-3 inhibitors also protects against hair cell loss, hearing loss, or vestibular dysfunction from cisplatin or aminoglycoside ototoxicity. However, systemic administration of caspase inhibitors provides less protection than intracochlear perfusion, ostensibly because of reduced uptake of the inhibitor from the bloodstream into the inner ear (Matsui et al. 2003; Wang et al. 2004). The improved vestibular and auditory function associated with caspase inhibition suggests that sublethally damaged cells remain functional and can undergo self-repair. Local application of caspase inhibitors to the inner ear would provide optimal protection and avoid potential toxic effects associated with systemic administration.

Drugs that block stress signaling pathways before caspase activation can also protect the inner ear. Small molecules such as CEP 1347, CEP 1104, and D-JNK-1, which inhibit MAPK/JNK stress pathways, protect against cochlear damage and hearing loss induced by aminoglycosides, electrode-insertion trauma, and noise-induced cochlear damage, but not cisplatin ototoxicity (Pirvola et al. 2000; Wang et al. 2003; Matsui et al. 2004; Eshraghi et al. 2007). The protective effect of CEP 1347 against aminoglycoside ototoxicity was less when the drug was administered systemically than when applied in organotypic cultures, possibly due to reduced drug uptake in vivo (Pirvola et al. 2000; Ylikoski et al. 2002). The differences may be due to the fact that the cell death pathways involved in acute aminoglycoside ototoxicity in culture are different from those seen with chronic aminoglycoside treatment in vivo (Jiang et al. 2006).

6. Future Directions

The rediscovery of hair cell regeneration and sound conditioning to prevent hearing loss in the late 1980s generated a groundswell of interest in the biochemical, cellular, and genetic events responsible for cell proliferation, differentiation, cellular repair, and cell death. The interest in regeneration, repair, and protection came at a time when there were rapid scientific and technological advances that provided investigators with the new tools and a more sophisticated scientific framework in genomics and proteomics that could be used to identify the biological processes involved in regeneration, repair, and cell death (Rivolta and Holley, Chapter 7).

6.1 Gene Arrays

Subjects with induced and natural gene mutations have provided valuable insights into the role specific genes have on the structure and function of the inner ear (Ernfors et al. 1994; Liberman et al. 2002). Another approach for studying the roles genes play in development, proliferation, differentiation, and repair involves the use of high-throughput gene microarrays that contain hundreds or thousands of genes. Gene arrays are composed of a series of microscopic spots attached to a membrane, polymer, or glass slide; each spot contains identical single-stranded polymeric molecules of oligonucleotides or cDNAs coding for a unique fragment of a particular gene. The pool of RNA or DNA obtained from a cell line or tissue is labeled and afterwards washed onto the gene array, allowing the labeled RNA or DNA to hybridize to its complimentary oligonucleotide or cDNA on the array. The signal from each spot is measured to determine if the gene is present and to estimate its abundance. To determine if the results are valid, it is common to follow up with reverse transcriptase-polymerase chain reaction (RT-PCR), in situ hybridization, or immunolabeling to corroborate the results and to identify the location of the gene or the protein coded by the gene.

Gene profiling can be used to assess which genes are expressed and the level of expression in a cell line or in a heterogeneous tissue such as the ear. Analysis of a heterogeneous tissue is complicated by the fact that a high level of expression of a gene in one cell type may be canceled out by low expression in another cell type. Further, if the proportion of cells of a given type (e.g., hair cells) is low, the genes unique to the low-abundance cell type may go undetected.

Gene profiling has been used to determine which genes are present in cochlear tissues or inner ear cell lines and to estimate their relative or absolute abundance (Germiller et al. 2004; Hildebrand et al. 2007). A comparison of gene profiles obtained at different developmental stages can be used to identify genes that are expressed at low or high levels during periods of cell proliferation and differentiation (Chen and Corey 2002a; Pompeia et al. 2004; Powles et al. 2004). A comparison of gene profiles obtained from mouse cochleas between P2 and P32 revealed an upregulation of many genes associated with ion channels, calcium binding, and transporters (Chen and Corey 2002b). Comparison of gene profiles from the mouse utricle identified the *Rb1* gene, which plays an important role in hair cell proliferation (Chen and Corey 2002a; Sage et al. 2005).

Gene profiles from the inner ear can be compared to other tissues to identify those that are unique to the inner ear versus those that are common to other tissues (Liu et al. 2004). This information can be used to create cDNA libraries for the inner ear and to identify genes unique to specific tissues within the ear (Beisel et al. 2004; Morris et al. 2005). One very elegant, but labor-intensive study identified some of the hair cell and support cell specific genes in the mammalian cristae ampullaris. The laser capture dissection technique was used to separate hair cells from supporting cells, and the mRNA from each cell population was amplified and gene profiled (Cristobal et al. 2005). More than 400 genes were identified that showed a significant difference in expression between hair cells and supporting cells.

Gene arrays from normal and noise-exposed chicks have been used to study gene expression changes associated with the early stage of damage and later stages of repair and regeneration (Lomax et al. 2000; Lomax et al. 2001). Noise-exposure upregulated genes associated with actin signaling, Rho GTPase, and protein degradation. In mammals, the early stages of noise-induced temporary threshold shift were correlated with a time-dependent upregulation of genes associated with protein synthesis, metabolism, cytoskeletal proteins, and calcium binding proteins (Taggart et al. 2001). Noise exposures associated with permanent threshold shift in mammals increased expression of immediate early genes involved with transcription (e.g., c-*FOS*) and cytokines (e.g., *LIF*) (Cho et al. 2004). Comparison of gene arrays from control versus short-term amino-glycoside treatment revealed a downregulation of genes that would reduce stress from reactive oxygen species and N-methyl-D-aspartate (NMDA) receptor.

6.2 Inner Ear Cell Lines

The diverse cell types and complex morphological organization of the inner ear, particularly the cochlea, makes it especially difficult to identify the biological signals that control the temporal and spatial features of proliferation and differentiation in this heterogeneous cell population. To reduce the complexity of the problem, cell lines have been established from the cochlea and vestibular system. Most of inner ear cell lines have been developed from a transgenic Immorto-mouse that contains a temperature sensitive, immortalizing protein that causes cells to proliferate at 33°C in the presence of γ-interferon, but not at normal body temperature (Noble et al. 1992). Individual cells derived from the inner ear at different stages of development and maintained at 33°C halt their program of differentiation at a particular developmental stage and proliferate to form a cell line with characteristics of the developmental stage and anatomical location from which it was derived. Inner ear cell lines have been derived from several locations and developmental stages including the embryonic otocyst, neuroblasts, and auditory epithelium and postnatal maculae of the utricle and organ of Corti (Barald et al. 1997; Rivolta et al. 1998; Lawlor et al. 1999; Lawoko-Kerali et al. 2004).

6.3 Gene Networks

Gene microarrays have been used to study the temporal pattern of gene expression associated with differentiation of an Immortomouse cell line derived from the organ of Corti at the early stage of hair cell differentiation (Rivolta et al. 1998; Rivolta and Holley 2002). As the clonal cells differentiated in synchrony, expression of hair cell markers Brn3c, myosin VIIa, and α9AChR increased. Analysis of temporal profiles has the potential to identify genes that are part of a functionally related program that drives development. To determine if genes that cluster together temporally are part of a signaling network, expression of a key regulatory gene can be knocked down with antisense oligonucleotides, antisense morpholinos, or siRNAs to determine its effects on the signaling cascade and subsequent differentiation (Rivolta and Holley, Chapter 7).

6.4 Cell Death

Cell lines derived from the inner ear have been used to study the mechanisms of ototoxicity and identify otoprotective compounds (Bertolaso et al. 2001; Devarajan et al. 2002; Jeong et al. 2005a, b). Inner ear cell lines provide a convenient and consistent source of tissues that can be used to study cell death pathways that may be unique to specific cell types within the inner ear and they can be used to evaluate the efficacy of otoprotective compounds. Cochlear OC-k3 cells treated with cisplatin or gentamicin died by apoptosis in a dose- and time-dependent manner. Cell death was accompanied by an increase in reactive oxygen species and a decrease in glutathione antioxidant enzyme. Treatments with exogenous antioxidants N-acetylcysteine, glutathione, and vitamin C were effective in reducing cell death (Bertolaso et al. 2001). Gene microarray analysis performed on OC-k3 cells treated with cisplatin showed an upregulation of many genes involved with cellular respiration, detoxification, arachidonate mobilization, and lipid, peroxidation. A temporal profile of the expression changes would be useful in identifying the proapoptotic and antiapoptotic gene networks associated with cisplatin and other forms of ototoxicity.

6.5 Differentiation Programs

Inner ear cell lines derived from embryonic or neonatal tissues could prove useful in identifying intrinsic or extrinsic signals that determine cell fate or lineage.

Cells of the clonal cell line UB/UE-1 derived from supporting cells from the postnatal macula of the utricle are multipotent and differentiate into supporting cells or hair cells. Two hair cell phenotypes were identified on the basis of their physiological characteristics (Lawlor et al. 1999). When UB/UE-1 clones were transfected with E-cadherin it inhibited the expression of some, but not all, hair cell specific proteins (e.g., myosin VIIa and acetylcholine receptors). Other exogenous and intrinsic factors that guide differentiation need to be explored. Cell lines derived from the embryonic brain of Immortomouse have been transplanted into the damaged brain, where they integrate and differentiate in the host tissue (Gray et al. 2000). Clonal cell lines derived from different developmental stages and different regions of the inner ear have yet to be transplanted into the ear.

7. Stem Cells and Therapy

In the mid-1960s, Altman and Das presented the first autoradiographic evidence of cell proliferation and neurogenesis in the adult mammalian brain (Altman and Das 1965a, b). This finding contradicted the long-held dogma that neurogenesis does not occur in the adult brain. Publications followed that contested Altman's remarkable findings, and unfortunately this discovery languished for many years only to be rediscovered (Reynolds et al. 1992). During the past decade, research on stem cells and their potential use in regenerative medicine has exploded and raised expectations. Stem cells come in many different "flavors" and are "more or less flexible" depending on the stage of development and the tissue from which they were derived (Rivolta and Holley, Chapter 7). Totipotent cells derived from the fusion of sperm and egg cells can develop into any cell type. Pluripotent stem cells, progeny of totipotent cells, can generate endoderm, mesoderm, and ectoderm. Multipotent cells produce progeny within the same germ line.

   Embryonic stem cells derived from mammalian embryos are pluripotent and retain undifferentiated features. Coaxing these cells to differentiate into specific cell types is an area of great scientific interest because of potential clinical applications (Lovell-Badge 2001). Multipotent stem cells from fetal and adult animals have been identified in many different tissues; those from hematopoietic, neural, and mesenchymal sources have been most thoroughly studied. Purified hematopoietic stem cells have been used to repair liver damage; injection of purified hematopoietic stem cells gave rise to new, donor-derived hepatocytes in the liver and restored biochemical function (Lagasse et al. 2000). Multipotent adult progenitor cells derived from a subpopulation of mesenchymal stem cells are extremely plastic. In vitro, they can differentiate into mesoderm, endoderm, and neuroectoderm. When injected into the tail vein of mice, they engraft, proliferate, and differentiate into the epithelium of lung, gut, and liver and the hematopoietic system (Jiang et al. 2002). A major advantage of mesenchymal and hematopoietic stem cells is that they can be derived from the person requiring therapy, thereby minimizing rejection by the immune system. The major sources of neural stem cells are the olfactory bulb, hippocampus, and subventricular zone

(Doetsch et al. 1997; Kempermann and Gage 2000; Pagano et al. 2000). Neural stem cells can differentiate into neurons and glia and under certain conditions into mesoderm and endoderm (Bjornson et al. 1999; Galli et al. 2000; Hsieh et al. 2004). Neural stem cells have shown therapeutic value in a number of models. For example, intravenous injection of human neural stem cells into a rat model of ischemic stroke led to engraftment of stem in the damaged region, differentiation into neural phenotypes, and improved behavioral performance (Chu et al. 2004).

7.1 Stem Cells from the Inner Ear

There is no evidence for stem cells in the adult mammalian cochlea; however, a limited number of new hair cells have been found in damaged vestibular organs, suggesting the presence of a stem cell pool. Pluripotent stem cells, dereveved from the utricular epithelium of adult mammals, can give rise to mesoderm, endoderm, and ectoderm. The self-renewing stem cells cluster into spheres and express genes and proteins characteristic of cells in the developing inner ear and nervous system (Li ct al. 2003; Lou et al. 2007). In postnatal mammals, self-renewing stem cell spheres have been found from tissue isolated from the organ of Corti, vestibular sensory organs, stria vascularis, and spiral ganglion (Oshima et al. 2007). Stem cells from the sensory epithelium generate progenitors with physiological and immunocytochemical features of immature hair cells. Cochlear stem cell spheres and progenitor cell markers decline rapildy in the first few postnatal weeks; this decline is more gradual in the verstibular system (Oshima et al. 2007). An important area of future study is the identification of factors that promote the decline of the stem cell pools. Developmental studies with mice lacking fibroblast growth factor receptor 1 suggest that this receptor is needed to generate the prosensory cells that give rise to auditory hair cells (Pirvola et al. 2002). The prosensory cells express the *SOX2* gene, which maintains the proliferative potential of cells in the stem cell pool (Ellis et al. 2004; Bani-Yaghoub et al. 2006).

7.2 Transplantation of Stem Cells

Transplantation of stem cells into the inner ear is still in its infancy, although success with stem cell transplantation in the brain has generated optimism that this approach can be applied to the ear. Labeled, autologous bone marrow cells transplanted into the gentamicin-damaged inner ear engrafted throughout much of the cochlea and some engrafted cells expressed neuron-specific proteins (Naito et al. 2004) Mouse bone marrow mesenchymal stem cells transplanted into embryonic chick ear organ cultures integrated into the epithelium and some expressed hair cell markers (Jeon et al. 2007). This suggests that mesenchymal stem cells may have potential for autologous stem cell transplantation.

8. Summary

While there is considerable optimism regarding the therapeutic potential of stem cell transplantation and gene therapy to replace missing hair cells and spiral ganglion neurons, numerous obstacles are likely to lie ahead. The inner ear has a complex morphological architecture, and getting stems or progenitor cells to engraft into the region where they are needed, to differentiate into the correct cell type, and to establish normal function will be challenging. Hair cells must insert into the apical surface of the epithelium and develop stereocilia with the proper orientation and attachments to the overlying tectorial membrane. The basal pole of each OHC rests in the Deiters' cell cup. If the Deiters' cells are missing, as is often the case in profound hearing loss, regeneration and stem engraftment are likely to occur incorrectly and do more harm than good. Even if the hair cells could be replaced, they would need to form afferent and efferent connections with afferent spiral ganglion neurons and efferent neurons in the brain. The guidance cues available in an ear that has been deafened for many years may no longer be available or may function improperly. Prolonged deafness and loss of neurotrophic support leads to atrophy and loss of spiral ganglion neurons and vestibular ganglion neurons. Effective gene therapy or stem cell therapy may require pharmacologic interventions during the early stages of deafness to prevent the loss of neurons or combined approaches to replace both hair cells and neurons. Many obstacles and challenges lie ahead, but one need only look backwards in the footsteps of time to recall an era in which hair cell regeneration did not occur and stem cells did not exist in the adult nervous system or inner ear. Our imagination, combined with a strong scientific foundation, will one day lead to new cures for deafness.

Acknowledgment. This work was supported by NIH grant R01 DC 00630.

References

Adler HJ, Raphael Y (1996) New hair cells arise from supporting cell conversion in the acoustically damaged chick inner ear. Neurosci Lett 205:17–20.

Altman J, Das GD (1965a) Autoradiographic and histological evidence of postnatal hippocampal neurogenesis in rats. J Comp Neurol 124:319–335.

Altman J, Das GD (1965b) Post-natal origin of microneurones in the rat brain. Nature 207:953–956.

Baird RA, Torres MA, Schuff NR (1993) Hair cell regeneration in the bullfrog vestibular otolith organs following aminoglycoside toxicity. Hear Res 65:164–174.

Baird RA, Burton MD, Fashena DS, Naeger RA (2000) Hair cell recovery in mitotically blocked cultures of the bullfrog saccule. Proc Natl Acad Sci USA 97:11722–11729.

Balak KJ, Corwin JT, Jones JE (1990) Regenerated hair cells can originate from supporting cell progeny: evidence from photoxicity and laser ablation experiments in the lateral line system. J Neurosci 10:2502–2512.

Balkany TJ, Eshraghi AA, Jiao H, Polak M, Mou C, Dietrich DW, Van De Water TR (2005) Mild hypothermia protects auditory function during cochlear implant surgery. Laryngoscope 115:1543–1547.

Bani-Yaghoub M, Tremblay RG, Lei JX, Zhang D, Zurakowski B, Sandhu JK, Smith B, Ribecco-Lutkiewicz M, Kennedy J, Walker PR, Sikorska M (2006) Role of Sox2 in the development of the mouse neocortex. Dev Biol 295:52–66.

Barald KF, Lindberg KH, Hardiman K, Kavka AI, Lewis JE, Victor JC, Gardner CA, Poniatowski A (1997) Immortalized cell lines from embryonic avian and murine otocysts: tools for molecular studies of the developing inner ear. Int J Dev Neurosci 15:523–540.

Bartus RT, Elliott PJ, Hayward NJ, Dean RL, Harbeson S, Straub JA, Li Z, Powers JC (1995) Calpain as a novel target for treating acute neurodegenerative disorders. Neurol Res 17:249–258.

Beisel KW, Shiraki T, Morris KA, Pompeia C, Kachar B, Arakawa T, Bono H, Kawai J, Hayashizaki Y, Carninci P (2004) Identification of unique transcripts from a mouse full-length, subtracted inner ear cDNA library. Genomics 83:1012–1023.

Bermingham-McDonogh O, Stone JS, Reh TA, Rubel EW (2001) FGFR3 expression during development and regeneration of the chick inner ear sensory epithelia. Dev Biol 238:247–259.

Bertolaso L, Martini A, Bindini D, Lanzoni I, Parmeggiani A, Vitali C, Kalinec G, Kalinec F, Capitani S, Previati M (2001) Apoptosis in the OC-k3 immortalized cell line treated with different agents. Audiology 40:327–335.

Bhave SA, Stone JS, Rubel EW, Coltrera MD (1995) Cell cycle progression in gentamicin-damaged avian cochleas. J Neurosci 15:4618–4628.

Bitner-Glindzicz M (2002) Hereditary deafness and phenotyping in humans. Br Med Bull 63:73–94.

Bjornson CR, Rietze RL, Reynolds BA, Magli MC, Vescovi AL (1999) Turning brain into blood: a hematopoietic fate adopted by adult neural stem cells in vivo. Science 283:534–537.

Boettcher FA, Spongr VP, Salvi RJ (1992) Physiological and histological changes associated with the reduction in threshold shift during interrupted noise exposure. Hear Res 62:217–236.

Boettger T, Hubner CA, Maier H, Rust MB, Beck FX, Jentsch TJ (2002) Deafness and renal tubular acidosis in mice lacking the K-Cl co-transporter Kcc4. Nature 416:874–878.

Cafaro J, Lee GS, Stone JS (2007) Atoh1 expression defines activated progenitors and differentiating hair cells during avian hair cell regeneration. Dev Dyn 236:156–170.

Campbell KC, Rybak LP, Meech RP, Hughes L (1996) D-methionine provides excellent protection from cisplatin ototoxicity in the rat. Hear Res 102:90–98.

Canlon B (1997) Protection against noise trauma by sound conditioning. Ear Nose Throat J 76:248–250, 253–255.

Canlon B, Fransson A (1998) Reducing noise damage by using a mid-frequency sound conditioning stimulus. NeuroReport 9:269–274.

Canlon B, Borg E, Flock A (1988) Protection against noise trauma by pre-exposure to a low level acoustic stimulus. Hear Res 34:197–200.

Canlon B, Meltser I, Johansson P, Tahera Y (2007) Glucocorticoid receptors modulate auditory sensitivity to acoustic trauma. Hear Res 226:61–69.

Capranica RR (1978) Sound communication and auditory physiology in anurans. J Acoust Soc Am 64:S2.

Chen L, Salvi RJ, Hashino E (1993) Recovery of CAP threshold and amplitude in chickens following kanamycin ototoxicity. Hear Res 69:15–24.

Chen L, Trautwein PG, Miller K, Salvi RJ (1995) Effects of kanamycin ototoxicity and hair cell regeneration on the DC endocochlear potential in adult chickens. Hear Res 89:28–34.

Chen L, Trautwein PG, Shero M, Salvi RJ (1996) Tuning, spontaneous activity and tonotopic map in chicken cochlear ganglion neurons following sound-induced hair cell loss and regeneration. Hear Res 98:152–164.

Chen L, Sun W, Salvi RJ (2001) Electrically evoked otoacoustic emissions from the chicken ear. Hear Res 161:54–64.

Chen P, Segil N (1999) p27(Kip1) links cell proliferation to morphogenesis in the developing organ of Corti. Development 126:1581–1590.

Chen P, Zindy F, Abdala C, Liu F, Li X, Roussel MF, Segil N (2003) Progressive hearing loss in mice lacking the cyclin-dependent kinase inhibitor Ink4d. Nat Cell Biol 5:422–426.

Chen ZY, Corey DP (2002a) Understanding inner ear development with gene expression profiling. J Neurobiol 53:276–285.

Chen ZY, Corey DP (2002b) An inner ear gene expression database. J Assoc Res Otolaryngol 3:140–148.

Cho Y, Gong TW, Kanicki A, Altschuler RA, Lomax MI (2004) Noise overstimulation induces immediate early genes in the rat cochlea. Brain Res Mol Brain Res 130:134–148.

Chu K, Kim M, Park KI, Jeong SW, Park HK, Jung KH, Lee ST, Kang L, Lee K, Park DK, Kim SU, Roh JK (2004) Human neural stem cells improve sensorimotor deficits in the adult rat brain with experimental focal ischemia. Brain Res 1016:145–153.

Clark WW, Bohne BA, Boettcher FA (1987) Effect of periodic rest on hearing loss and cochlear damage following exposure to noise. J Acoust Soc Am 82:1253–1264.

Clerici WJ, Hensley K, DiMartino DL, Butterfield DA (1996) Direct detection of ototoxicant-induced reactive oxygen species generation in cochlear explants. Hear Res 98:116–124.

Coling DE, Yu KC, Somand D, Satar B, Bai U, Huang TT, Seidman MD, Epstein CJ, Mhatre AN, Lalwani AK (2003) Effect of *SOD1* overexpression on age- and noise-related hearing loss. Free Radic Biol Med 34:873–880.

Coling DE, Ding D, Young R, Lis M, Stofko E, Blumenthal KM, Salvi RJ (2007) Proteomic analysis of cisplatin-induced cochlear damage: methods and early changes in protein expression. Hear Res 226:140–156.

Corwin JT (1981) Postembryonic production and aging of inner ear hair cells in sharks. J Comp Neurol 201:541–553.

Corwin JT (1985) Perpetual production of hair cell and maturation changes in hair cells and maturation changes in hair cell ultrastructure accompany postembryoic growth in an amphibian ear. Proc Nat Acad Sci USA 82:3911–3915.

Corwin JT, Cotanche D (1988) Regeneration of sensory hair cells after acoustic trauma. Science 240:1771–1774.

Cotanche DA (1987) Regeneration of the tectorial membrane in the chick cochlea following severe acoustic trauma. Hear Res 30:197–206.

Cristobal R, Wackym PA, Cioffi JA, Erbe CB, Roche JP, Popper P (2005) Assessment of differential gene expression in vestibular epithelial cell types using microarray analysis. Brain Res Mol Brain Res 133:19–36.

Cruz RM, Lambert PR, Rubel EW (1987) Light microscopic evidence of hair cell regeneration after gentamicin toxicity in chick cochlea. Arch Otolaryngol Head Neck Surg 113:1058–1062.

Cullen KE, Sarge KD (1997) Characterization of hypothermia-induced cellular stress response in mouse tissues. J Biol Chem 272:1742–1746.

Cunningham LL, Brandon CS (2006) Heat shock inhibits both aminoglycoside- and cisplatin-induced sensory hair cell death. J Assoc Res Otolaryngol 7:299–307.

Cunningham LL, Cheng AG, Rubel EW (2002) Caspase activation in hair cells of the mouse utricle exposed to neomycin. J Neurosci 22:8532–8540.

Danial NN, Korsmeyer SJ (2004) Cell death: critical control points. Cell 116:205–219.

Davis RR, Newlander JK, Ling X, Cortopassi GA, Krieg EF, Erway LC (2001) Genetic basis for susceptibility to noise-induced hearing loss in mice. Hear Res 155:82–90.

Debnath J, Baehrecke EH, Kroemer G (2005) Does autophagy contribute to cell death? Autophagy 1:66–74.

Devarajan P, Savoca M, Castaneda MP, Park MS, Esteban-Cruciani N, Kalinec G, Kalinec F (2002) Cisplatin-induced apoptosis in auditory cells: role of death receptor and mitochondrial pathways. Hear Res 174:45–54.

Ding D, McFadden SL, Salvi RJ (2001) Calpain activation and morphological damage in chinchilla inner ears after carboplatin. J Assoc Res Otolaryngol 3:68–79.

Doetsch F, Garcia-Verdugo JM, Alvarez-Buylla A (1997) Cellular composition and three-dimensional organization of the subventricular germinal zone in the adult mammalian brain. J Neurosci 17:5046–5061.

Dooling RJ, Ryals BM, Manabe K (1997) Recovery of hearing and vocal behavior after hair-cell regeneration. Proc Natl Acad Sci USA 94:14206–14210.

Dooling RJ, Ryals BM, Dent ML, Reid TL (2006) Perception of complex sounds in budgerigars (*Melopsittacus undulatus*) with temporary hearing loss. J Acoust Soc Am 119:2524–2532.

Drescher DG (1974) Noise-induced reduction of inner-ear microphonic response: dependence on body temperature. Science 185:273–274.

Duckert LG, Rubel EW (1990) Ultrastructural observations on regenerating hair cells in the chick basilar papilla. Hear Res 48:161–182.

Duckert LG, Rubel EW (1993) Morphological correlates of functional recovery in the chicken inner ear after gentamicin treatment. J Comp Neurol 331:75–96.

Eddison M, Le Roux I, Lewis J (2000) Notch signaling in the development of the inner ear: lessons from *Drosophila*. Proc Natl Acad Sci USA 97:11692–11699.

Ellis P, Fagan BM, Magness ST, Hutton S, Taranova O, Hayashi S, McMahon A, Rao M, Pevny L (2004) SOX2, a persistent marker for multipotential neural stem cells derived from embryonic stem cells, the embryo or the adult. Dev Neurosci 26:148–165.

Ernfors P, Lee KF, Kucera J, Jaenisch R (1994) Lack of neurotrophin-3 leads to deficiencies in the peripheral nervous system and loss of limb proprioceptive afferents. Cell 77:503–512.

Ernfors P, Van De Water T, Loring J, Jaenisch R (1995) Complementary roles of BDNF and NT-3 in vestibular and auditory development. Neuron 14:1153–1164.

Eshraghi AA, Wang J, Adil E, He J, Zine A, Bublik M, Bonny C, Puel JL, Balkany TJ, Van De Water TR (2007) Blocking c-Jun-N-terminal kinase signaling can prevent hearing loss induced by both electrode insertion trauma and neomycin ototoxicity. Hear Res 226:168–177.

Fairfield DA, Lomax MI, Dootz GA, Chen S, Galecki AT, Benjamin IJ, Dolan DF, Altschuler RA (2005) Heat shock factor 1-deficient mice exhibit decreased recovery of hearing following noise overstimulation. J Neurosci Res 81:589–596.

Fekete DM, Wu DK (2002) Revisiting cell fate specification in the inner ear. Curr Opin Neurobiol 12:35–42.

Fetoni AR, Sergi B, Ferraresi A, Paludetti G, Troiani D (2004) Protective effects of alpha-tocopherol and tiopronin against cisplatin-induced ototoxicity. Acta Otolaryngol 124:421–426.

Forge A, Li L (2000) Apoptotic death of hair cells in mammalian vestibular sensory epithelia. Hear Res 139:97–115.

Forge A, Li L, Corwin JT, Nevill G (1993) Ultrastructural evidence for hair cell regeneration in the mammalian inner ear. Science 259:1616–1619.

Forge A, Li L, Nevill G (1998) Hair cell recovery in the vestibular sensory epithelia of mature guinea pigs. J Comp Neurol 397:69–88.

Forge A, Becker D, Casalotti S, Edwards J, Marziano N, Nevill G (2003) Gap junctions in the inner ear: comparison of distribution patterns in different vertebrates and assessment of connexin composition in mammals. J Comp Neurol 467:207–231.

Fritzsch B, Silos-Santiago I, Bianchi LM, Farinas I (1997) The role of neurotrophic factors in regulating the development of inner ear innervation. Trends Neurosci 20:159–164.

Fritzsch B, Tessarollo L, Coppola E, Reichardt LF (2004) Neurotrophins in the ear: their roles in sensory neuron survival and fiber guidance. Prog Brain Res 146:265–278.

Fuchs J, Huflejt ME, Rothfuss LM, Wilson DS, Carcamo G, Packer L (1989) Acute effects of near ultraviolet and visible light on the cutaneous antioxidant defense system. Photochem Photobiol 50:739–744.

Furman AC, Avissar M, Saunders JC (2006) The effects of intense sound exposure on phase locking in the chick (*Gallus domesticus*) cochlear nerve. Eur J Neurosci 24:2003–2010.

Galli R, Borello U, Gritti A, Minasi MG, Bjornson C, Coletta M, Mora M, De Angelis MG, Fiocco R, Cossu G, Vescovi AL (2000) Skeletal myogenic potential of human and mouse neural stem cells. Nat Neurosci 3:986–991.

Germiller JA, Smiley EC, Ellis AD, Hoff JS, Deshmukh I, Allen SJ, Barald KF (2004) Molecular characterization of conditionally immortalized cell lines derived from mouse early embryonic inner ear. Dev Dyn 231:815–827.

Girod DA, Duckert LG, Rubel EW (1989) Possible precursors of regenerated hair cells in the avian cochlea following acoustic trauma. Hear Res 42:175–194.

Girod DA, Tucci DL, Rubel EW (1991) Anatomical correlates of functional recovery in the avian inner ear following aminoglycoside ototoxicity. Laryngoscope 101:1139–1149.

Gitter AH (1990) A short history of hearing research. I. Antiquity. Laryngorhinootologie 69:442–445.

Gleich O, Dooling RJ, Manley GA (1994) Inner-ear abnormalities and their functional consequences in Belgian Waterslager canaries (*Serinus canarius*). Hear Res 79:123–136.

Goodyear R, Killick R, Legan PK, Richardson GP (1996) Distribution of beta-tectorin mRNA in the early posthatch and developing avian inner ear. Hear Res 96:167–178.

Gray JA, Grigoryan G, Virley D, Patel S, Sinden JD, Hodges H (2000) Conditionally immortalized, multipotential and multifunctional neural stem cell lines as an approach to clinical transplantation. Cell Transplant 9:153–168.

Grifa A, Wagner CA, D'Ambrosio L, Melchionda S, Bernardi F, Lopez-Bigas N, Rabionet R, Arbones M, Monica MD, Estivill X, Zelante L, Lang F, Gasparini P (1999) Mutations in *GJB6* cause nonsyndromic autosomal dominant deafness at DFNA3 locus. Nat Genet 23:16–18.

Guan MX, Fischel-Ghodsian N, Attardi G (2000) A biochemical basis for the inherited susceptibility to aminoglycoside ototoxicity. Hum Mol Genet 9:1787–1793.

Halliwell B, Gutteridge JM, Cross CE (1992) Free radicals, antioxidants, and human disease: where are we now? J Lab Clin Med 119:598–620.

Hashino E, Salvi RJ (1993) Changing spatial patterns of DNA replication in the noise-damaged chick cochlea. J Cell Sci 105:23–31.

Hashino E, Sokabe M, Miyamoto K (1988) Frequency specific susceptibility to acoustic trauma in the budgerigar (*Melopsittacus undulatus*). J Acoust Soc Am 83:2450–2453.

Hashino E, Tanaka Y, Salvi RJ, Sokabe M (1992) Hair cell regeneration in the adult budgerigar after kanamycin ototoxicity. Hear Res 59:46–58.

Hashino E, TinHan EK, Salvi RJ (1995) Base-to-apex gradient of cell proliferation in the chick cochlea following kanamycin-induced hair cell loss. Hear Res 88:156–168.

Hasson T, Gillespie PG, Garcia JA, MacDonald RB, Zhao Y, Yee AG, Mooseker MS, Corey DP (1997) Unconventional myosins in inner-ear sensory epithelia. J Cell Biol 137:1287–1307.

He DZ, Beisel KW, Chen L, Ding DL, Jia S, Fritzsch B, Salvi R (2003) Chick hair cells do not exhibit voltage-dependent somatic motility. J Physiol 546:511–520.

Henry KR, Chole RA (1984) Hypothermia protects the cochlea from noise damage. Hear Res 16:225–230.

Henry WJ, Makaretz M, Saunders JC, Schneider ME, Vrettakos P (1988) Hair cell loss and regeneration after exposure to intense sound in neonatal chicks. Otolaryngol Head Neck Surg 98:607–611.

Hildebrand MS, de Silva MG, Klockars T, Campbell CA, Smith RJ, Dahl HH (2007) Gene expression profiling analysis of the inner ear. Hear Res 225:1–10.

Hofstetter P, Ding D, Powers N, Salvi RJ (1997) Quantitative relationship of carboplatin dose to magnitude of inner and outer hair cell loss and the reduction in distortion product otoacoustic emission amplitude in chinchillas. Hear Res 112:199–215.

Housby JN, Cahill CM, Chu B, Prevelige R, Bickford K, Stevenson MA, Calderwood SK (1999) Non-steroidal anti-inflammatory drugs inhibit the expression of cytokines and induce HSP70 in human monocytes. Cytokine 11:347–358.

Hsieh J, Aimone JB, Kaspar BK, Kuwabara T, Nakashima K, Gage FH (2004) IGF-I instructs multipotent adult neural progenitor cells to become oligodendrocytes. J Cell Biol 164:111–122.

Hu BH, Zheng XY, McFadden SL, Kopke RD, Henderson D (1997) R-phenylisop-ropyladenosine attenuates noise-induced hearing loss in the chinchilla. Hear Res 113:198–206.

Hu BH, Henderson D, Nicotera TM (2006) Extremely rapid induction of outer hair cell apoptosis in the chinchilla cochlea following exposure to impulse noise. Hear Res 211:16–25.

Huard JM, Youngentob SL, Goldstein BJ, Luskin MB, Schwob JE (1998) Adult olfactory epithelium contains multipotent progenitors that give rise to neurons and non-neural cells. J Comp Neurol 400:469–486.

Hudspeth AJ, Choe Y, Mehta AD, Martin P (2000) Putting ion channels to work: mecha-noelectrical transduction, adaptation, and amplification by hair cells. Proc Natl Acad Sci USA 97:11765–11772.

Ipakchi R, Kyin T, Saunders JC (2005) Loss and recovery of sound-evoked otoacoustic emissions in young chicks following acoustic trauma. Audiol Neurootol 10:209–219.

Izumikawa M, Minoda R, Kawamoto K, Abrashkin KA, Swiderski DL, Dolan DF, Brough DE, Raphael Y (2005) Auditory hair cell replacement and hearing improvement by *Atoh1* gene therapy in deaf mammals. Nat Med 11:271–276.

Jacono AA, Hu B, Kopke RD, Henderson D, Van De Water TR, Steinman HM (1998) Changes in cochlear antioxidant enzyme activity after sound conditioning and noise exposure in the chinchilla. Hear Res 117:31–38.

Jeon SJ, Oshima K, Heller S, Edge AS (2007) Bone marrow mesenchymal stem cells are progenitors in vitro for inner ear hair cells. Mol Cell Neurosci 34:59–68.

Jeong HJ, Hong SH, Park RK, Shin T, An NH, Kim HM (2005a) Hypoxia-induced IL-6 production is associated with activation of MAP kinase, HIF-1, and NF-kappaB on HEI-OC1 cells. Hear Res 207:59–67.

Jeong HJ, Kim JB, Hong SH, An NH, Kim MS, Park BR, Park RK, Kim HM (2005b) Vascular endothelial growth factor is regulated by hypoxic stress via MAPK and HIF-1 alpha in the inner ear. J Neuroimmunol 163:84–91.

Jiang H, Sha SH, Forge A, Schacht J (2006) Caspase-independent pathways of hair cell death induced by kanamycin in vivo. Cell Death Differ 13:20–30.

Jiang Y, Jahagirdar BN, Reinhardt RL, Schwartz RE, Keene CD, Ortiz-Gonzalez XR, Reyes M, Lenvik T, Lund T, Blackstad M, Du J, Aldrich S, Lisberg A, Low WC, Largaespada DA, Verfaillie CM (2002) Pluripotency of mesenchymal stem cells derived from adult marrow. Nature 418:41–49.

Johnson KR, Erway LC, Cook SA, Willott JF, Zheng QY (1997) A major gene affecting age-related hearing loss in C57BL/6J mice. Hear Res 114:83–92.

Jones JE, Corwin JT (1993) Replacement of lateral line sensory organs during tail regeneration in salamanders: identification of progenitor cells and analysis of leukocyte activity. J Neurosci 13:1022–1034.

Jones JE, Corwin JT (1996) Regeneration of sensory cells after laser ablation in the lateral line system: hair cell lineage and macrophage behavior revealed by time-lapsed video microscopy. J Neurosci 16:649–662.

Jargensen JM, Mathiesen C (1988) The avian inner ear. Continuous production of hair cells in vestibular sensory organs, but not in the auditory papilla. Naturwissenschaften 75:319–320.

Kageyama R, Ohtsuka T (1999) The Notch-Hes pathway in mammalian neural development. Cell Res 9:179–188.

Kageyama R, Ohtsuka T, Hatakeyama J, Ohsawa R (2005) Roles of bHLH genes in neural stem cell differentiation. Exp Cell Res 306:343–348.

Katayama A, Corwin JT (1989) Cell production in the chicken cochlea. J Comp Neurol 281:129–135.

Katoh I, Tomimori Y, Ikawa Y, Kurata S (2004) Dimerization and processing of procaspase-9 by redox stress in mitochondria. J Biol Chem 279:15515–15523.

Keithley EM, Canto C, Zheng QY, Fischel-Ghodsian N, Johnson KR (2004) Age-related hearing loss and the ahl locus in mice. Hear Res 188:21–28.

Kelley MW (2003) Cell adhesion molecules during inner ear and hair cell development, including notch and its ligands. Curr Top Dev Biol 57:321–356.

Kempermann G, Gage FH (2000) Neurogenesis in the adult hippocampus. Novartis Found Symp 231:220–235; discussion 235–241, 302–306.

Kil J, Warchol ME, Corwin JT (1997) Cell death, cell proliferation, and estimates of hair cell life spans in the vestibular organs of chicks. Hear Res 114:117–126.

Kopke RD, Weisskopf PA, Boone JL, Jackson RL, Wester DC, Hoffer ME, Lambert DC, Charon CC, Ding DL, McBride D (2000) Reduction of noise-induced hearing loss using L-NAC and salicylate in the chinchilla. Hear Res 149:138–146.

Kopke RD, Coleman JK, Liu J, Campbell KC, Riffenburgh RH (2002) Candidate's thesis: enhancing intrinsic cochlear stress defenses to reduce noise-induced hearing loss. Laryngoscope 112:1515–1532.

Lagasse E, Connors H, Al-Dhalimy M, Reitsma M, Dohse M, Osborne L, Wang X, Finegold M, Weissman IL, Grompe M (2000) Purified hematopoietic stem cells can differentiate into hepatocytes in vivo. Nat Med 6:1229–1234.

Lawlor P, Marcotti W, Rivolta MN, Kros CJ, Holley MC (1999) Differentiation of mammalian vestibular hair cells from conditionally immortal, postnatal supporting cells. J Neurosci 19:9445–9458.

Lawoko-Kerali G, Rivolta MN, Lawlor P, Cacciabue-Rivolta DI, Langton-Hewer C, van Doorninck JH, Holley MC (2004) GATA3 and NeuroD distinguish auditory and vestibular neurons during development of the mammalian inner ear. Mech Dev 121:287–299.

Lee KH, Cotanche DA (1996) Localization of the hair-cell-specific protein fimbrin during regeneration in the chicken cochlea. Audiol Neurootol 1:41–53.

Lewis ER, Li CW (1973) Evidence concerning the morphogenesis of saccular receptors in the bullfrog (*Rana catesbeiana*). J Morphol 139:351–361.

Li CW, Lewis ER (1979) Structure and development of vestibular hair cells in the larval bullfrog. Ann Otol Rhinol Laryngol 88:427–437.

Li H, Liu H, Heller S (2003) Pluripotent stem cells from the adult mouse inner ear. Nat Med 9:1293–1299.

Li S, Price SM, Cahill H, Ryugo DK, Shen MM, Xiang M (2002) Hearing loss caused by progressive degeneration of cochlear hair cells in mice deficient for the *Barhl1* homeobox gene. Development 129:3523–3532.

Liberman MC, Gao J, He DZ, Wu X, Jia S, Zuo J (2002) Prestin is required for electro-motility of the outer hair cell and for the cochlear amplifier. Nature 419:300–304.

Lifshitz J, Furman AC, Altman KW, Saunders JC (2004) Spatial tuning curves along the chick basilar papilla in normal and sound-exposed ears. J Assoc Res Otolaryngol 5:171–184.

Lim HH, Jenkins OH, Myers MW, Miller JM, Altschuler RA (1993) Detection of HSP 72 synthesis after acoustic overstimulation in rat cochlea. Hear Res 69:146–150.

Liu W, Staecker H, Stupak H, Malgrange B, Lefebvre P, Van De Water TR (1998) Caspase inhibitors prevent cisplatin-induced apoptosis of auditory sensory cells. NeuroReport 9:2609–2614.

Liu X, Mohamed JA, Ruan R (2004) Analysis of differential gene expression in the cochlea and kidney of mouse by cDNA microarrays. Hear Res 197:35–43.

Lomax MI, Huang L, Cho Y, Gong TL, Altschuler RA (2000) Differential display and gene arrays to examine auditory plasticity. Hear Res 147:293–302.

Lomax MI, Gong TW, Cho Y, Huang L, Oh SH, Adler HJ, Raphael Y, Altschuler RA (2001) Differential gene expression following noise trauma in birds and mammals. Noise Health 3:19–35.

Lombarte A, Popper AN (1994) Quantitative analyses of postembryonic hair cell addition in the otolithic endorgans of the inner ear of the European hake, *Merluccius merluccius* (Gadiformes, Teleostei). J Comp Neurol 345:419–428.

Lou X, Zhang Y, Yuan C (2007) Multipotent stem cells from the young rat inner ear. Neurosci Lett 416:28–33.

Lovell-Badge R (2001) The future for stem cell research. Nature 414:88–91.

Lynch ED, Gu R, Pierce C, Kil J (2004) Ebselen-mediated protection from single and repeated noise exposure in rat. Laryngoscope 114:333–337.

Lynch ED, Gu R, Pierce C, Kil J (2005) Reduction of acute cisplatin ototoxicity and nephrotoxicity in rats by oral administration of allopurinol and ebselen. Hear Res 201:81–89.

Manabe K, Sadr EI, Dooling RJ (1998) Control of vocal intensity in budgerigars (*Melopsittacus undulatus*): differential reinforcement of vocal intensity and the Lombard effect. J Acoust Soc Am 103:1190–1198.

Marean GC, Burt JM, Beecher MD, Rubel EW (1993) Hair cell regeneration in the European starling (*Sturnus vulgaris*): recovery of pure-tone detection thresholds. Hear Res 71:125–136.

Marean GC, Burt JM, Beecher MD, Rubel EW (1998) Auditory perception following hair cell regeneration in European starling (*Sturnus vulgaris*): frequency and temporal resolution. J Acoust Soc Am 103:3567–3580.

Matsui JI, Ogilvie JM, Warchol ME (2002) Inhibition of caspases prevents ototoxic and ongoing hair cell death. J Neurosci 22:1218–1227.

Matsui JI, Haque A, Huss D, Messana EP, Alosi JA, Roberson DW, Cotanche DA, Dickman JD, Warchol ME (2003) Caspase inhibitors promote vestibular hair cell survival and function after aminoglycoside treatment in vivo. J Neurosci 23:6111–6122.

Matsui JI, Gale JE, Warchol ME (2004) Critical signaling events during the aminoglycoside-induced death of sensory hair cells in vitro. J Neurobiol 61:250–266.

McFadden SL, Henderson D, Shen YH (1997) Low-frequency 'conditioning' provides long-term protection from noise-induced threshold shifts in chinchillas. Hear Res 103:142–150.

McFadden SL, Ding D, Burkard RF, Jiang H, Reaume AG, Flood DG, Salvi RJ (1999) Cu/Zn SOD deficiency potentiates hearing loss and cochlear pathology in aged 129,CD-1 mice. J Comp Neurol 413:101–112.

McFadden SL, Ding D, Salvemini D, Salvi RJ (2003) M40403, a superoxide dismutase mimetic, protects cochlear hair cells from gentamicin, but not cisplatin toxicity. Toxicol Appl Pharmacol 186:46–54.

McFadden SL, Ding D, Jiang H, Salvi RJ (2004) Time course of efferent fiber and spiral ganglion cell degeneration following complete hair cell loss in the chinchilla. Brain Res 997:40–51.

Michils A, Redivo M, Zegers de Beyl V, de Maertelaer V, Jacobovitz D, Rocmans P, Duchateau J (2001) Increased expression of high but not low molecular weight heat shock proteins in resectable lung carcinoma. Lung Cancer 33:59–67.

Miller JD, Watson C, Covell W (1963) Deafening effects of noise on the cat. Acta Otolaryngologica (Stockholm) Suppl 176:1–91.

Monks J, Rosner D, Geske FJ, Lehman L, Hanson L, Neville MC, Fadok VA (2005) Epithelial cells as phagocytes: apoptotic epithelial cells are engulfed by mammary alveolar epithelial cells and repress inflammatory mediator release. Cell Death Differ 12:107–114.

Montcouquiol M, Corwin JT (2001a) Intracellular signals that control cell proliferation in mammalian balance epithelia: key roles for phosphatidylinositol-3 kinase, mammalian target of rapamycin, and S6 kinases in preference to calcium, protein kinase C, and mitogen-activated protein kinase. J Neurosci 21:570–580.

Montcouquiol M, Corwin JT (2001b) Brief treatments with forskolin enhance s-phase entry in balance epithelia from the ears of rats. J Neurosci 21:974–982.

Morris KA, Snir E, Pompeia C, Koroleva IV, Kachar B, Hayashizaki Y, Carninci P, Soares MB, Beisel KW (2005) Differential expression of genes within the cochlea as defined by a custom mouse inner ear microarray. J Assoc Res Otolaryngol 6:75–89.

Muller M, Smolders JW, Ding-Pfennigdorff D, Klinke R (1996) Regeneration after tall hair cell damage following severe acoustic trauma in adult pigeons: correlation between cochlear morphology, compound action potential responses and single fiber properties in single animals. Hear Res 102:133–154.

Muller M, Smolders JW, Ding-Pfennigdorff D, Klinke R (1997) Discharge properties of pigeon single auditory nerve fibers after recovery from severe acoustic trauma. Int J Dev Neurosci 15:401–416.

Murray RC, Calof AL (1999) Neuronal regeneration: lessons from the olfactory system. Semin Cell Dev Biol 10:421–431.

Naito Y, Nakamura T, Nakagawa T, Iguchi F, Endo T, Fujino K, Kim TS, Hiratsuka Y, Tamura T, Kanemaru S, Shimizu Y, Ito J (2004) Transplantation of bone marrow stromal cells into the cochlea of chinchillas. NeuroReport 15:1–4.

Navaratnam DS, Su HS, Scott SP, Oberholtzer JC (1996) Proliferation in the auditory receptor epithelium mediated by a cyclic AMP-dependent signaling pathway. Nat Med 2:1136–1139.

Ni YQ, Tang H, Fu WS (2005) Expression of heat shock protein 70 mRNA in guinea pig cochlea with ototoxicity of gentamicin. Sheng Li Xue Bao 57:328–332.

Niemiec AJ, Raphael Y, Moody DB (1994a) Return of auditory function following structural regeneration after acoustic trauma: behavioral measures from quail. Hear Res 75:209–224.

Niemiec AJ, Raphael Y, Moody DB (1994b) Return of auditory function following structural regeneration after acoustic trauma: behavioral measures from quail. Hear Res 79:1–16.

Noble M, Groves AK, Ataliotis P, Jat PS (1992) From chance to choice in the generation of neural cell lines. Brain Pathol 2:39–46.

Oesterle EC, Tsue TT, Rubel EW (1997) Induction of cell proliferation in avian inner ear sensory epithelia by insulin-like growth factor-I and insulin. J Comp Neurol 380:262–274.

Oesterle EC, Bhave SA, Coltrera MD (2000) Basic fibroblast growth factor inhibits cell proliferation in cultured avian inner ear sensory epithelia. J Comp Neurol 424:307–326.

Ohlemiller KK, Wright JS, Dugan LL (1999a) Early elevation of cochlear reactive oxygen species following noise exposure. Audiol Neurootol 4:229–236.

Ohlemiller KK, McFadden SL, Ding DL, Flood DG, Reaume AG, Hoffman EK, Scott RW, Wright JS, Putcha GV, Salvi RJ (1999b) Targeted deletion of the cytosolic Cu/Zn-superoxide dismutase gene (Sod1) increases susceptibility to noise-induced hearing loss. Audiol Neuro-Otol 4:237–246.

Oshima K, Grimm CM, Corrales CE, Senn P, Martinez Monedero R, Geleoc GS, Edge A, Holt JR, Heller S (2007) Differential distribution of stem cells in the auditory and vestibular organs of the inner ear. J Assoc Res Otolaryngol 8:18–31.

Pagano SF, Impagnatiello F, Girelli M, Cova L, Grioni E, Onofri M, Cavallaro M, Etteri S, Vitello F, Giombini S, Solero CL, Parati EA (2000) Isolation and characterization of neural stem cells from the adult human olfactory bulb. Stem Cells 18:295–300.

Pirvola U, Xing-Qun L, Virkkala J, Saarma M, Murakata C, Camoratto AM, Walton KM, Ylikoski J (2000) Rescue of hearing, auditory hair cells, and neurons by CEP-1347/KT7515, an inhibitor of c-Jun N-terminal kinase activation. J Neurosci 20:43–50.

Pirvola U, Ylikoski J, Trokovic R, Hebert JM, McConnell SK, Partanen J (2002) FGFR1 is required for the development of the auditory sensory epithelium. Neuron 35:671–680.

Poje CP, Sewell DA, Saunders JC (1995) The effects of exposure to intense sounds on the DC endocochlear potential in the chick. Hear Res 82:197–204.

Pompeia C, Hurle B, Belyantseva IA, Noben-Trauth K, Beisel K, Gao J, Buchoff P, Wistow G, Kachar B (2004) Gene expression profile of the mouse organ of Corti at the onset of hearing. Genomics 83:1000–1011.

Popper AN, Hoxter B (1984) Growth of a fish ear: 1. Quantitative analysis of hair cell and ganglion cell proliferation. Hear Res 15:133–142.

Powles N, Babbs C, Ficker M, Schimmang T, Maconochie M (2004) Identification and analysis of genes from the mouse otic vesicle and their association with developmental subprocesses through in situ hybridization. Dev Biol 268:24–38.

Presson JC (1994) Immunocytochemical reactivities of precursor cells and their progeny in the ear of a cichlid fish. Hear Res 80:1–9.

Presson JC, Smith T, Mentz L (1995) Proliferating hair cell precursors in the ear of a postembryonic fish are replaced after elimination by cytosine arabinoside. J Neurobiol 26:579–584.

Prezant TR, Agaian JV, Bohlman MC (1993) Mitochondrial ribosomal RNA mutation associated with both antibiotic-induced and non-syndromic deafness. Nat Genet 4:289–293.

Reggiori F, Klionsky DJ (2005) Autophagosomes: biogenesis from scratch? Curr Opin Cell Biol 17:415–422.

Reynolds BA, Tetzlaff W, Weiss S (1992) A multipotent EGF-responsive striatal embryonic progenitor cell produces neurons and astrocytes. J Neurosci 12:4565–4574.

Richardson GP, Russell IJ (1991) Cochlear cultures as a model system for studying aminoglycoside induced ototoxicity. Hear Res 53:293–311.

Ritossa F (1996) Discovery of the heat shock response. Cell Stress Chaperones 1:97–98.

Rivolta MN, Holley MC (2002) Cell lines in inner ear research. J Neurobiol 53:306–318.

Rivolta MN, Grix N, Lawlor P, Ashmore JF, Jagger DJ, Holley MC (1998) Auditory hair cell precursors immortalized from the mammalian inner ear. Proc Biol Sci 265:1595–1603.

Roberson DW, Kreig CS, Rubel EW (1996) Light microscopic evidence that direct trans-differentiation gives rise to new hair cells in regenerating avian auditory epithelium. Audit Neurosci 2:195–205.

Ruben RJ (1967) Development of the inner ear of the mouse: a radioautographic study of terminal mitoses. Acta Otolaryngol Suppl. 220:1–44.

Ryals BM, Rubel EW (1988) Hair cell regeneration after acoustic trauma in adult Coturnix quail. Science 240:1774–1776.

Ryals BM, Westbrook EW (1994) TEM analysis of neural terminals on autoradiographi-cally identified regenerated hair cells. Hear Res 72:81–88.

Ryals BM, Ten Eyck B, Westbrook EW (1989) Ganglion cell loss continues during hair cell regeneration. Hear Res 43:81–90.

Ryals BM, Dooling RJ, Westbrook E, Dent ML, MacKenzie A, Larsen ON (1999) Avian species differences in susceptibility to noise exposure. Hear Res 131:71–88.

Rzadzinska AK, Schneider ME, Davies C, Riordan GP, Kachar B (2004) An actin molecular treadmill and myosins maintain stereocilia functional architecture and self-renewal. J Cell Biol 164:887–897.

Sage C, Huang M, Karimi K, Gutierrez G, Vollrath MA, Zhang DS, Garcia-Anoveros J, Hinds PW, Corwin JT, Corey DP, Chen ZY (2005) Proliferation of functional hair cells in vivo in the absence of the retinoblastoma protein. Science 307:1114–1118.

Salvi RJ, Henderson D, Hamernik RP, Ahroon WA (1983) Neural correlates of sensorineural hearing loss. Ear Hear 4:115–129.

Saunders J, Dooling R (1974) Noise-induced threshold shift in the parakeet (*Melopsittacus undulatus*). Proc Natl Acad Sci USA 71:1962–1965.

Saunders JC, Doan DE, Poje CP, Fisher KA (1996) Cochlear nerve activity after intense sound exposure in neonatal chicks. J Neurophysiol 76:770–787.

Saunders SS, Salvi RJ (1993) Psychoacoustics of normal adult chickens: thresholds and temporal integration. J Acoust Soc Am 94:83–90.

Saunders SS, Salvi RJ (1995) Pure tone masking patterns in adult chickens before and after recovery from acoustic trauma. J Acoust Soc Am 98:1365–1371.

Saunders SS, Salvi RJ, Miller KM (1995) Recovery of thresholds and temporal integration in adult chickens after high-level 525-Hz pure-tone exposure. J Acoust Soc Am 97:1150–1164.

Savill J, Fadok V, Henson P, Haslett C (1993) Phagocyte recognition of cells undergoing apoptosis. Immunol Today 14:131–136.

Schneider ME, Belyantseva IA, Azevedo RB, Kachar B (2002) Rapid renewal of auditory hair bundles. Nature 418:837–838.

Schwob JE (2002) Neural regeneration and the peripheral olfactory system. Anat Rec 269:33–49.

Sente M (2004) The history of audiology. Med Pregl 57:611–616.

Severinsen S, Jørgensen J, Nyengaard J (2003) Structure and growth of the utricular macula in the inner ear of the slider turtle *Trachemys scripta*. J Assoc Res Otolaryngol 4:505–520.

Sha SH, Schacht J (1999) Salicylate attenuates gentamicin-induced ototoxicity. Lab Invest 79:807–813.

Sha SH, Zajic G, Epstein CJ, Schacht J (2001a) Overexpression of copper/zinc-superoxide dismutase protects from kanamycin-induced hearing loss. Audiol Neurootol 6:117–123.

Sha SH, Taylor R, Forge A, Schacht J (2001b) Differential vulnerability of basal and apical hair cells is based on intrinsic susceptibility to free radicals. Hear Res 155:1–8.

Shou J, Zheng JL, Gao WQ (2003) Robust generation of new hair cells in the mature mammalian inner ear by adenoviral expression of Hath1. Mol Cell Neurosci 23:169–179.

Sliwinska-Kowalska M, Rzadzinska A, Jedlinska U, Rajkowska E (2000) Hair cell regeneration in the chick basilar papilla after exposure to wide-band noise: evidence for ganglion cell involvement. Hear Res 148:197–212.

Sobkowicz HM, Slapnick SM, August BK (1995) The kinocilium of auditory hair cells and evidence for its morphogenetic role during the regeneration of stereocilia and cuticular plates. J Neurocytol 24:633–653.

Sobkowicz HM, August BK, Slapnick SM(1996) Post-traumatic survival and recovery of the auditory sensory cells in culture. Acta Oto-Laryngol 116:257–262.

Speidel C (1947) Correlated studies of sense organs and nerves of the lateral-line in living frog tadpoles I. Regeneration of denervated organs. J Comp Neurol 87: 29–55.

Steyger PS, Burton M, Hawkins JR, Schuff NR, Baird RA (1997) Calbindin and parvalbumin are early markers of non-mitotically regenerating hair cells in the bullfrog vestibular otolith organs. Int J Dev Neurosci 15:417–432.

Stone JS, Rubel EW (2000) Temporal, spatial, and morphologic features of hair cell regeneration in the avian basilar papilla. J Comp Neurol 417:1–16.

Stone JS, Leano SG, Baker LP, Rubel EW (1996) Hair cell differentiation in chick cochlear epithelium after aminoglycoside toxicity: in vivo and in vitro observations. J Neurosci 16:6157–6174.

Stone LS (1933) The development of lateral-line sense organs in amphibians observed in living and vital-stained preparations. J Comp Neurol 57:507–540.

Stone LS (1937) Further experimental studies of the development of lateral-line sense organs in amphibians observed in living preparations. J Comp Neurol 68:83–115.

Subramaniam M, Henderson D, Spongr V (1991) Frequency differences in the development of protection against NIHL by low level "toughening" exposure. J Acoust Soc Am 89:1865.

Sun H, Hashino E, Ding DL, Salvi RJ (2001) Reversible and irreversible damage to cochlear afferent neurons by kainic acid excitotoxicity. J Comp Neurol 430:172–181.

Taggart RT, McFadden SL, Ding DL, Henderson D, Jin X, Sun W, Salvi R (2001) Gene expression changes in chinchilla cochlea from noise-induced temporary threshold shift. Noise Health 3:1–18.

Tahera Y, Meltser I, Johansson P, Bian Z, Stierna P, Hansson AC, Canlon B (2006) NF-kappaB mediated glucocorticoid response in the inner ear after acoustic trauma. J Neurosci Res 83:1066–1076.

Tang LS, Alger HM, Pereira FA (2006) COUP-TFI controls Notch regulation of hair cell and support cell differentiation. Development 133:3683–3693.

Taylor RR, Forge A (2005) Hair cell regeneration in sensory epithelia from the inner ear of a urodele amphibian. J Comp Neurol 484:105–120.

Tilney LG, Tilney MS, Saunders JS, DeRosier DJ (1986) Actin filaments, stereocilia, and hair cells of the bird cochlea. III. The development and differentiation of hair cells and stereocilia. Dev Biol 116:100–118.

Trautwein P, Salvi RJ, Miller K, Shero M, Hashino E (1996) Incomplete recovery of chicken distortion product otoacoustic emissions following acoustic overstimulation. Audiol Neurootol 1:86–103.

Trautwein PG, Chen L, Salvi RJ (1997) Steady state EP is not responsible for hearing loss in adult chickens following acoustic trauma. Hear Res 110:266–270.

Tucci DL, Rubel EW (1990) Physiologic status of regenerated hair cells in the avian inner ear following aminoglycoside ototoxicity. Otolaryngol Head Neck Surg 103:443–450.

Vahava O, Morell R, Lynch ED, Weiss S, Kagan ME, Ahituv N, Morrow JE, Lee MK, Skvorak AB, Morton CC, Blumenfeld A, Frydman M, Friedman TB, King MC, Avraham KB (1998) Mutation in transcription factor POU4F3 associated with inherited progressive hearing loss in humans. Science 279:1950–1954.

Wang J, Ding D, Shulman A, Stracher A, Salvi RJ (1999) Leupeptin protects sensory hair cells from acoustic trauma. NeuroReport 10:811–816.

Wang J, Van De Water TR, Bonny C, de Ribaupierre F, Puel JL, Zine A (2003) A peptide inhibitor of c-Jun N-terminal kinase protects against both aminoglycoside and acoustic trauma-induced auditory hair cell death and hearing loss. J Neurosci 23:8596–8607.

Wang J, Ladrech S, Pujol R, Brabet P, Van De Water TR, Puel JL (2004) Caspase inhibitors, but not c-Jun NH_2-terminal kinase inhibitor treatment, prevent cisplatin-induced hearing loss. Cancer Res 64:9217–9224.

Warchol ME (1999) Immune cytokines and dexamethasone influence sensory regeneration in the avian vestibular periphery. J Neurocytol 28:889–900.

Warchol ME (2002) Cell density and N-cadherin interactions regulate cell proliferation in the sensory epithelia of the inner ear. J Neurosci 22:2607–2616.

Warchol ME, Lambert PR, Goldstein BJ, Forge A, Corwin JT (1993) Regenerative proliferation in inner ear sensory epithelia from adult guinea pigs and humans. Science 259:1619–1622.

Watanabe F, Koga K, Hakuba N, Gyo K (2001) Hypothermia prevents hearing loss and progressive hair cell loss after transient cochlear ischemia in gerbils. Neurosci 102:639–645.

Watson GM, Graugnard EM, Mire P (2007) The involvement of Arl-5b in the Repair of Hair Cells in Sea Anemones. J Assoc Res Otolaryngol 8:183–193.

Williams JA, Holder N (2000) Cell turnover in neuromasts of zebrafish larvae. Hear Res 143:171–181.

Winter J, Jakob U (2004) Beyond transcription—new mechanisms for the regulation of molecular chaperones. Crit Rev Biochem Mol Biol 39:297–317.

Witte MC, Montcouquiol M, Corwin JT (2001) Regeneration in avian hair cell epithelia: identification of intracellular signals required for S-phase entry. Eur J Neurosci 14:829–838.

Wright M (1947) Regeneration and degeneration experiments on lateral line nerves and sense organs in anurans. J Exp Zool 105:221–257.

Wu WJ, Sha SH, McLaren JD, Kawamoto K, Raphael Y, Schacht J (2001) Aminoglycoside ototoxicity in adult CBA, C57BL and BALB mice and the Sprague-Dawley rat. Hear Res 158:165–178.

Xia JH, Liu CY, Tang BS, Pan Q, Huang L, Dai HP, Zhang BR, Xie W, Hu DX, Zheng D, Shi XL, Wang DA, Xia K, Yu KP, Liao XD, Feng Y, Yang YF, Xiao JY, Xie DH, Huang JZ (1998) Mutations in the gene encoding gap junction protein beta-3 associated with autosomal dominant hearing impairment. Nat Genet 20:370–373.

Xu SA, Shepherd RK, Chen Y, Clark GM (1993) Profound hearing loss in the cat following the single co-administration of kanamycin and ethacrynic acid. Hear Res 70:205–215.

Yamamoto N, Tanigaki K, Tsuji M, Yabe D, Ito J, Honjo T (2006) Inhibition of Notch/RBP-J signaling induces hair cell formation in neonate mouse cochleas. J Mol Med 84:37–45.

Yamasoba T, Dolan DF, Miller JM (1999) Acquired resistance to acoustic trauma by sound conditioning is primarily mediated by changes restricted to the cochlea, not by systemic responses. Hear Res 127:31–40.

Ylikoski J, Xing-Qun L, Virkkala J, Pirvola U (2002) Blockade of c-Jun N-terminal kinase pathway attenuates gentamicin-induced cochlear and vestibular hair cell death. Hear Res 166:33–43.

Yoshida N, Kristiansen A, Liberman MC (1999) Heat stress and protection from permanent acoustic injury in mice. J Neurosci 19:10116–10124.

Zhao Y, Yamoah EN, Gillespie PG (1996) Regeneration of broken tip links and restoration of mechanical transduction in hair cells. Proc Natl Acad Sci USA 93:15469–15474.

Zheng J, Shen W, He DZ, Long KB, Madison LD, Dallos P (2000a) Prestin is the motor protein of cochlear outer hair cells. Nature 405:149–155.

Zheng JL, Gao WQ (1996) Differential damage to auditory neurons and hair cells by ototoxins and neuroprotection by specific neurotrophins in rat cochlear organotypic cultures. Eur J Neurosci 8:1897–1905.

Zheng JL, Gao WQ (2000) Overexpression of Math1 induces robust production of extra hair cells in postnatal rat inner ears. Nat Neurosci 3:580–586.

Zheng JL, Shou J, Guillemot F, Kageyama R, Gao W (2000b) Hes1 is a negative regulator of inner ear hair cell differentiation. Development 127:4551–4560.

Zine A, Aubert A, Qiu J, Therianos S, Guillemot F, Kageyama R, de Ribaupierre F (2001) Hes1 and Hes5 activities are required for the normal development of the hair cells in the mammalian inner ear. J Neurosci 21:4712–4720.

2
Morphological Correlates of Regeneration and Repair in the Inner Ear

Jason R. Meyers and Jeffrey T. Corwin

1. Introduction

The loss of hair cells is a major cause of disabling hearing impairments that affect approximately 250 million persons worldwide and a contributor to inner ear balance disorders that can lead to falls late in life. Normal healthy human inner ears contain approximately 16,000 hair cells in the sound sensing cochlea, around 8000 hair cells in each of the three rotation sensitive semicircular canal cristae, and 18,000–33,000 hair cells each in the gravity sensing utriculus and sacculus (Fig. 2.1; Rosenhall 1972; Wright et al. 1987). The majority of people will lose some of those hair cells as they mature and age and many will develop age-related deficits of hearing called presbycusis (Fig. 2.2; Bredberg 1968; Rosenhall 1973). Presbycusis and age-related deterioration of vestibular reflexes correlate with declines in the numbers of cochlear and vestibular hair cells, respectively (Fig. 2.2; Bredberg 1968; Rosenhall 1973; Paige 1992).

Hair cells can be damaged and killed by loud sounds, infections, head trauma, and autoimmune disorders. The clinical use of aminoglycoside antibiotics and certain chemotherapeutic agents such as cisplatin can also cause hair cell loss. As life expectancies have lengthened, the occurrence of hearing and balance disabilities has grown, because hair cell losses in humans and other mammals are permanent and cumulative.

In contrast to mammals, many nonmammalian vertebrates produce hair cells throughout life and regenerate replacements for hair cells that have been lost. The replacement cells become innervated, which leads to the restoration of hearing and balance sensitivity, usually within a matter of weeks. Hair cell regeneration also occurs in the lateral line neuromasts of fish and amphibians, which share many aspects of tissue structure and organization with the hearing and balance end organs of the ear (Fig. 2.3).

This chapter reviews and discusses current understanding and highlights unresolved questions pertaining to morphological aspects of the mechanisms that

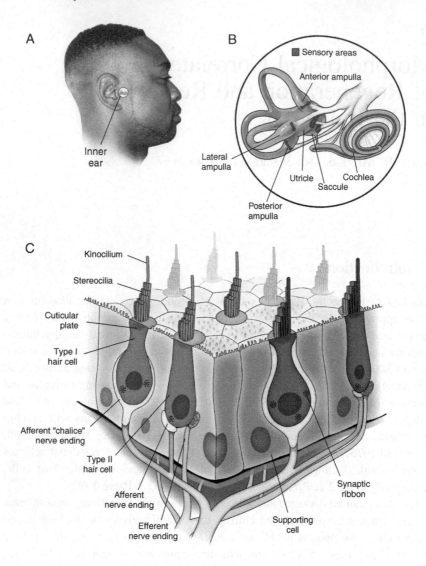

FIGURE 2.1. Schematic diagrams of the human inner ear. (**A**) Approximate size and location of the sensory organs of the inner ear. (**B**) Schematic drawing of the auditory and vestibular labyrinth. The sensory patches within each organ are indicated by the dark coloration. (**C**) Schematic representation of a section of the sensory epithelium from a vestibular organ to demonstrate the general structure and organization of the sensory epithelium.

underlie hair cell regeneration and repair in vertebrates. It also reviews investigations that have explored the potential for hair cell regeneration in the ears of mammals.

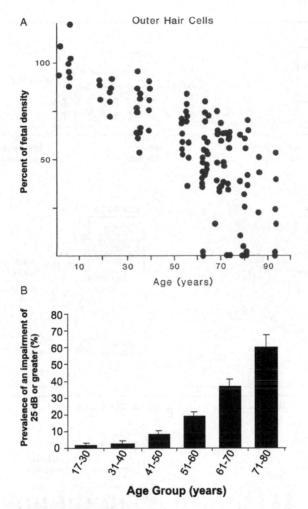

FIGURE 2.2. Loss of hair cell number and function in the human cochlea with increasing age. (**A**) Density of outer hair cells in the cochlea decreases with age, expressed as a percentage of the fetal density (100%). (**B**) Prevalence of hearing impairment of 25 dB or greater in the better hearing ear among individuals of different ages. The increasing prevalence of hearing loss at increased age correlates strongly with the loss of cochlear hair cells. (**A**) Modified from Bredberg (1968). (**B**) Modified from Davis (1989).

2. Ongoing Production of Hair Cells in Nonmammalian Vertebrates

In many nonmammalian vertebrates, the production of hair cells is not limited to embryonic development (Corwin 1981, 1983, 1985; Popper and Hoxter 1984; Lanford et al. 1996). For example, the ears of sharks and rays add hundreds of thousands of hair cells in the ear and increase in sensitivity as juveniles grow into

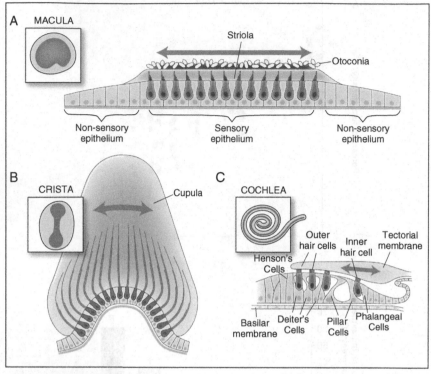

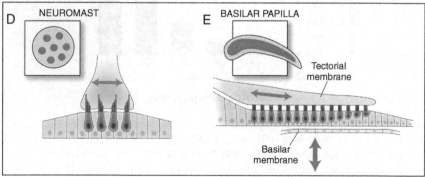

FIGURE 2.3. Schematic diagrams showing general structural organization of sensory epithelia. (**A**) Organization of otolith macular organs. The sensory patch, indicated by the darkened color in the inset, is comprised of hair cells and supporting cells in a mosaic pattern. The hair cells insert their hair bundles into a gelatinous matrix covered by otoconial crystals that provide inertial weight to acceleration of the head (arrow). The sensory epithelium is surrounded by a nonsensory epithelium, made up of supporting-like cells. (**B**) Organization of the semicircular canal cristae. The barbell-shaped sensory patch is organized very similar to that of the otolith macular organs. The hair bundles, however, are inserted into a large cupula, which transduces rotational movement of fluid through the canal to the hair bundles (arrow). (**C**) Organization of the mammalian cochlea. In contrast to the generally uniform population of supporting cells in vestibular organs,

adults (Fig. 2.4a; Corwin 1981, 1983). The ears of bony fish and amphibians also add thousands of hair cells during postembryonic growth (Lewis and Li 1973; Li and Lewis 1979; Popper and Hoxter 1984; Corwin 1985; Lombarte and Popper 1994; Lanford et al. 1996).

2.1 Postembryonic Hair Cell Production

These new hair cells are the product of postnatal proliferation, as [³H]thymidine labels both newly produced hair cells and supporting cells (Corwin 1981). The distribution of proliferating cells and newly produced hair cells in elasmobranchs and amphibians is strongly biased toward appositional growth, with cells added at the outer edge of the sensory epithelium (Fig. 2.4b,c), but some interstitial proliferation and addition of new hair cells occurs in the central regions of the maculae as well (Lewis and Li 1973; Li and Lewis 1979; Corwin 1983, 1985).

Appositional expansion is not the only pattern for ongoing growth in the ear. In teleost fish, interstitial addition outweighs appositional growth at the outer margin (Popper and Hoxter 1990; Lombarte and Popper 1994; Lanford et al. 1996). There is also ongoing proliferation and production of hair cells within the vestibular maculae of chickens occurring throughout the epithelium (Jørgensen and Mathiesen 1988; Roberson et al. 1992; Kil et al. 1997).

Thus, the ears of many nonmammalian vertebrate classes show signs of normal continued production of hair cells throughout life. The permanent loss of hair cells that occurs in mammalian ears appears to be unique, as hair cell epithelia retain progenitors in nonmammalian vertebrates that can produce hair cells.

3. Generation and Regeneration of Lateral Line Hair Cells

Hair cells are also continually generated in the sensory organs of the lateral line. Lateral line neuromasts are small collections of hair cells and supporting cells on the heads and bodies of fish and aquatic amphibians. The hair cells in the lateral line are morphologically similar to hair cells from the inner ear. They show expression of the same molecular markers as inner ear hair cells,

◄───

FIGURE 2.3. the cochlear supporting cells (e.g., Deiters', pillar, and phalangeal cells) show discrete morphological specializations to form the unique structure of the organ of Corti. (**D**) The lateral line neuromasts found along the body of fish and amphibians is organized similar to vestibular organs, though there are many fewer hair cells per organ. (**E**) The basilar papilla of birds shows some similarity to the vertebrate cochlea, though the supporting cells do not form morphologically discrete subtypes or show high levels of specialization.

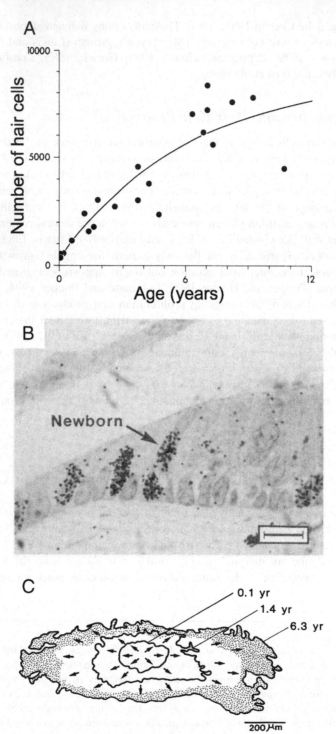

FIGURE 2.4.

including homologs of atonal (Itoh and Chitnis 2001) and myosin VIIa (Ernest et al. 2000), and are thought to be homologous. During the life of the animal, the lateral line organs bud off accessory neuromasts that lie dorsal or ventral to the primary organ and gradually develop into collections of up to 30 neuromasts (Stone 1937; Ledent 2002). The accessory organs begin as a few supporting cells around a single hair cell, but grow to include several hair cells, indicating that the cells of the neuromasts continue to produce new sensory structures and hair cells (Ledent 2002).

3.1 Regeneration of Hair Cell Epithelia After Amputation

In amphibians, neuromasts also can be replaced during regeneration after amputation of the tail. Cells in the last neuromast on the tail stump divide to give rise to a regenerative placode that is morphologically similar to the embryonic placode that first produces the lateral line organs in embryos and larvae. The regenerative placode migrates along the regenerating tail producing replacement neuromasts similar in number to those lost through amputation (Stone 1933, 1937; Speidel 1947; Wright 1947). These replacement organs develop hair cells with normal morphology and become reinnervated (Stone 1933, 1937; Speidel 1947; Wright 1947; Jørgensen and Flock 1976). Normally, the regenerating placode forms from the posterior edge of the last neuromast, but grafting experiments that reversed the orientation of the neuromasts demonstrated that production of a regenerating placode and neuromasts could develop from the anterior edge as well (Stone 1937). This suggests that the marginal cells closest to the wound are able to respond to some signal from the wound to initiate the regenerative process.

FIGURE 2.4. Ongoing production of hair cells continues throughout life in the inner ear of many vertebrate classes. (**A**) A plot of the number of hair cells within the macula neglecta of a ray versus age of the animal. The slope indicates that approximately three hair cells are added per day in the first years of life and approximately one hair cell is added per day beyond 6 years of age. (**B**) A micrograph of the edge of the sensory epithelium of a toad's saccule. A [^{3}H]thymidine labeled newborn hair cell is indicated at the edge of the epithelium, and three labeled supporting cells lie just beyond the last hair cell. The transition from the sensory epithelium to the nonsensory epithelium is the primary location for addition of new hair cells in cartilaginous fish and amphibians. (**C**) Superimposed outlines of the sensory epithelium from the macula neglecta of rays at different ages showing the increase in size of the sensory patch. The arrows indicate the orientation of the hair bundles pointing away from a central point of symmetry. The shaded portion indicates the region of hair cells with immature hair bundles at the outer edge of the epithelium. (**A, C**) From Corwin (1983); (**B**) from Corwin (1985).

3.2 Identifying the Progenitors of Replacement Hair Cell Epithelia

It was suggested decades ago that for both neuromast budding and regenerating placodes, the likely source of the cells was the supporting cells of the neuromast (Stone 1933, 1937). Time-lapse microscopy confirmed the identity of progenitor cells of the regenerating placode as mantle-type supporting cells that reside at the edge of the neuromast, adjacent to the internal supporting cells and hair cells (Jones and Corwin 1993). The location of the regenerating cells at the border between the sensory epithelium and surrounding epidermis is similar to the location of the proliferative cells that produce new hair cells throughout the lives of elasmobranches and amphibians (see Section 2.2). Neuromasts formed from a regenerative placode have the ability to form a new regenerating placode in response to reamputation of the tail (Speidel 1947). The progenitor cells, therefore, can make all of the differentiated cell types within the neuromast, and can reproduce latent precursors satisfying two defining characteristics of stem cells: self-renewal and multipotency.

3.3 Regeneration of Lateral Line Hair Cells In Situ

The lateral line can also generate replacement hair cells within individual neuromasts after selective destruction of individual hair cells (Balak et al. 1990; Song et al. 1995). In response to loss of hair cells within a neuromast, the surrounding supporting cells become proliferative and can generate both replacement hair cells and supporting cells (Jones and Corwin 1996). Thus, within the lateral line sensory organs, regeneration can replace lost hair cells or develop multiple new neuromasts, with all of their sensory and nonsensory cells, during tail regeneration.

4. Hair Cell Regeneration in the Inner Ear

Similar to the findings of ongoing addition of hair cells in anamniotes and the regenerative capacity of the lateral line, nonmammalian vertebrates such as fish, amphibians, reptiles, and birds are capable of significant regeneration of hair cells after damage to the inner ear.

4.1 Morphological Recovery in the Avian Hearing Organ

The full complement of hair cells in the chicken basilar papilla sensory epithelium is produced embryonically and the cells all become quiescent after embryogenesis (Tilney et al. 1986; Katayama and Corwin 1989). In response to tonal acoustic overstimulation, hair cells within the region of the basilar papilla tuned to the stimulating tone are lost or suffer significant damage, while hair cells in regions tuned to other frequencies are spared (Fig. 2.5a; Cotanche 1987).

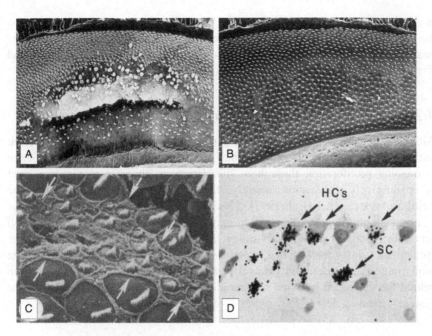

FIGURE 2.5. Hair cell regeneration in the chicken basilar papilla. (**A**) After 48 h of acoustic overstimulation at 1.5 kHz, hair cells within the region of the cochlea tuned to 1.5 kHz are lost and can be seen extruding from the epithelium. (**B**) After a 10-day recovery after acoustic overstimulation, the epithelium has been repaired. Hair cells can be seen throughout the formerly damaged regions. (**C**) Close-up micrograph of hair cells in the damaged region of a chick cochlea 6 days after damage shows a number of new hair cells with small bundles and small apical surfaces. (**D**) Injection of the birds with tritiated thymidine during the 10-day recovery period reveals labeling of both hair cells (HC) and supporting cells (SC), indicating that the new hair cells are progeny of dividing supporting cells. (**A–D**) From Corwin and Cotanche (1988).

This is followed by proliferation at the lesion site, and over 6–10 days there is a dramatic recovery in the number of hair cells (Fig. 2.5b; Cotanche 1987; Corwin and Cotanche 1988). Many of the hair cells within the damaged region exhibit small hair bundles reminiscent of immature hair bundles on developing hair cells (Fig. 2.5c). The ability to stimulate proliferation and regeneration of new hair cells was similar in young chickens and adult quail (Corwin and Cotanche 1988; Ryals and Rubel 1988; Ryals and Westbrook 1990). Thus, cells within the avian cochlea retain the capacity to reenter the cell cycle and generate new hair cells throughout life, even though they normally become mitotically quiescent after embryogenesis. The newly differentiating cochlear hair cells take on the distinct positional morphologies of the cells they replace (Cotanche 1987), suggesting that positional identity such as tonotopy can be passed on to the regenerating cells. Regeneration of hair cells also occurs in birds in response to aminoglycoside toxicity in both the auditory and vestibular systems (Lippe et al. 1991; Weisleder and Rubel 1993). After regeneration of hair cells in birds, the replacement cells become innervated and there is significant recovery of hearing and vestibular function (see discussion by Dooling, Dent, Lauer, and Ryals, Chapter 4).

4.2 Ongoing Proliferation in the Avian Vestibular System

Although the cells in the avian basilar papilla are normally mitotically quiescent until damage induces cell cycle reentry, in the avian vestibular system there is ongoing proliferation throughout life. However, there is also a low level of ongoing cell death, approximately 90% of which are hair cells within the vestibular system (Kil et al. 1997). This suggests that the continual production of new cells in the vestibular system may be in response to the dying cells and serves to replace lost hair cells rather than promote continued growth of the epithelium. Consistent with that hypothesis, inhibition of cell death in the utricle limits ongoing proliferation (Matsui et al. 2002). The mechanism and reason behind the continued loss and regeneration of hair cells in the avian vestibular system remain unclear.

4.3 Regeneration in Fish and Amphibians

The sensory epithelia from fish and amphibians also show substantial damage-induced proliferation and regeneration of hair cells. New hair cells are produced after aminoglycoside treatment in fish (Lombarte et al. 1993) and in bullfrogs (Baird et al. 1993), and after destruction of hair cells in the lateral line (Balak et al. 1990; Song et al. 1995). This generation of hair cells is accompanied by increased proliferation (Balak et al. 1990; Presson and Popper 1990; Baird et al. 1993; Presson et al. 1996; Avallone et al. 2003), similar to that seen in the avian organs after damage. In summary, fish, amphibians, and birds are all capable of significant regeneration of hair cells after damage to the inner ear. The replacement hair cells in these species occurs via cell divisions that occur at the sites of damage.

4.4 Regenerative Responses in Mammalian Ears

The hair cells and supporting cells in the vestibular system of mammals are similar to their counterparts in birds, fish, and amphibians (Fig. 2.3). In contrast, hair cells and supporting cells in mammalian cochleae are highly specialized, without direct counterparts in the auditory organs of other vertebrates. This high degree of differentiation and specialization suggests that regeneration may be more limited in the mammalian cochlea. However,, the similarity between mammalian and nonmammalian vestibular cells paired with the robust regeneration in nonmammalian ears suggests that hair cell regeneration may be more likely in mammalian vestibular epithelia.

4.4.1 Morphological Recovery in the Mammalian Vestibular Organs

After aminoglycoside-induced hair cell loss, guinea pigs allowed to recover for 4 weeks developed small immature hair bundles in the damaged regions of the utricular epithelium (Fig. 2.6; Forge et al. 1993, 1998). These cells were reminiscent of the appearance of regenerating hair cells in chickens (Cotanche 1987). Utricles from adult guinea pigs maintained in vitro after aminoglycoside treatment had small numbers of [^{3}H]thymidine-labeled cells throughout the sensory epithelium, showing that hair cell damage can trigger proliferation in vestibular organs of mammals (Warchol et al. 1993). Profiles of putative [^{3}H]thymidine-labeled hair cells were also found in those epithelia. When utricles obtained during otologic surgery on adult human patients were treated similarly, that resulted in at least 100 [^{3}H]thymidine-labeled cells in each utricle (Warchol et al. 1993). Thus even supporting cells from the sensory epithelium of adult humans can proliferate after damage.

However, the tens to hundreds of dividing cells in mammalian utricles after damage are considerably less than the thousands that would be seen in avian

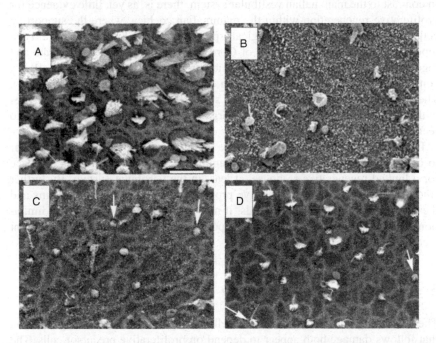

FIGURE 2.6. Regeneration in the mammalian vestibular epithelium. (A) Scanning electron micrograph of the striola region of a control utricle from a mature guinea pig. (B–D) Damage and recovery of hair bundles in guinea pig utricles after aminoglycoside toxicity. (B) The striola of a utricle 1 week after the end of aminoglycoside treatment, showing loss of most of the hair bundles. (C) The striola of a utricle 2 weeks after the end of aminoglycoside treatment showing initial recovery of hair bundles (arrows). (D) The striola of a utricle 4 weeks after the end of aminoglycoside treatment showing continued replacement of hair bundles (arrows). (A–D) From Forge et al. (1998).

utricles after similar damage (Warchol et al. 1993; Matsui et al. 2000). There was also a notable quantitative mismatch between the large numbers of recovering hair bundles reported by Forge and colleagues (1993, 1998) and the limited number of proliferating cells found by Warchol et al. (1993). There may be a substantial difference in the level of damage to the utricles in vivo versus in vitro, which may make it difficult to compare the response between these studies. The studies showing substantial recovery of hair bundle numbers were done in vivo, wherein aminoglycoside nephrotoxicity limits the concentrations that can be used without severe systemic repercussions, while the investigations of proliferation were performed in vitro, wherein high levels of aminoglycoside ensure the death of most of the hair cells. Thus, whether the difference in the magnitude of proliferation versus hair bundle recovery is due to the level of damage, or whether it reflects a nonproliferative mechanism of regenerating hair cells remains an open question.

4.4.2 Minimal Hair Cell Regeneration in the Mammalian Cochlea

In contrast to the mammalian vestibular system, there is, as yet, little evidence for proliferative regeneration within the mammalian cochlea, where the supporting cells show significant morphological specialization. While there have been reports that the cells of the organ of Corti can be stimulated to proliferate and regenerate hair cells after aminoglycoside treatments (Lefebvre et al. 1993), confirmation of those results has proven difficult (Chardin and Romand 1995). The specialization and differentiation of supporting cells into discrete cell types with distinct morphologies within the organ of Corti may limit their ability to reenter the cell cycle.

In summary, while regeneration in the mammalian inner ear does not appear to occur at the robust level as in nonmammalian vertebrates, there is evidence for a limited regenerative response, at least in the vestibular system. By further studying both the strong regenerative processes in nonmammalian systems and the limited regeneration in mammals, it may be possible to override the limitations in the mammalian regenerative response and promote clinically significant recovery from hair cell loss.

5. The Source of Regenerating Hair Cells

The continual production of new hair cells throughout life and the regeneration that follows damage both appear to depend on proliferative precursor cells. The morphological identity of the cells that produce regenerating hair cells within damaged epithelia has been a key question in establishing a mechanism for the observed recovery of functional hair cells. Proposed sources have included the hyaline cells that lie at the margin of the chick basilar papilla (the same position as the cells that result in appositional growth in elasmobranchs and neuromast regeneration in amphibians), the differentiated supporting cells within the sensory organs, or a specialized reserve pool of distinct stem cells.

During damage-induced regeneration, proliferating cells can be labeled with tritiated thymidine or 5-bromo-2-deoxyuridine (BrdU). In the avian ear, supporting cells within the sensory epithelium are the first cells to be labeled by mitotic markers, and this is followed by the production of new, labeled, hair cells (Fig. 2.5d; Corwin and Cotanche 1988; Ryals and Rubel 1988; Raphael 1992; Roberson et al. 1992; Hashino and Salvi 1993; Stone and Cotanche 1994). Hyaline cells, which reside at the abneural edge of the sensory epithelium, will also migrate into the wounded region and proliferate after extensive damage to the basilar papilla (Girod et al. 1989; Cotanche et al. 1995). However, proliferation of hyaline cells has not been associated with hair cell regeneration (Corwin et al. 1991; Raphael 1992; Cotanche et al. 1995; Warchol and Corwin 1996). In addition to the indirect evidence of BrdU- or [^{3}H]thymidine-labeled supporting cells pointing to them as the source of new hair cells, direct observation via time-lapse microscopy has shown that supporting cells within salamander lateral line neuromasts divide and the progeny differentiate into replacement hair cells and replacement supporting cells (Jones and Corwin 1996). Thus, in response to damage, cells within the epithelium that show the morphological characteristics of supporting cells appear to be the primary cells that reenter the cell cycle to produce replacement hair cells and supporting cells (Fig. 2.7).

5.1 Reserve Stem Cells versus Dedifferentiation of Supporting Cells?

Although cells classified as supporting cells have been identified as the primary source for regenerating hair cells, there are two hypotheses about the nature of those cells. They may be normal functional supporting cells that dedifferentiate, return to the cell cycle, and produce replacement hair cells and supporting cells, or they may be a discrete population of reserve stem cells that are morphologically similar to supporting cells.

The possibility that there may be a discrete population of proliferative cells within the sensory epithelium separate from the supporting cells has been considered (Presson and Popper 1990), though more recent studies have

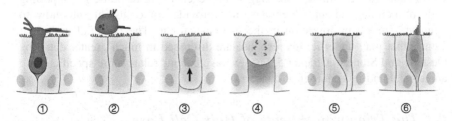

FIGURE 2.7. Schematic diagram of regenerative proliferation by supporting cells. In response to loss of a hair cell (1) the supporting cell nucleus migrates toward the epithelial surface (2–3) where it undergoes mitosis (4). The progeny of that division then go on to form replacement hair cells or supporting cells (5–6).

concluded that dividing precursor cells are morphologically indistinguishable from other supporting cells in the fish inner ear (Presson et al. 1996). Further, destruction of the proliferating precursor cells with cytosine arabinoside (Ara-C) is followed by a new wave of proliferating cells that replace the lost cells and generate new hair cells (Presson et al. 1995). After damage in the chick inner ear, the majority of supporting cells downregulate expression of G_0 markers and begin expressing markers consistent with passage to late G_1 phase (Bhave et al. 1995). These data suggest that regeneration does not depend on a discrete population of reserve cells, but rather originates from a larger collection of supporting cells that have the capacity to dedifferentiate, reenter the cell cycle, and produce replacement hair cells.

The heavy bias of proliferation and addition of hair cells at the lateral edge of the epithelium in elasmobranches and amphibians strongly suggests that a class of stem cells resides at the moving junction between the sensory epithelium and the nonsensory epithelium in these animals. A similar population of stem cells is found in the ciliary marginal zone in the retinas of fish, amphibians, and birds, contributing to growth and regeneration of the retina (Straznicky and Gaze 1971; Johns 1977; Wetts and Fraser 1988; Fischer and Reh 2000). Growing knowledge about these stem cells may be applicable to questions about the proliferative cells at the edges of inner ear epithelia. Although they do not have similar ongoing growth of the organs, the auditory organs of chickens and the vestibular organs of mammals initially develop in a pattern similar to the radial addition of cells in many anamniote ears. Proliferation and cell differentiation occur first near the center of the sensory epithelium and progress radially outward from there until the whole sensory epithelium has been produced (Sans and Chat 1982; Mbiene and Sans 1986; Katayama and Corwin 1989). Perhaps the mechanisms that control ongoing hair cell production at the margin of sensory epithelia in anamniotes may also be active in amniotes during development, but normally suppressed after maturity.

6. What Triggers the Regeneration Response?

Understanding the stimulus that triggers the regenerative response in supporting cells of nonmammalian vertebrates may provide some insight into how to stimulate similar proliferation in the mammalian ear. The potential molecular signals that may underlie this response are discussed in much greater detail by Oesterle and Stone in Chapter 5, but several elements relate more specifically to anatomical and morphological concerns.

6.1 The Triggering Effects of Hair Cell Loss

The loss of hair cells is an important stimulus for regeneration of new hair cells. In undamaged avian cochleae, there is little or no supporting cell proliferation (Corwin and Cotanche 1988; Ryals and Rubel 1988; Katayama and

Corwin 1989), but individual laser ablation of a patch of hair cells is sufficient to stimulate regenerative proliferation in the area around the wound (Warchol and Corwin 1996). The necessity for hair cell loss and the restriction of proliferation to the damaged area suggests that hair cells may suppress proliferation in supporting cells (Corwin et al. 1991). Similarly, along the outer margin of an anamniote's sensory epithelium, the outermost row of cells may not contact hair cells and thus may be relieved from this suppression of proliferation. One hypothesis is that hair cells express proteins along their basolateral membrane that prevent adjacent supporting cells from reentering the cell cycle (Corwin et al. 1991). Interaction of cadherins, transmembrane proteins that mediate cell–cell adhesion at adherens junctions, can suppress proliferation (Caveda et al. 1996), and is thought to mediate contact-inhibition of proliferation through sequestration of β-catenin (Fagotto and Gumbiner 1996). Changes in cadherin-mediated cell–cell contact, such as loss of hair cell-supporting cell junctions, could be a trigger for reinitiation of proliferation. Another cell–cell signaling pathway that may play a role in maintaining inhibition of supporting cell proliferation is the Notch pathway.

Notably, however, while the proliferative response in damaged chick basilar papillae is restricted to the area around the lesion, dividing supporting cells can be seen up to 180 μm away from the edge of the wound, in apparently undamaged regions (Warchol and Corwin 1996). Thus, while loss of hair cells is necessary to stimulate regenerative proliferation, it is not yet established whether loss of direct cell–cell signaling, such as through cadherins or notch–delta, is a necessary trigger for proliferation of supporting cells, or whether broader, more diffusible signals play the critical role.

6.2 The Potential Influence of Immune Responses to Damage in the Inner Ear

Several studies suggested the possibility that immune responses may influence regeneration within hair cell sensory epithelia. After tail amputation in axolotls, macrophages are recruited to the posterior side of the posteriormost neuromast, which will give rise to the regenerative placode (Jones and Corwin 1993). This suggests that they may play a role in the regenerative placode formation, for example, breaking down the glycocalyx that surrounds the neuromast, allowing mantle supporting cells to migrate out to form the migratory placode. Similarly, macrophages and microglia are recruited to sites of damage in the avian utricle and basilar papilla (Warchol 1997, 1999; Bhave et al. 1998) and in rat organ of Corti (Wang and Li 2000). The exact role of macrophages and microglia in repair and regeneration of damaged sensory epithelia is unknown, but they are known to have significant effects during epidermal wound healing (Gailit and Clark 1994; Martin 1997). Among their reported roles are removal of dead or dying cells and cellular detritus, production of mitogenic cytokines, and modulation of extracellular matrix composition. In hair cell epithelia, macrophages will scavenge and remove debris, dead or damaged hair cells, and even progeny from recent cell divisions. Macrophages likely also play a role in production of growth factors. Treatment of aminoglycoside-damaged chick utricles with dexamethasone to block macrophage cytokine production significantly reduces the number of proliferating supporting cells without significantly altering the number of macrophages (Warchol 1999). This indicates that chemical signaling from macrophages may be more important than their direct reparative effects on the epithelium. An important open question is whether the immune response to damage in the mammalian inner ear is similar to that in the avian inner ear or axolotl lateral line, as the extent of macrophage invasion into the damaged mammalian organ of Corti may be limited (Fredelius 1988; Fredelius and Rask-Andersen 1990; Hirose et al. 2005).

6.3 The Potential Importance of Cellular Shape Change

Cell shape has a direct impact on the ability of cells to proliferate (Folkman and Moscona 1978). In cultured endothelial cells, cells allowed to spread out on micropatterned substrates were able to proliferate, while cells forced to maintain compact cell shapes did not proliferate (Chen et al. 1997; Huang and Ingber 1999). Several experiments have suggested that there may be an important role for cell spreading in control of proliferation in hair cell sensory epithelia. In isolated cultures of chick utricular sensory epithelium, cells at the edge of the epithelium that spread out proliferate well while central cells that remain columnar and tight packed do not proliferate (Warchol 1995, 2002; Witte et al. 2001). Within the inner ear, one of the initial responses to loss of hair cells is expansion of the supporting cell surface to cover the space of the missing hair cells (Forge 1985; Cotanche 1987; Cotanche and Dopyera 1990; Marsh et al. 1990; Li et al. 1995). The significant expansion of the supporting cells in the avian basilar papilla coincides with the time during which they return to the cell cycle, and their return to normal surface dimensions occurs as replacement hair cells and supporting cells differentiate (Corwin and Cotanche 1988; Marsh et al. 1990). In mammalian utricular epithelia, cellular shape following damage is strongly correlated with cell-cycle entry, with tall, columnar cells rarely entering the cell cycle while most flattened, spread cells returned to the cell cycle (Meyers and Corwin 2007). These data are all consistent with the hypothesis that morphological shape change in supporting cells may be an important trigger in stimulation of regeneration, though further experimentation is necessary to clarify the specific role of such shape change on initiating proliferation.

7. Nonproliferative Restoration of Hair Cells

Although much of the morphological recovery from damage in nonmammalian vertebrates appears to be from proliferation of supporting cells, there is growing evidence that not all of the recovery is due to proliferation. The first hair cells to appear following damage to the chick are not labeled with BrdU, though subsequent generation of hair cells is predominantly from proliferative means (Roberson et al. 2004). Similarly, morphological recovery in the newt (*Notophthalmus viridescens*) appears to occur in the absence of proliferation, as new hair cells are not BrdU labeled and still arise in the presence of mitotic inhibitors (Taylor and Forge 2005). In the mammalian vestibular system, there is a quantitative mismatch between the large number of hair bundles that reappear at the epithelial surface and the small number of dividing cells that can be observed (Warchol et al. 1993; Forge et al. 1998; Berggren et al. 2003). Further, differentiated hair bundles in bullfrog (*Rana catesbeiana*) and chick organs reappear after damage even in the presence of mitotic inhibitors (Adler and Raphael 1996; Adler et al. 1997; Steyger et al. 1997; Baird et al. 2000; Gale et al. 2002). Two potential mechanisms that have been proposed for this nonproliferative recovery are repair of damaged hair cells and phenotypic conversion of a supporting cell directly into a hair cell without intervening mitosis.

7.1 Sublethal Hair Cell Damage and Repair

Acoustic overstimulation, aminoglycoside antibiotics, platinum-based chemotherapeutics, loop-diuretics, head trauma, or infection can all lead to damage or death of hair cells. In many cases, the specific manner in which these insults lead to damage or death of hair cells is unknown, but many recent advances have been made in understanding the mechanisms of damage and to develop potential strategies to protect hair cells from such damage (see Forge and Van De Water, Chapter 6).

7.1.1 Hair Cell Responses to Damage

Generally, the initial response of hair cells to low levels of furosemide, aminoglycosides, cisplatin, or mechanical overstimulation involves disruption of the tip-links connecting adjacent stereocilia, followed by splaying of the bundle and/or fusion of the stereocilia (Fig. 2.8a,b; Engstrom et al. 1983; Pickles et al. 1987a, b; Osborne and Comis 1990b; Clark and Pickles 1996). More extensive insult can lead to loss of stereocilia at their point of insertion into the cuticular plate and loss of the hair bundle (Fig. 2.8c; Pickles et al. 1987a; Osborne and Comis 1990b; Gale et al. 2002). At higher levels of insult, hair cells are killed and lost from the epithelium, either via physical extrusion of the entire cell, or apoptosis within the epithelium (Fig. 2.8d,e; Li et al. 1995; Nakagawa et al. 1997). Any of these levels of damage will produce a loss-of-function, either from disruption of mechanosensation in surviving cells or from loss of the sensory cells themselves.

Because hair cells can exhibit a range of morphological effects depending on the level of insult, careful analysis of the level of damage is necessary to ensure that later regenerative steps are appropriately identified. For example, many studies have used hair bundle counts as the primary assay for loss of hair cells, but hair cells may be able to lose their stereociliary bundles and survive. In fact,

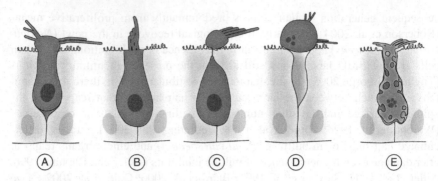

FIGURE 2.8. Hair cells can have a number of morphologic responses to insult. (**A**) In response to light to moderate acoustic overstimulation or ototoxic insult, the hair bundle may fragment and the stereocilia may become splayed. (**B**) At higher levels of damage, stereocilia may fuse together and portions of the hair bundle may be lost. This is often accompanied by swelling or blebbing of the apical surface of the hair cell. (**C**) Hair cells may also completely pinch off their damaged hair bundle along with a bit of apical cytoplasm, resulting in a bundleless hair cell. In response to higher levels of damage, hair cells may be fully extruded from the epithelium (**D**) or become vacuolated and degenerate within the epithelium (**E**).

alterations to the hair bundle and stereociliary loss are often the initial responses to traumatic insult, and may in fact serve to protect the cell from further insult. Thus a critical step in analysis of regeneration is solid determination of the level of damage to the epithelium.

7.1.2 Repair and Replacement of Damaged Hair Bundles

The mechanosensory function of hair cells is dependent on the presence of a functioning hair bundle, but often the initial response to insult is damage to the hair bundle. Tip-links are extracellular filaments that connect stereocilia and are proposed to act as the gating spring that opens sensory transduction channels at the tips of the stereocilia in response to movement of the hair bundle (Pickles et al. 1984; Ricci et al. 2006). Tip-links can be damaged by moderate acoustic overstimulation (Clark and Pickles 1996; Husbands et al. 1999), treatment with calcium chelators (Assad et al. 1991), or elastase (Osborne and Comis 1990a). Breakage of tip-links, such as by calcium chelation, leads to a loss of mechanotransduction consistent with the hypothesis that the tip-link gates the mechanotransduction channels in response to hair bundle movement (Assad et al. 1991). There is rapid morphological recovery of tip-links after calcium-chelation-induced breakage, accompanied by a physiological recovery of mechanotransduction (Zhao et al. 1996). There is similar recovery in the number of tip-links after acoustic overstimulation in tall hair cells of the chick basilar papilla within 1–4 days after the damage (Husbands et al. 1999). Thus, the fine structure of the hair bundle can be readily repaired after light damage, and this repair returns functionality to the hair cells.

In cases where damage to the hair bundle has been more extensive, resulting in fusion or loss of stereocilia or even loss of the entire hair bundle, regeneration of the damaged bundle would be necessary to permit functional recovery (Fig. 2.9a). In bullfrog (*Rana catesbeiana*) saccules treated with low doses of aminoglycoside antibiotics, a number of cells lose a large portion of the bundle, but retain some of their stereocilia, often fused together. Within a week after the damage, some of these hair cells rebuild a hair bundle with the traditional staircase pattern adjacent to the remnants of their damaged bundle (Gale et al. 2002). This is suggestive of reconstruction of a hair bundle in mature hair cells. Hair cells can also completely lose their hair bundles and survive as bundle-less hair cells within the epithelium for days in both bullfrogs and rats (Zheng et al. 1999; Gale et al. 2002). There also can be recovery of hair bundle number, though the number of hair cell bodies remains constant, in the presence of mitotic inhibitors to prevent new cell birth (Zheng et al. 1999; Gale et al. 2002). Thus, even after relatively significant damage, such as complete loss of their bundle, hair cells are able to recover and repair, which may lead to recovery of function within the damaged organ even in the absence of proliferation.

Such plasticity of the hair bundle may be a normal part of its physiology. Studies examining the actin composition of hair bundles have found that actin is continually added to the tips of stereocilia and removed at the base, resulting

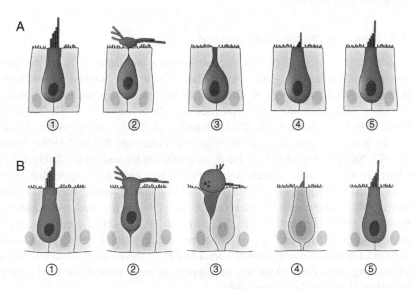

FIGURE 2.9. Schematic diagrams of two proposed mechanisms of nonproliferative regeneration. (**A**) Repair of damaged hair cells. In response to damage, the hair cell loses its hair bundle and apical cytoplasm (1–2), but over time is able to regenerate a new bundle that matures into a normal hair bundle (3–5). (**B**) Direct conversion of a supporting cell into a hair cell without an intervening mitosis. When a hair cell is lost (1–2), one of the neighboring supporting cells, no longer suppressed by an adjacent hair cell (3), is able to redifferentiate into a hair cell and develop a morphologically mature hair bundle (4-5).

in a complete turnover of the actin in each stereocilia every 48 hours (Schneider et al. 2002). Thus hair bundles are not static, but rather are dynamic with the cytoskeletal elements of the stereocilia continually recycled. It is not unreasonable, then, to suggest that after damage to the bundle, alterations in the actin dynamics may lead to repair or reconstruction of stereocilia. Mouse hair cells that have their bundles mechanically disrupted are able to rebuild a kinocilium, leading to reorganization of the cuticular plate and reinitiation of stereocilia formation (Sobkowicz et al. 1995). Together, these data suggest that if hair cells survive the insult, they may have a significant capacity for repair that could lead to morphological, and possibly functional, recovery. Such repair may account for some of the nonproliferative regeneration observed in many systems.

7.2 Phenotypic Conversion into Hair Cells

A second hypothesis that has been put forth to account for the nonproliferative generation of new hair cells, is that supporting cells are able to convert directly into hair cells without undergoing an intervening mitosis (Fig. 2.9b). This process has been termed either transdifferentiation or phenotypic conversion. The term phenotypic conversion will be used herein to eliminate any confusion with transdifferentiation, as that term has also been used to describe the conversion of one tissue type to another.

7.2.1 Evidence for Phenotypic Conversion

The evidence in support of this hypothesis includes the disparity in the number of dividing cells after injury and the number of new hair bundles or hair cells seen in the epithelium (Forge et al. 1998), a lack of incorporation of S-phase markers (e.g., BrdU) in hair cells after recovery (Roberson et al. 1996, 2004; Li and Forge 1997; Berggren et al. 2003; Matsui et al. 2003), and recovery of hair cells even in the presence of a mitotic inhibitor (Adler and Raphael 1996; Adler et al. 1997; Steyger et al. 1997; Baird et al. 2000; Taylor and Forge 2005). Cells that have an intermediate morphology between hair cells and supporting cells (i.e., nuclei in the basal layer, contact with the basal lamina, apical specializations and microvilli similar to hair bundles) can be found during recovery from damage (Adler et al. 1997; Li and Forge 1997; Matsui et al. 2003), consistent with supporting cells in the process of redifferentiating as hair cells. Time-lapse recordings of regeneration in axolotl lateral line have shown occasional hair cells differentiating from cells identified as supporting cells without an intervening cell division (Jones and Corwin 1996).

7.2.2 Developmental Window for Conversion in the Mammalian Cochlea

In the embryonic mammalian cochlea, shortly after terminal mitosis, newly differentiating supporting cells can convert into hair cells after ablation of adjacent hair cells, but the supporting cells rapidly lose this capacity as they mature (Kelley et al. 1993). Thus, there may be a limited window during differentiation when mammalian cochlear supporting cells are plastic enough to convert into hair cells, though mature cochlear supporting cells with their morphological specializations are unlikely to do so.

In the mature mammalian organ of Corti, with its highly differentiated supporting cells, there have been several reports of morphological changes in supporting cells after hair cell loss. Deiters' cells, which surround the three rows of outer hair cells, develop dense bundles of microvilli that resemble immature hair bundles, and are occasionally contacted by efferent and afferent fibers after loss of the outer hair cells (Romand et al. 1996; Lenoir and Vago 1997; Daudet et al. 1998, 2002). While they are able to develop actin-rich microvilli bundles, the bundles never mature into stereocilia, the cells do not develop a tall kinocilium, they do not acquire morphological or molecular features of hair cells, and they retain features of supporting cells including gap junctions between adjacent cells (Daudet et al. 1998, 2002). Over time, most of the atypical cells and neighboring supporting cells are lost from the sensory epithelium and replaced by cells from the outer sulcus (Daudet et al. 1998). Thus, mature Deiters' cells attempt to undergo morphological change, but are unable to complete the process and are removed from the epithelium.

7.2.3 Conversion into Hair Cells by Gene Manipulation

There is evidence that the cells that lie just outside the organ of Corti, including the Hensen's cells and tectal cells adjacent to the outer hair cells, and cells of the greater epithelial ridge adjacent to the inner hair cells, though not in the sensory epithelium, are nonetheless capable of being converted into hair cells. Application of retinoic acid to the mouse cochlea causes the production of supernumerary hair cells next to both the inner and outer hair cells without any additional cell divisions, but once again only if applied during a short developmental window between E14 and E18 (Kelley et al. 1993). Similar results have been reported for treatment with epidermal growth factor (EGF) and transforming growth factor-β (TGF-β) in neonatal rat cochleae (Chardin and Romand 1997; Lefebvre et al. 2000). The nonsensory epithelial cells adjacent to the organ of Corti thus retain the ability to convert into sensory hair cells when pushed by application of exogenous growth factors, though this capacity seems to be lost as the cells mature.

In cochlear explants from neonatal rats, exogenous expression of *Atoh1* (the atonal homolog also known as *Math1*) in cells of the greater epithelial ridge (GER) is sufficient to convert nonsensory epithelial cells there into a hair cell phenotype: expressing myosin VIIa, a molecular marker of hair cells, taking on the morphology of hair cells, and developing hair bundles (Zheng and Gao 2000). Similarly, the use of adenovirus to transfect *Math1* into mammalian inner ear organs leads to formation of ectopic hair cells in the nonsensory epithelium around the organ of Corti in vivo in mature guinea pigs and may convert supporting cells from damaged adult organs of Corti and adult utricles into hair cells (Kawamoto et al. 2003; Shou et al. 2003). More recently, there is suggestion that transfection with *Atoh1* may be able to promote a reconstitution of the organ of Corti after aminoglycoside damage by converting surviving supporting cells into hair cells, allowing a limited recovery of cochlear function (Izumikawa et al. 2005). If the surprisingly accurate reconstitution of the epithelium after damage by *Atoh1* can be replicated, this would suggest that the supporting cells of the cochlea are not only capable of being converted into hair cells by *Atoh1*, but also that the positional cues that specify the identity of the newly differentiating cells are maintained. Such genetic manipulations should continue to provide information about control of cell fate in the inner ear, even if it remains unlikely that genetic therapy will be a viable approach for inducing clinical regeneration.

7.2.4 Evaluating the Current Evidence for Supporting Cell Conversion

The role that phenotypic conversion plays in normal regeneration is unclear at present, as much of the indirect evidence supporting conversion could also be consistent with repair of damaged hair cells, such as recovery in the presence of mitotic inhibitors and recovery of hair bundle numbers. Supporting cell conversion in the absence of proliferation notably requires a decrease of supporting cell number as they convert into hair cells. However, decreases in the number of supporting cells are smaller than the number of cells that develop new hair bundles (Forge et al. 1998; Zheng et al. 1999). To establish whether significant supporting cell conversion occurs, it will be necessary to do specific and conclusive experiments that demonstrate conversion and exclude other mechanisms such as repair. Nonetheless, supporting cells or a subpopulation of less differentiated (e.g., recently produced) cells can convert into hair cells and contribute to morphological recovery, particularly in epithelia where ongoing cell addition means that cells at different states of differentiation may be present.

7.3 Mitotic Contribution of G_2 Cells

An additional mechanism that may contribute to apparent nonproliferative regeneration is mitosis and differentiation of cells that have already passed S-phase. Such cells will not be blocked by many mitotic inhibitors, which often block passage into S-phase, nor will they label with mitotic labels, which are incorporated in S-phase. In bullfrog (*Rana catesbeiana*) saccular cultures blocked with aphidicolin, occasional pairs of immature bundles were found throughout the epithelium, consistent with mitotic production in the absence of the cells passing through S-phase (Gale et al. 2002). In particular, in epithelia where there is constant addition of cells, there may be a fraction of cells residing in G_2-phase that can contribute to regenerative recovery. G_2-arrest of precursors is another potential source for mitosis in the absence of S-phase passage. During salamander limb regeneration, myotubes arrest in G_2 (Tanaka et al. 1997), and during differentiation of *Drosophila* eye and wing development, cells arrest in G_2 in a Wnt/Notch-dependent manner (Kimura et al. 1997; Johnston and

Edgar 1998). The potential contribution of G_2-phase cells to nonproliferative recover seems worthy of further investigation, such as screening for cells that have twice the DNA content of neighboring cells.

7.4 Nonproliferative versus Proliferative Recovery

There is a strong possibility that conversion, repair, and proliferative regeneration all contribute to morphological recovery of hair cells after damage. Importantly, only in mammals is regenerative proliferation strictly limited, and only in mammals are sensory deficits attributable to loss of hair cells permanent. Nonproliferative mechanisms of regeneration, while they may contribute significantly to morphological recovery from low levels of damage, may not be sufficient to bring about functional recovery in response to significant hair cell loss in the mammalian ear. Also, without proliferation, the number of supporting cells that can be converted while still maintaining sufficient supporting cell number for epithelial function will be limited. Generation of clinically significant numbers of replacement hair cells may therefore require returning mature mammalian supporting cells to a proliferative state, as occurs in other vertebrates.

8. Can Regeneration Restore Hair Cells in Mammals?

As discussed in the preceding text, the number of hair cells in the mammalian inner ear decreases with age, few cells are regenerated in the vestibular and auditory system after damage, and clinically, loss of hair cells in humans leads to permanent sensory deficit. The ongoing proliferation and robust regenerative proliferation that occur in nonmammalian vertebrates therefore do not occur in mammals at sufficient levels to be physiologically relevant. However, there may be low levels of repair and regenerative processes that can be stimulated to lead to functional regeneration in mammals.

8.1 Developmental Production of Hair Cells in the Mammalian Inner Ear

In mammals, the hair cells in the auditory and vestibular organs become fully established during embryonic development. In rodents, nearly all of the progenitor cells in the organ of Corti and vestibular system undergo terminal mitosis during mid to late embryonic development, followed by hair cell differentiation producing functional hair cells by birth (Ruben 1967; Kaltenbach and Falzarano 1994; Kaltenbach et al. 1994; Geleoc and Holt 2003). While a full cohort of cells is produced embryonically, and the cells are thought to become postmitotic, low ongoing rates of proliferation could occur and remain well below the sensitivity limits of the tests used to look for production in postembryonic mammalian ears. In fact, examination of hair bundles within the utricle maculae from mature guinea pigs has shown on average one cell per thousand that has a small, immature bundle characteristic of newly differentiating hair cells, and six to seven cells per thousand that have intermediate size hair bundles (Lambert et al. 1997). Occasional immature-looking hair cells have also been reported in the sensory epithelia of the bat (Kirkegaard and Jørgensen 2001). If there is even a small level of ongoing proliferative production of cells, as occurs in most nonmammalian vertebrates examined, such a mechanism may provide an entry point into initiating a large proliferative response in mammals.

8.2 Reserve Stem Cells versus Supporting Cell Proliferation in Mammals

In contrast to the robust supporting cell proliferation in nonmammalian verte-brates previously discussed, studies of proliferation in response to damage in mammals suggest that few cells contribute to the regenerative response. Does this represent a distinct mechanism of replacing lost hair cells between mammals and other vertebrates, such as the presence of a few stem cells rather than a large population of potential progenitors?

Consistent with this hypothesis, a small fraction of cells from the adult mouse utricular epithelium act as colony-forming stem cells when dissociated, and the progeny of these cells can differentiate into many cell types including cells that express proteins in common with hair cells (Li et al. 2003). These colonies each have one to three cells that can be subcloned to form their own spheres in cultures. The demonstrated multipotency and self-renewal of these progenitors are the hallmarks of a stem cell population, suggesting that there may be a handful of isolated stem cells within mature mammalian vestibular epithelia. At this point, the identity or other characteristics of these potential stem cells are unknown.

What, then, of the capacity for the many supporting cells within the epithelium to reenter the cell cycle as occurs in nonmammalian vertebrates? The majority of supporting cells from perinatal rodents can be stimulated to proliferate in culture with growth factor and pharmacologic treatment, in a manner similar to, and using the same intracellular cascades as, avian supporting cells (Gu et al. 1996; Montcouquiol and Corwin 2001a, b; Witte et al. 2001). Thus the proliferative capacity of perinatal utricular epithelia is consistent with the hypothesis that most supporting cells are capable of proliferating. This is in contrast to the limited proliferation seen in mature organs in response to damage. However, when mature supporting cells are stimulated to spread and flatten following a large lesion, nearly all spread cells reenter the cell cycle, suggesting that most supporting cells retain the capacity to reenter the cell cycle, but are restricted from doing so under normal conditions (Meyers and Corwin 2007). Further experimentation will be necessary to determine whether a few reserve stem cells exist in the mature mammalian inner ear, or whether all mature supporting cells can serve as a potential, though only rarely stimulated, progenitor. It seems unlikely that mammals would retain only a few stem cells, in particular too few to bring about functional repair, while all other vertebrates can utilize any supporting cell as a regenerative precursor. Rather, it may be that the cellular control of proliferation has been significantly tightened in mammals and that only rarely can a mature supporting cell overcome the strong inhibition.

8.3 Loss of Mammalian Proliferative Capacity with Age

Notably, the capacity to stimulate supporting cells to reenter the cell cycle is critically dependent on the age of the animal. While large numbers of supporting cells from perinatal animals can be stimulated to proliferate, few to no cells enter S-phase in cultures from mature animals (Gu et al. 1997; Hume et al. 2003; Gu et al. 2007). Thus, the proliferative mechanisms that operate in cultures from the neonatal inner ear are turned off or suppressed in more mature epithelia, consistent with the limited proliferation found in mature guinea pig or human utricles after damage. Coincident with this change in proliferation, there are changes in the components of the basal lamina and cytoskeleton that affect the ability of supporting cells to change shape and reenter the cell cycle (Davies et al. 2007). Pharmacological treatments that alter the cytoskeleton show promise in promoting supporting cell proliferation in mature utricles (Davies et al. 2007; Meyers and Corwin 2007). The fundamental nature of the cytoskeletal and other changes that underlie this change in proliferative capacity are unknown, but future studies should help elucidate the mechanisms that repress proliferation in mature epithelia (see also Oesterle and Stone, Chapter 5).

8.4 Challenges and Promises of Mammalian Regeneration

The restricted proliferative response to damage in the mammalian inner ear suggests that significant numbers of new cells are not produced as they are in nonmammalian vertebrates, consistent with the clinical permanence of auditory and vestibular deficits associated with hair cells loss and the continual decrease in hair cell number throughout the life of mammals. There may be some non-proliferative mechanisms that allow morphological recovery, and more work needs to be done to assess whether this could contribute to functional recovery. While mammalian supporting cells appear to lose their capacity to reenter the cell cycle as they mature, the presence of some supporting cells that reenter the cell cycle and immature hair bundles in the adult mammalian vestibular system indicate that the cellular processes necessary for regeneration can occur in mammalian sensory epithelia. Work has begun to point to genetic changes and cellular pathways that may underlie the loss of proliferative capacity, but the fundamental changes that restrict supporting cell proliferation in mammalian epithelia remain unclear. Further investigation into the control of proliferation in the mammalian inner ear and the mechanisms that underlie the robust proliferation in nonmammalian vertebrates may lead to strategies that can upregulate proliferative regeneration, bringing about morphological and functional recovery from damage.

9. Levels of Cell Differentiation

One recurring theme in this chapter has been that cells that become too differentiated become refractory to regenerative processes, whether that is cell cycle reentry or conversion of cell fate. Framed more broadly, the plasticity of the inner ear to respond to damage may be tied into the level of specialization and differentiation of the cells within the sensory organ (Fig. 2.10). In particular, the most specialized cell type within the ear, the hair cells, are terminally differentiated and under normal conditions do not reenter the cell cycle. Genetic manipulation of *retinoblastoma* control of the G_1–S transition can force differentiating hair cells to stay in the cell cycle (Sage et al. 2005), but normal hair cells are not thought to contribute to regeneration, except via self-repair. In the mammalian cochlea, where supporting cells show the highest degree of morphological specialization and differentiation, the normal proliferative response is minimal, and capacity for cell fate conversion is minimal outside a narrow developmental window. Supporting cells in the mammalian vestibular organs show fewer specializations than those in the cochlea, and are occasionally able to reenter the cell cycle and may be capable of redifferentiating into a hair cell by phenotypic conversion. Supporting cells in nonmammalian sensory organs also show few morphological specializations and appear generally capable of reentry into the cell cycle as

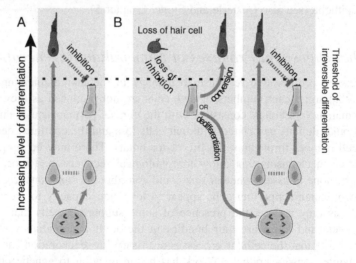

FIGURE 2.10. Schematic of how levels of differentiation may affect hair cell regeneration. (A) Multipotent progenitors (bottom) have no differentiated characteristics and can divide to produce both hair cells and supporting cells. As the hair cell differentiates, it crosses a threshold beyond which it is incapable of returning to the cell cycle or respecifying its fate. An inhibitory cue from the hair cell holds the supporting cell in a less morphologically differentiated state. (B) When the hair cell is lost, the inhibition of the supporting cell is relieved, allowing it to either directly differentiate into a hair cell (conversion) or dedifferentiate into a multipotent progenitor that returns to the cell cycle.

well as potentially able to convert directly into hair cells. Notably, in sensory epithelia that have ongoing proliferation, such as the lateral line or the avian, fish, or amphibian vestibular system, a sizable population of recently produced, partially differentiated supporting cells may reside and be capable of phenotypic plasticity or cell cycle reentry.

As the supporting cells in the mammalian cochlea and vestibular organs differentiate, they become less capable of reentry into the cell cycle as well as less capable of converting into a hair cell in response to hair cell ablation or pharmacological treatment. One could therefore imagine a hierarchy of differentiation, with the original progenitor cells for the hair cells and supporting cells serving as an undifferentiated, multipotent cell (Fig. 2.10). The progeny of that cell begin to differentiate, with the hair cell quickly crossing a threshold of differentiation beyond which the cell cannot reenter the cell cycle or change its fate (potentially controlled by the *retinoblastoma* pathway among other signaling mechanisms). In nonmammalian vertebrates, the supporting cell differentiates, but retains enough plasticity either to alter its fate directly or to return to a multipotential progenitor state. In elasmobranchs and amphibians, multipotent progenitors may also be retained at the margin between the sensory and nonsensory epithelium. In the mammalian vestibular system, the supporting cells retain plasticity for a short time after they are produced, but gradually become refractory to reentering the cell cycle or converting their fate to become hair cells. In the mammalian cochlea, the supporting cells again initially show some plasticity, but quickly specialize to a point where few can return to a multipotent progenitor or change their fate without direct genetic manipulation. As more is discovered about the genes involved in supporting cell differentiation and maintenance of a multipotential progenitor fate, several aspects of this hypothesis can be more completely tested.

10. Reinnervation of Hair Cells

In addition to the challenges faced in recovering an appropriate number of hair cells after damage, these hair cells must become integrated with the nervous system for there to be restoration of function. The recovery of hearing and vestibular function in birds in the days to weeks after loss of hair cells indicates that once hair cells regenerate, sufficient reinnervation occurs to enable communication between the sensory cells and the central nervous system (see Saunders and Salvi, Chapter 3). Ten days after acoustic overstimulation in adult quail, both efferent and afferent terminals with defined synaptic specializations were seen on regenerated hair cells (Ryals and Westbrook 1994), lagging about 3 days behind the regeneration of hair cells (Wang and Raphael 1996). In contrast, in studies of gentamicin-treated chicks, few synapses were present on regenerating hair cells after 10 days, though terminals gradually formed over several months, matured, and developed specializations, though not matching the complexity of normal synaptic terminals (Hennig and Cotanche 1998; Zakir and Dickman 2006). The difference in speed of neural reconnection may be due to the more complete lesion of hair cells and the broader systemic effects from aminoglycoside antibiotics, such as kidney damage, compared to acoustic overstimulation, which would be restricted to the basilar papilla. Nonetheless, reinnervation in the avian inner ear can recapitulate the original innervation and lead to functional recovery, suggesting that substantial plasticity is retained in the regenerating afferents and efferents. Importantly, such reinnervation requires healthy neurons. Studies in the chicken found that low doses of kainic acid would damage the glutamatergic cochlear afferent synapse but allow normal reinnervation and return of cochlear function, while high doses killed the cochlear neurons and led to irreversible loss of the synapses and cochlear function (Sun et al. 2000, 2001).

10.1 Reinnervation in Mammals

Although significant spontaneous hair cell regeneration does not occur in the mammalian cochlea, the neural fibers do show signs of plasticity that suggest that reinnervation is possible. After acoustic overstimulation, the fibers innervating the organ of Corti degenerate, though it is unclear whether the loss of the fibers is due to direct excitotoxicity from the overstimulation (Puel et al. 1994) or is secondary to the loss of hair cells (Lawner et al. 1997). While some neurons in the spiral ganglion die after overstimulation, after 1 year, nerve fibers can be seen throughout the damaged regions of the cochlea, doubling back toward the spiral ganglion, traveling laterally along the basilar membrane, or spiraling around the basilar membrane (Bohne and Harding 1992). Some of the regenerating fibers terminate on cuboidal or squamous epithelial cells while others migrate into remnants of the organ of Corti and terminate on supporting cells or surviving hair cells (Bohne and Harding 1992).

The regenerating fibers within the cochlea appear to be exclusively afferent neurites (Strominger et al. 1995), though neurons within the cochlear nucleus will sprout small axonal processes between 2 and 8 months after overstimulation-induced degeneration (Bilak et al. 1997). Thus, while hair cells are not replaced after loss in the mammalian organ of Corti, the innervating neurons can regenerate. Paired with the finding that functional recovery follows reinnervation of regenerated hair cells in the avian inner ear, this suggests that if new hair cells could be produced in mammals, they are likely to be reinnervated, at least by afferent neurons, which may lead to functional recovery.

11. Summary

Morphological evidence of regeneration and repair in the lateral line and inner ear sensory organs of vertebrates has accumulated since the 1930s. After trauma to their lateral lines, aquatic amphibians regenerate entire sensory epithelia and the individual hair cells within them. This regeneration begins when supporting cells reenter the cell cycle, replicate their chromosomes, divide, and produce new cellular progeny that differentiate as replacement hair cells and supporting cells.

Supporting cells in the ears of fish, amphibians, and birds also divide and produce new hair cells and supporting cells throughout life, contributing to significant growth of the hair cell populations in the ears of some species, continual turnover of hair cells in others, and most significantly to the regenerative replacement of hair cells that have been killed by trauma or toxicity. Reptiles have not been widely investigated, but at present it appears that mammals alone lack the capacity for effective replacement of hair cells.

Research aimed at the discovery of treatments that may stimulate regeneration of hair cells in mammals has yielded considerable progress. In mammals, there is strong evidence for recovery in the number of hair bundles after damage to the vestibular system, indicating that some process of repair or regeneration can occur in vivo. While the proliferative responses are limited in mature animals, some supporting cells divide in response to damage, even in epithelia from 50-year-old humans.

Mammalian supporting cells share many morphological characteristics with the supporting cells that are the key to the replacement of hair cells in nonmammals, but the mammalian cells become quiescent early in the postnatal maturation. Before this, the proliferation of supporting cells in neonatal mammalian vestibular epithelia can be greatly stimulated by appropriate growth factor treatments and by direct activation of intracellular signal cascades. In addition, it has been shown that the forced expression of the transcription factor *Atoh1* will result in the differentiation of new hair cells in mammalian ears. Thus, the machinery for the regeneration of mammalian hair cells is present and appears to be capable of functioning under experimental conditions.

11.1 Open Questions and Future Directions

While much progress has been made in identifying the underlying mechanisms for regeneration in the inner ear, many open questions remain. We would like to briefly summarize a few of what we feel are important morphological questions.

1. Is there a population of stem cells at the margins of the elasmobranch and amphibian sensory epithelia? Do these cells recapitulate the original developmental processes, or are the cells distinct from the original progenitors that lay out the sensory epithelium? Do the cells at the margin of the mammalian epithelia retain any of the properties of these ongoing progenitors?
2. What is the trigger for stimulating regeneration? Is this trigger missing or decreased in mammals or is it the response to the trigger that is decreased?
3. What processes underlie the dedifferentiation of supporting cells allowing them to return to the cell cycle?
4. What is the contribution of nonproliferative mechanisms to recovery following damage? Are there limits to how much hair cells can repair damage? Can protective treatments preserve hair cells and allow sufficient post-damage repair to bring about return of function? Does supporting cell conversion contribute significantly to recovery? In addition, because these mechanisms are believed to occur in mammals, but mammals do not have significant

functional recovery from significant damage, what physiological relevance do these mechanisms have in mammals versus nonmammalian vertebrates? As the mechanisms underlying nonproliferative repair are elucidated, can these mechanisms be augmented to stimulate functional recovery in mammals?

5. What is the primary limiting factor for stimulating regenerative proliferation in mammals? Is it that mammalian supporting cells are too highly differentiated to return to the cell cycle, and only a handful of stem cells remain? Or do all supporting cells retain the capacity to reenter the cell cycle if appropriately stimulated?

11.2 Conclusions

Continued research into the mechanisms of hair cell regeneration at all levels from species that produce hair cells throughout life to those where proliferation is activated only in response to damage, to mature mammalian cochleae where little regeneration occurs, is likely to provide further insight into the factors that limit regeneration. Four steps must be achieved to turn hopes of regeneration in mammalian into reality. Quiescence of supporting cells must be reversed or suspended, supporting cell proliferation must produce new cells, the new cells must be induced to differentiate as replacement hair cells, and those cells must become reinnervated and functionally integrated into the nervous system. Progress has been made toward the achievement of each of those steps, and continued research ultimately holds the promise of developing treatments to help the millions of people affected by hearing loss and vestibular dysfunctions.

References

Adler HJ, Raphael Y (1996) New hair cells arise from supporting cell conversion in the acoustically damaged chick inner ear. Neurosci Lett 205:17–20.

Adler HJ, Komeda M, Raphael Y (1997) Further evidence for supporting cell conversion in the damaged avian basilar papilla. Int J Dev Neurosci 15:375–385.

Assad JA, Shepherd GM, Corey DP (1991) Tip-link integrity and mechanical transduction in vertebrate hair cells. Neuron 7:985–994.

Avallone B, Porritiello M, Esposito D, Mutone R, Balsamo G, Marmo F (2003) Evidence for hair cell regeneration in the crista ampullaris of the lizard *Podarcis sicula*. Hear Res 178:79–88.

Baird RA, Torres MA, Schuff NR (1993) Hair cell regeneration in the bullfrog vestibular otolith organs following aminoglycoside toxicity. Hear Res 65:164–174.

Baird RA, Burton MD, Fashena DS, Naeger RA (2000) Hair cell recovery in mitotically blocked cultures of the bullfrog saccule. Proc Natl Acad Sci USA 97:11722–11729.

Balak KJ, Corwin JT, Jones JE (1990) Regenerated hair cells can originate from supporting cell progeny: evidence from phototoxicity and laser ablation experiments in the lateral line system. J Neurosci 10:2502–2512.

Berggren D, Liu W, Frenz D, Van De Water T (2003) Spontaneous hair-cell renewal following gentamicin exposure in postnatal rat utricular explants. Hear Res 180:114–125.

Bhave SA, Stone JS, Rubel EW, Coltrera MD (1995) Cell cycle progression in gentamicin-damaged avian cochleas. J Neurosci 15:4618–4628.

Bhave SA, Oestele EC, Coltrera MD (1998) Macrophage and microglia-like cells in the avian inner ear. J Comp Neurol 398:241–256.

Bilak M, Kim J, Potashner SJ, Bohne BA, Morest DK (1997) New growth of axons in the cochlear nucleus of adult chinchillas after acoustic trauma. Exp Neurol 147:256–268.

Bohne BA, Harding GW (1992) Neural regeneration in the noise-damaged chinchilla cochlea. Laryngoscope 102:693–703.

Bredberg G (1968) Cellular pattern and nerve supply of the human organ of Corti. Acta Otolaryngol Suppl 236:1–135.

Caveda L, Martin-Padura I, Navarro P, Breviario F, Corada M, Gulino D, Lampugnani MG, Dejana E (1996) Inhibition of cultured cell growth by vascular endothelial cadherin (cadherin-5/VE-cadherin). J Clin Invest 98:886–893.

Chardin S, Romand R (1995) Regeneration and mammalian auditory hair cells. Science 267:707–711.

Chardin S, Romand R (1997) Factors modulating supernumerary hair cell production in the postnatal rat cochlea in vitro. Int J Dev Neurosci 15:497–507.

Chen CS, Mrksich M, Huang S, Whitesides GM, Ingber DE (1997) Geometric control of cell life and death. Science 276:1425–1428.

Clark JA, Pickles JO (1996) The effects of moderate and low levels of acoustic over-stimulation on stereocilia and their tip links in the guinea pig. Hear Res 99:119–128.

Corwin JT (1981) Postembryonic production and aging in inner ear hair cells in sharks. J Comp Neurol 201:541–553.

Corwin JT (1983) Postembryonic growth of the macula neglecta auditory detector in the ray, *Raja clavata*: continual increases in hair cell number, neural convergence, and physiological sensitivity. J Comp Neurol 217:345–356.

Corwin JT (1985) Perpetual production of hair cells and maturational changes in hair cell ultrastructure accompany postembryonic growth in an amphibian ear. Proc Natl Acad Sci USA 82:3911–3915.

Corwin JT, Cotanche DA (1988) Regeneration of sensory hair cells after acoustic trauma. Science 240:1772–1774.

Corwin JT, Jones JE, Katayama A, Kelley MW, Warchol ME (1991) Hair cell regeneration: the identities of progenitor cells, potential triggers and instructive cues. Ciba Found Symp 160:103–120; discussion 120–130.

Cotanche DA (1987) Regeneration of hair cell stereociliary bundles in the chick cochlea following severe acoustic trauma. Hear Res 30:181–195.

Cotanche DA, Dopyera CE (1990) Hair cell and supporting cell response to acoustic trauma in the chick cochlea. Hear Res 46:29–40.

Cotanche DA, Saunders JC, Tilney LG (1987) Hair cell damage produced by acoustic trauma in the chick cochlea. Hear Res 25:267–286.

Cotanche DA, Messana EP, Ofsie MS (1995) Migration of hyaline cells into the chick basilar papilla during severe noise damage. Hear Res 91:148–159.

Daudet N, Vago P, Ripoll C, Humbert G, Pujol R, Lenoir M (1998) Characterization of atypical cells in the juvenile rat organ of Corti after aminoglycoside ototoxicity. J Comp Neurol 401:145–162.

Daudet N, Ripoll C, Lenoir M (2002) Transforming growth factor-alpha-induced cellular changes in organotypic cultures of juvenile, amikacin-treated rat organ of corti. J Comp Neurol 442:6–22.

Davies D, Magnus C, Corwin JT (2007) Developmental changes in cell-extracellular matrix interactions limit proliferation in the mammalian inner ear. Eur J Neurosci 25:985–998.

Engstrom B, Flock A, Borg E (1983) Ultrastructural studies of stereocilia in noise-exposed rabbits. Hear Res 12:251–264.

Ernest S, Rauch GJ, Haffter P, Geisler R, Petit C, Nicolson T (2000) Mariner is defective in myosin VIIA: a zebrafish model for human hereditary deafness. Hum Mol Genet 9:2189–2196.

Fagotto F, Gumbiner BM (1996) Cell contact-dependent signaling. Dev Biol 180:445–454.

Fischer AJ, Reh TA (2000) Identification of a proliferating marginal zone of retinal progenitors in postnatal chickens. Dev Biol 220:197–210.

Folkman J, Moscona A (1978) Role of cell shape in growth control. Nature 273:345–349.

Forge A (1985) Outer hair cell loss and supporting cell expansion following chronic gentamicin treatment. Hear Res 19:171–182.

Forge A, Li L, Corwin JT, Nevill G (1993) Ultrastructural evidence for hair cell regeneration in the mammalian inner ear. Science 259:1616–1619.

Forge A, Li L, Nevill G (1998) Hair cell recovery in the vestibular sensory epithelia of mature guinea pigs. J Comp Neurol 397:69–88.

Fredelius L (1988) Time sequence of degeneration pattern of the organ of Corti after acoustic overstimulation. A transmission electron microscopy study. Acta Otolaryngol 106:373–385.

Fredelius L, Rask-Andersen H (1990) The role of macrophages in the disposal of degeneration products within the organ of corti after acoustic overstimulation. Acta Otolaryngol 109:76–82.

Gailit J, Clark RA (1994) Wound repair in the context of extracellular matrix. Curr Opin Cell Biol 6:717–725.

Gale JE, Meyers JR, Periasamy A, Corwin JT (2002) Survival of bundleless hair cells and subsequent bundle replacement in the bullfrog's saccule. J Neurobiol 50:81–92.

Geleoc GS, Holt JR (2003) Developmental acquisition of sensory transduction in hair cells of the mouse inner ear. Nat Neurosci 6:1019–1020.

Girod DA, Duckert LG, Rubel EW (1989) Possible precursors of regenerated hair cells in the avian cochlea following acoustic trauma. Hear Res 42:175–194.

Gu R, Marchionni M, Corwin JT (1996) Glial growth factor enhances supporting cell proliferation in rodent vestibular epithelia cultured in isolation. Soc Neurosci Abstr 21:520.

Gu R, Marchionni M, Corwin JT (1997) Age-related decreases in proliferation within isolated mammalian vestibular epithelia cultured in control and glial growth factor 2 medium. Assoc Res Otolaryngol Abstr 20:98.

Gu R, Montcouquiol M, Marchionni M, Corwin JT (2007) Proliferative responses to growth factors decline rapidly during postnatal maturation of mammalian hair cell epithelia. Eur J Neurosci 25:1363–1372.

Hashino E, Salvi RJ (1993) Changing spatial patterns of DNA replication in the noise-damaged chick cochlea. J Cell Sci 105 (Pt 1):23–31.

Hennig AK, Cotanche DA (1998) Regeneration of cochlear efferent nerve terminals after gentamycin damage. J Neurosci 18:3282–3296.

Hirose K, Discolo CM, Keasler JR, Ransohoff R (2005) Mononuclear phagocytes migrate into the murine cochlea after acoustic trauma. J Comp Neurol 489:180–194.

Huang S, Ingber DE (1999) The structural and mechanical complexity of cell-growth control. Nat Cell Biol 1:E131–138.

Hume CR, Kirkegaard M, Oesterle EC (2003) ErbB expression: the mouse inner ear and maturation of the mitogenic response to heregulin. J Assoc Res Otolaryngol 4:422–443.

Husbands JM, Steinberg SA, Kurian R, Saunders JC (1999) Tip-link integrity on chick tall hair cell stereocilia following intense sound exposure. Hear Res 135:135–145.

Itoh M, Chitnis AB (2001) Expression of proneural and neurogenic genes in the zebrafish lateral line primordium correlates with selection of hair cell fate in neuromasts. Mech Dev 102:263–266.

Izumikawa M, Minoda R, Kawamoto K, Abrashkin KA, Swiderski DL, Dolan DF, Brough DE, Raphael Y (2005) Auditory hair cell replacement and hearing improvement by Atoh1 gene therapy in deaf mammals. Nat Med 11:271–276.

Johns PR (1977) Growth of the adult goldfish eye. III. Source of the new retinal cells. J Comp Neurol 176:343–357.

Johnston LA, Edgar BA (1998) Wingless and Notch regulate cell-cycle arrest in the developing Drosophila wing. Nature 394:82–84.

Jones JE, Corwin JT (1993) Replacement of lateral line sensory organs during tail regeneration in salamanders: identification of progenitor cells and analysis of leukocyte activity. J Neurosci 13:1022–1034.

Jones JE, Corwin JT (1996) Regeneration of sensory cells after laser ablation in the lateral line system: hair cell lineage and macrophage behavior revealed by time-lapse video microscopy. J Neurosci 16:649–662.

Jørgensen JM, Mathiesen C (1988) The avian inner ear. Continuous production of hair cells in vestibular sensory organs, but not in the auditory papilla. Naturwissenschaften 75:319–320.

Jørgensen JM, Flock Å (1976) Non-innervated sense organs of the lateral line: development in the regenerating tail of the salamander Ambystoma mexicanum. J Neurocytol 5:33–41.

Kaltenbach JA, Falzarano PR (1994) Postnatal development of the hamster cochlea. I. Growth of hair cells and the organ of Corti. J Comp Neurol 340:87–97.

Kaltenbach JA, Falzarano PR, Simpson TH (1994) Postnatal development of the hamster cochlea. II. Growth and differentiation of stereocilia bundles. J Comp Neurol 350:187–198.

Katayama A, Corwin JT (1989) Cell production in the chicken cochlea. J Comp Neurol 281:129–135.

Kawamoto K, Ishimoto S, Minoda R, Brough DE, Raphael Y (2003) Math1 gene transfer generates new cochlear hair cells in mature guinea pigs in vivo. J Neurosci 23:4395–4400.

Kelley MW, Xu XM, Wagner MA, Warchol ME, Corwin JT (1993) The developing organ of Corti contains retinoic acid and forms supernumerary hair cells in response to exogenous retinoic acid in culture. Development 119:1041–1053.

Kil J, Warchol ME, Corwin JT (1997) Cell death, cell proliferation, and estimates of hair cell life spans in the vestibular organs of chicks. Hear Res 114:117–126.

Kimura K, Usui-Ishihara A, Usui K (1997) G2 arrest of cell cycle ensures a determination process of sensory mother cell formation in g. Dev Genes Evol 207:199–202.

Kirkegaard M, Jørgensen JM (2001) The inner ear macular sensory epithelia of the Daubenton's bat. J Comp Neurol 438:433–444.

Lambert PR, Gu R, Corwin JT (1997) Analysis of small hair bundles in the utricles of mature guinea pigs. Am J Otol 18:637–643.

Lanford PJ, Presson JC, Popper AN (1996) Cell proliferation and hair cell addition in the ear of the goldfish, Carassius auratus. Hear Res 100:1–9.

Lawner BE, Harding GW, Bohne BA (1997) Time course of nerve-fiber regeneration in the noise-damaged mammalian cochlea. Int J Dev Neurosci 15:601–617.

Ledent V (2002) Postembryonic development of the posterior lateral line in zebrafish. Development 129:597–604.

Lefebvre PP, Malgrange B, Staecker H, Moonen G, Van de Water TR (1993) Retinoic acid stimulates regeneration of mammalian auditory hair cells. Science 260:692–695.

Lefebvre PP, Malgrange B, Thiry M, Van De Water TR, Moonen G (2000) Epidermal growth factor upregulates production of supernumerary hair cells in neonatal rat organ of corti explants. Acta Otolaryngol 120:142–145.

Lenoir M, Vago P (1997) Does the organ of Corti attempt to differentiate new hair cells after antibiotic intoxication in rat pups? Int J Dev Neurosci 15:487–495.

Lewis ER, Li CW (1973) Evidence concerning the morphogenesis of saccular receptors in the bullfrog (Rana catesbeiana). J Morphol 139:351–361.

Li CW, Lewis ER (1979) Structure and development of vestibular hair cells in the larval bullfrog. Ann Otol Rhinol Laryngol 88:427–437.

Li H, Liu H, Heller S (2003) Pluripotent stem cells from the adult mouse inner ear. Nat Med 9:1293–1299.

Li L, Forge A (1997) Morphological evidence for supporting cell to hair cell conversion in the mammalian utricular macula. Int J Dev Neurosci 15:433–446.

Li L, Nevill G, Forge A (1995) Two modes of hair cell loss from the vestibular sensory epithelia of the guinea pig inner ear. J Comp Neurol 355:405–417.

Lippe WR, Westbrook EW, Ryals BM (1991) Hair cell regeneration in the chicken cochlea following aminoglycoside toxicity. Hear Res 56:203–210.

Lombarte A, Popper AN (1994) Quantitative analyses of postembryonic hair cell addition in the otolithic endorgans of the inner ear of the European hake, Merluccius merluccius (Gadiformes, Teleostei). J Comp Neurol 345:419–428.

Lombarte A, Yan HY, Popper AN, Chang JS, Platt C (1993) Damage and regeneration of hair cell ciliary bundles in a fish ear following treatment with gentamicin. Hear Res 64:166–174.

Marsh RR, Xu LR, Moy JP, Saunders JC (1990) Recovery of the basilar papilla following intense sound exposure in the chick. Hear Res 46:229–237.

Martin P (1997) Wound healing—aiming for perfect skin regeneration. Science 276:75–81.

Matsui JI, Oesterle EC, Stone JS, Rubel EW (2000) Characterization of damage and regeneration in cultured avian utricles. J Assoc Res Otolaryngol 1:46–63.

Matsui JI, Ogilvie JM, Warchol ME (2002) Inhibition of caspases prevents ototoxic and ongoing hair cell death. J Neurosci 22:1218–1227.

Matsui JI, Haque A, Huss D, Messana EP, Alosi JA, Roberson DW, Cotanche DA, Dickman JD, Warchol ME (2003) Caspase inhibitors promote vestibular hair cell survival and function after aminoglycoside treatment in vivo. J Neurosci 23:6111–6122.

Mbiene JP, Sans A (1986) Differentiation and maturation of the sensory hair bundles in the fetal and postnatal vestibular receptors of the mouse: a scanning electron microscopy study. J Comp Neurol 254:271–278.

Meyers JR, Corwin JT (2007) Shape change controls supporting cell proliferation in lesioned mammalian balance epithelium. J Neurosci 27:4313–4325.

Montcouquiol M, Corwin JT (2001a) Brief treatments with forskolin enhance s-phase entry in balance epithelia from the ears of rats. J Neurosci 21:974–982.

Montcouquiol M, Corwin JT (2001b) Intracellular signals that control cell proliferation in mammalian balance epithelia: key roles for phosphatidylinositol-3 kinase, mammalian target of rapamycin, and S6 kinases in preference to calcium, protein kinase C, and mitogen-activated protein kinase. J Neurosci 21:570–580.

Nakagawa T, Yamane H, Shibata S, Takayama M, Sunami K, Nakai Y (1997) Two modes of auditory hair cell loss following acoustic overstimulation in the avian inner ear. ORL J Otorhinolaryngol Relat Spec 59:303–310.

Osborne MP, Comis SD (1990a) Action of elastase, collagenase and other enzymes upon linkages between stereocilia in the guinea-pig cochlea. Acta Otolaryngol 110:37–45.

Osborne MP, Comis SD (1990b) High resolution scanning electron microscopy of stereocilia in the cochlea of normal, postmortem, and drug-treated guinea pigs. J Electron Microsc Tech 15:245–260.

Paige GD (1992) Senescence of human visual-vestibular interactions. 1. Vestibulo-ocular reflex and adaptive plasticity with aging. J Vestib Res 2:133–151.

Pickles JO, Comis SD, Osborne MP (1984) Cross-links between stereocilia in the guinea pig organ of Corti, and their possible relation to sensory transduction. Hear Res 15:103–112.

Pickles JO, Comis SD, Osborne MP (1987a) The effect of chronic application of kanamycin on stereocilia and their tip links in hair cells of the guinea pig cochlea. Hear Res 29:237–244.

Pickles JO, Osborne MP, Comis SD (1987b) Vulnerability of tip links between stereocilia to acoustic trauma in the guinea pig. Hear Res 25:173–183.

Popper AN, Hoxter B (1984) Growth of a fish ear: 1. Quantitative analysis of hair cell and ganglion cell proliferation. Hear Res 15:133–142.

Popper AN, Hoxter B (1990) Growth of a fish ear. II. Locations of newly proliferated sensory hair cells in the saccular epithelium of Astronotus ocellatus. Hear Res 45:33–40.

Presson JC, Popper AN (1990) Possible precursors to new hair cells, support cells, and Schwann cells in the ear of a post-embryonic fish. Hear Res 46:9–21.

Presson JC, Smith T, Mentz L (1995) Proliferating hair cell precursors in the ear of a postembryonic fish are replaced after elimination by cytosine arabinoside. J Neurobiol 26:579–584.

Presson JC, Lanford PJ, Popper AN (1996) Hair cell precursors are ultrastructurally indistinguishable from mature support cells in the ear of a postembryonic fish. Hear Res 100:10–20.

Puel JL, Pujol R, Tribillac F, Ladrech S, Eybalin M (1994) Excitatory amino acid antagonists protect cochlear auditory neurons from excitotoxicity. J Comp Neurol 341:241–256.

Raphael Y (1992) Evidence for supporting cell mitosis in response to acoustic trauma in the avian inner ear. J Neurocytol 21:663–671.

Ricci AJ, Kachar B, Gale J, Van Netten SM (2006) Mechano-electrical transduction: new insights into old ideas.J Membr Biol 209:71–88.

Roberson DF, Weisleder P, Bohrer PS, Rubel EW (1992) Ongoing production of sensory cells in the vestibular epithelium of the chick. Hear Res 57:166–174.

Roberson DW, Kreig CS, Rubel EW (1996) Light microscopic evidence that direct transdifferentiation gives rise to new hair cells in regenerating avian auditory epithelium. Audit Neurosci 2:195–205.

Roberson DW, Alosi JA, Cotanche DA (2004) Direct transdifferentiation gives rise to the earliest new hair cells in regenerating avian auditory epithelium. J Neurosci Res 78:461–471.

Romand R, Chardin S, Le Calvez S (1996) The spontaneous appearance of hair cell-like cells in the mammalian cochlea following aminoglycoside ototoxicity. NeuroReport 8:133–137.

Rosenhall U (1972) Vestibular macular mapping in man. Ann Otol Rhinol Laryngol 81:339–351.

Rosenhall U (1973) Degenerative patterns in the aging human vestibular neuro-epithelia. Acta Otolaryngol 76:208–220.

Ruben RJ (1967) Development of the inner ear of the mouse: a radioautographic study of terminal mitoses. Acta Otolaryngol:Suppl 220:221–244.

Ryals BM, Rubel EW (1988) Hair cell regeneration after acoustic trauma in adult Coturnix quail. Science 240:1774–1776.

Ryals BM, Westbrook EW (1990) Hair cell regeneration in senescent quail. Hear Res 50:87–96.

Ryals BM, Westbrook EW (1994) TEM analysis of neural terminals on autoradiographically identified regenerated hair cells. Hear Res 72:81–88.

Sage C, Huang M, Karimi K, Gutierrez G, Vollrath MA, Zhang DS, Garcia-Anoveros J, Hinds PW, Corwin JT, Corey DP, Chen ZY (2005) Proliferation of functional hair cells in vivo in the absence of the retinoblastoma protein. Science 307:1114–1118.

Sans A, Chat M (1982) Analysis of temporal and spatial patterns of rat vestibular hair cell differentiation by tritiated thymidine radioautography. J Comp Neurol 206:1–8.

Schneider ME, Belyantseva IA, Azevedo RB, Kachar B (2002) Rapid renewal of auditory hair bundles. Nature 418:837–838.

Shou J, Zheng JL, Gao WQ (2003) Robust generation of new hair cells in the mature mammalian inner ear by adenoviral expression of Hath1. Mol Cell Neurosci 23:169–179.

Sobkowicz HM, Slapnick SM, August BK (1995) The kinocilium of auditory hair cells and evidence for its morphogenetic role during the regeneration of stereocilia and cuticular plates. J Neurocytol 24:633–653.

Song J, Yan HY, Popper AN (1995) Damage and recovery of hair cells in fish canal (but not superficial) neuromasts after gentamicin exposure. Hear Res 91:63–71.

Speidel C (1947) Correlated studies of sense organs and nerves of the lateral-line in living frog tadpoles I. Regeneration of denervated organs. J Comp Neurol 87:27–55.

Steyger PS, Burton M, Hawkins JR, Schuff NR, Baird RA (1997) Calbindin and parvalbumin are early markers of non-mitotically regenerating hair cells in the bullfrog vestibular otolith organs. Int J Dev Neurosci 15:417–432.

Stone JS, Cotanche DA (1994) Identification of the timing of S phase and the patterns of cell proliferation during hair cell regeneration in the chick cochlea. J Comp Neurol 341:50–67.

Stone LS (1933) The development of lateral-line sense organs in amphibians observed in living and vital-stained preparations. J Comp Neurol 57:507–540.

Stone LS (1937) Further experimental studies of the development of lateral-line sense organs in amphibians observed in living preparations. J Comp Neurol 68:83–115.

Straznicky K, Gaze RM (1971) The growth of the retina in Xenopus laevis: an autoradiographic study. J Embryol Exp Morphol 26:67–79.

Strominger RN, Bohne BA, Harding GW (1995) Regenerated nerve fibers in the noise-damaged chinchilla cochlea are not efferent. Hear Res 92:52–62.

Sun H, Salvi RJ, Ding DL, Hashino DE, Shero M, Zheng XY (2000) Excitotoxic effect of kainic acid on chicken otoacoustic emissions and cochlear potentials. J Acoust Soc Am 107:2136–2142.

Sun H, Hashino E, Ding DL, Salvi RJ (2001) Reversible and irreversible damage to cochlear afferent neurons by kainic acid excitotoxicity. J Comp Neurol 430:172–181.

Tanaka EM, Gann AA, Gates PB, Brockes JP (1997) Newt myotubes reenter the cell cycle by phosphorylation of the retinoblastoma protein. J Cell Biol 136:155–165.

Taylor RR, Forge A (2005) Hair cell regeneration in sensory epithelia from the inner ear of a urodele amphibian. J Comp Neurol 484:105–120.

Tilney LG, Tilney MS, Saunders JS, DeRosier DJ (1986) Actin filaments, stereocilia, and hair cells of the bird cochlea. III. The development and differentiation of hair cells and stereocilia. Dev Biol 116:100–118.

Wang Y, Raphael Y (1996) Re-innervation patterns of chick auditory sensory epithelium after acoustic overstimulation. Hear Res 97:11–18.

Wang Z, Li H (2000) Microglia-like cells in rat organ of Corti following aminoglycoside ototoxicity. NeuroReport 11:1389–1393.

Warchol ME (1995) Supporting cells in isolated sensory epithelia of avian utricles proliferate in serum-free culture. NeuroReport 6:981–984.

Warchol ME (1997) Macrophage activity in organ cultures of the avian cochlea: demonstration of a resident population and recruitment to sites of hair cell lesions. J Neurobiol 33:724–734.

Warchol ME (1999) Immune cytokines and dexamethasone influence sensory regeneration in the avian vestibular periphery. J Neurocytol 28:889–900.

Warchol ME (2002) Cell density and N-cadherin interactions regulate cell proliferation in the sensory epithelia of the inner ear. J Neurosci 22:2607–2616.

Warchol ME, Corwin JT (1996) Regenerative proliferation in organ cultures of the avian cochlea: identification of the initial progenitors and determination of the latency of the proliferative response. J Neurosci 16:5466–5477.

Warchol ME, Lambert PR, Goldstein BJ, Forge A, Corwin JT (1993) Regenerative proliferation in inner ear sensory epithelia from adult guinea pigs and humans. Science 259:1619–1622.

Weisleder P, Rubel EW (1993) Hair cell regeneration after streptomycin toxicity in the avian vestibular epithelium. J Comp Neurol 331:97–110.

Wetts R, Fraser SE (1988) Multipotent precursors can give rise to all major cell types of the frog retina. Science 239:1142–1145.

Witte MC, Montcouquiol M, Corwin JT (2001) Regeneration in avian hair cell epithelia: identification of intracellular signals required for S-phase entry. Eur J Neurosci 14:829–838.

Wright A, Davis A, Bredberg G, Ulehlova L, Spencer H (1987) Hair cell distributions in the normal human cochlea. Acta Otolaryngol Suppl 444:1–48.

Wright M (1947) Regeneration and degeneration experiments on lateral line nerves and sense organs in anurans. J Exp Zool 105:221–257.

Zakir M, Dickman JD (2006) Regeneration of vestibular otolith afferents after ototoxic damage. J Neurosci 26:2881–2893.

Zhao Y, Yamoah EN, Gillespie PG (1996) Regeneration of broken tip links and restoration of mechanical transduction in hair cells. Proc Natl Acad Sci USA 93:15469–15474.

Zheng JL, Gao WQ (2000) Overexpression of *Math1* induces robust production of extra hair cells in postnatal rat inner ears. Nat Neurosci 3:580–586.

Zheng JL, Keller G, Gao WQ (1999) Immunocytochemical and morphological evidence for intracellular self-repair as an important contributor to mammalian hair cell recovery. J Neurosci 19:2161–2170.

3
Recovery of Function in the Avian Auditory System After Ototrauma

James C. Saunders and Richard J. Salvi

1. Introduction

The auditory system of birds has been scrutinized in ever increasing detail since the discovery of hair cell regeneration in the late 1980s, and numerous reviews of hair cell regeneration have appeared in recent years (Cotanche et al. 1994; Forge 1996; Cotanche 1997, 1999; Smolders 1999; Stone and Rubel 2000). It was recognized early that if functional recovery in the peripheral auditory system accompanied the appearance of new hair cells, then the phenomenon of hair cell regeneration would be much more than a biologic curiosity. During the ensuing two and a half decades, it has become amply clear that the emergence of new hair cells is one aspect of a complicated array of structural repair processes in the avian inner ear (Cotanche 1999). Moreover, this repair is accompanied by nearly complete recovery of auditory capability (Smolders 1999). Nevertheless, the contribution of regenerated hair cells to the restoration of auditory function is not fully understood, and depends on the ototraumatic events that caused inner ear damage and hair cell destruction.

Peripheral auditory function might be considered from different perspectives. Using the patterns of cochlear (auditory) nerve discharge as an example, activity might be seen as a cipher in its own right, coding various properties of the acoustic stimulus. Another approach might consider the same activity as a proxy for inner ear processes. Tuning curves, for example, could provide an indication of the frequency analytic mechanisms in the cochlea; phase-locking and adaptation behavior might reflect membrane and synaptic mechanisms of the hair cell, while rate-level functions and two-tone suppression reflect nonlinear inner-ear processes. Other peripheral responses such as the endocochlear potential or otoacoustic and electrically evoked emissions provide additional insight to the health of the inner ear and its analytic mechanisms. These and other phenomena can be associated with particular tissue or cellular compartments, or analytic processes, within the inner ear. Abnormal behavior in any aspect of peripheral physiology provides a window into the pathophysiology of the inner ear induced by ototrauma.

Two procedures have evolved for damaging the chick basilar papilla. The first overstimulates the ear with intense sound exposures, and in birds pure tones have been used most frequently. Acoustic trauma in the chick produces two well defined lesions. The so called "patch" lesion, located over the basilar membrane (at the tonotopic location of the exposure frequency), damages principally abneural hair cells. Depending on exposure level and animal age, the hair cell destruction in the patch may be relatively modest to substantial. With the exception of tip-link loss, most neural hair cells survive the exposure unharmed. A second, much less understood area of damage is the "stripe" lesion, which occurs on the high-frequency side of the patch, and is seen as an area of hair cell destruction, perhaps several hair cells wide, lying along the midline of the papilla.

The second procedure uses ototoxic agents such as gentamicin or kanamycin to produce a massive lesion of hair cell destruction. The extent of papilla damage is limited by the nephrotoxic side effects, and animal survival depends on the dose level and number of daily injections. Ototoxic drugs have been applied systemically or directly through the round window of the labyrinth. All hair cells are completely destroyed over the basal half of the papilla. Trauma to other cells of the papilla or the tectorial membrane is minimized by this treatment.

In this chapter, the focus is mainly on the peripheral physiology of the avian ear, particularly that of the chicken, and when considering structural and functional loss from intense sound exposure, the discussion is confined to relatively mild exposures rather than those that create extreme levels of damage.

2. Peripheral Auditory Processes in Normal and Damaged Ears

2.1 The Endocochlear Potential

The endocochlear potential (EP) is a DC voltage that is positive in scala media, and is measured with respect to the adjacent fluid compartments of the inner ear. The fluid in the avian scala media (referred to as endolymph) is high in potassium and low in sodium (much like that of the intracellular environment), and nearly identical to that found in the mammalian scala media. Mammalian and avian perilymphs, in scala vestibule and tympani, have nearly identical ionic concentrations, and are much like that of extracellular fluid (e.g., high sodium and low potassium; Sterkers et al. 1988).

Specialized cells that secrete K^+ and other ions into scala media, and the permeability gradient of the tissue compartments lining scala media, determine the EP voltage level (Salt et al. 1987). In mammals, the EP is approximately 80 mV (von Békésy et al. 1952) while in the young chick it is between 8 and 16 mV, increasing in the adult to between 18 and 23 mV (Trautwein et al. 1997). These levels are typical of all avian species, and the question is, how do mammals and birds, with the same approximate ionic concentrations, produce such dramatic voltage differences? The most frequent explanation evokes the concept of "leakiness" in the cellular lining of scala media. If K^+ ions are more easily shunted across the endolymph/perilymph boundaries of scala media, then the resulting EP voltage should be lower.

Potassium is secreted into scala media via the dark cells of the tegmentum vasculosum (Hara et al. 2002). The tegmentum is a richly vascularized structure whose organization is homologous to that of the mammalian stria vascularis. However, the tegmentum forms a tissue boundary between scala vestibuli and scala media, occupying the same physical location as Reissners' membrane in the mammalian cochlea. The tegmentum is thus mechanically vibrated by the propagation of acoustic pressure waves through the cochlear fluids, and during intense sound stimulation there is the possibility of inducing tissue damage.

The EP plays an important role in hair cell transduction. Ion channels in the tips and shafts of the hair cell stereocilia carry hair cell transduction currents. These channels are gated mechanically by tension exerted through tip-links, fine threadlike structures interconnecting the tip and shaft of adjacent sensory hairs. (Markin and Hudspeth 1995; Kachar et al. 2000). The dominant extracellular ion is K^+, which is interesting because the open transduction channel presents an already high intracellular K^+ environment. The flow of K^+ into the hair cell is aided by the EP, and even at 15 mV provides a driving force that "pushes" K^+ ions into the opened transduction channel. Increases in intracellular K^+ concentration depolarize the hair cell membrane, triggering events that lead to exocytosis of neurotransmitter vesicles and activation of the auditory nerve. Reducing EP amplitude by injecting current into scala media attenuates sound-driven cochlear nerve unit activity, but causes little change in spontaneous activity (Vossieck et al. 1991).

Acoustic overstimulation in young chicks, sufficient to cause hair cell loss and substantial shifts in threshold, also produced dynamic changes in the EP. A 48-hour pure tone exposure, between 1 and 3 days of age, caused a 63% loss in EP shortly after removal from the exposure. Within 4 days postexposure the EP level recovered to normal. Figure 3.1 demonstrates that the time course of EP recovery mimics the rate of evoked response threshold recovery seen in the chick cochlear nucleus after a similar exposure. On the basis of this relationship, it was hypothesized that the loss and recovery of EP played a role in the loss and recovery of peripheral function (Saunders et al. 1996b). Similar recordings in adult chickens, after acoustic overstimulation sufficiently intense to cause hair cell loss and threshold shifts, failed to show any postexposure change in the positive EP (Trautwein et al. 1997). However, the acoustic trauma altered the negative EP recorded during anoxia. The negative EP was still reduced 4 months postexposure suggesting a long-term disruption of this potential. The difference between very young and mature animals was striking, but there is reason to believe that both observations are valid.

A histologic evaluation of the tegmentum vasculosum in the adult quail after an intense pure-tone exposure for 12 hours revealed considerable injury (Ryals et al. 1995). With 6 days of recovery, the tegmentum regained a normal

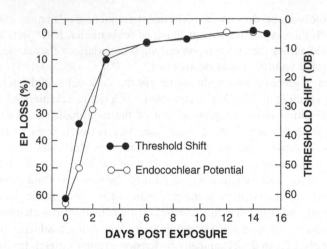

FIGURE 3.1. Endocochlear potential (EP) loss is plotted against postexposure time in young chicks. Also plotted is evoked response threshold shift (in dB), averaged for 0.9, 1.3, and 1.5 kHz, in similarly exposed chicks at the same age. Within 3 days, the EP and threshold shifts are nearly fully recovered. The correspondence between these data suggests that EP recovery might play a role in evoked response threshold recovery (Data from Poje et al. 1995, with permission.)

appearance. Endocochlear potential changes in the sound damaged quail ear have yet to be reported, though other measures of peripheral auditory function recover in the same time frame as the tegmentum repair. Similar postexposure morphologic damage and recovery of the tegmentum has been reported in the chick. The area occupied by dark cells was greatly reduced immediately after removal from overstimulation, and appeared normal 6 days later (Ramakrishna et al. 2004; Askew et al. 2006).

The postexposure loss of EP in the chick could arise from disruption of signaling pathways leading to K^+ secretion and modification in the number or types of ion pumps in the tegmentum, or from greater leakiness of K^+ in the walls of scala media due to acoustic damage. It is likely that the differences between the adult and young chicken are due to the immaturity of the chick inner ear at the time of exposure (Trautwein et al. 1997). Regardless, additional research is needed to understand more fully the role of ionic homeostasis, particularly the loss and recovery of tegmental structure and EP function after intense sound exposure.

High doses of aminoglycoside antibiotics such as kanamycin and gentamicin have also been used to damage the sensory cells in the avian ear, and over time, the hair cells exhibit regeneration. Interestingly, adult chickens treated with kanamycin for 10 days failed to show changes in the steady state EP (Chen et al. 1995). Nevertheless, other sound-driven functional deficits were apparent and these also recovered with the passage of time. This observation suggested that ototoxic damage to the epithelium does not injure the EP, perhaps because of the absence of any mechanical trauma to the papilla. It further suggested that impairment to sound-driven activity can arise from sources other than the driving force behind the hair cell transduction current.

The EP might be viewed as a "gross" potential within the cochlea. There are other gross potentials such as the cochlear microphonic (CM), summating potential, and the compound action potential (CAP). All these suffer changes as a consequence of ototrauma. Changes in CAP thresholds, for example, show deterioration after exposure to both intense sound and aminoglycoside treatment (Chen et al. 1993; Müller et al. 1996, 1997).

2.2 Innervation of the Avian Cochlea

The organization and innervation of the avian cochlear nerve are important for understanding and interpreting results from single-unit studies in the normal and damaged ear. Hair cell innervation and cochlear nerve organization are well understood on the chicken basilar papilla (Fischer et al. 1992; Fischer 1994). The avian cochlear nerve contains three neuronal components: the auditory afferents, lagenar afferents, and a small component of efferent fibers (Köppl et al. 2000). In the chick, there are approximately 12,400 auditory afferent fibers, all myelinated, and of a fairly uniform diameter (approximately 2.0 μm). Innervation density is greatest in the area located between 40% and 70% from the apex (approximately 0.9–3.0 kHz). There are estimated to be between 100 and 200 efferent fibers in the chick, and because all hair cells synapse with at least one efferent fiber, the small number implies considerable divergence of innervation. These efferent fibers are myelinated and exhibit a diameter similar to the afferent fibers. There is evidence of unmyelinated fibers in the cochlear nerve, but their numbers are few, and it is uncertain if they are autonomic nervous system fibers, traversing the cochlear nerve, or perhaps isolated efferent fibers. The interesting aspect of this innervation pattern is the absence of unmyelinated neurons similar to the type II afferents innervating the outer hair cells (OHCs) of the mammalian cochlea.

Three types of hair cells were described across the sensory sheet of the basilar papilla: tall, intermediate, and short. These names were derived from a consideration of the ratio of cell apical diameter to its length (Smith 1985). Short hair cells were thought to occupy the abneural half of the papilla, whereas tall hair cells were found on the neural half (Fig. 3.2), with intermediate cells in between. This description has been refined to take into account the innervation pattern of the hair cells rather than their morphology. Those hair cells on the far abneural edge of the papilla, innervated exclusively by efferent nerve fibers with large chalice-like synaptic boutons (see Fig. 3.2), are now considered the short hair cells (Manley et al. 1989). Throughout the remainder of the chapter we will refer to neural and abneural hair cells. The neural cells are equivalent to the tall hair cells lying over the superior fibrocartilagenous plate, while the abneural hair cells are situated over the basilar membrane.

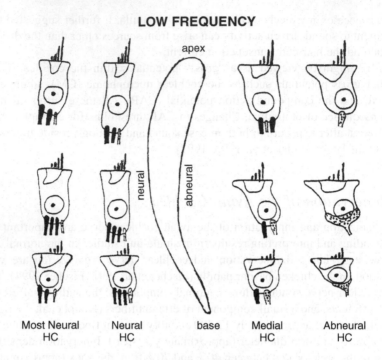

LOW FREQUENCY

HIGH FREQUENCY

apex

neural

abneural

base

Most Neural   Neural   Medial   Abneural
HC   HC   HC   HC

FIGURE 3.2. This cartoon depicts the innervation and organization of chick neural (left side) and abneural (right side) hair cells. The afferent fibers are colored black while efferent fibers are indicated by the speckles. (Modified from Fisher 1994, with permission.)

Changes occur in hair cell innervation from the abneural to neural edges of the papilla. The abneural hair cells have large efferent endings and minimal size afferent endings. Neural hair cells exhibit large afferent endings with small efferent terminals (Fig. 3.2). Depending on location, the number of afferent fibers that innervate each neural hair cell is between 1.4 and 2.4 (Fig. 3.2).

Ototrauma from loud sound or ototoxic drugs results in hair cell loss, which inevitably means that the neural connections with the hair cell are broken. These connections *must* repair themselves if auditory function is to return to normal (Ryals and Dooling 1996). Issues such as the postexposure fate of synaptic boutons, the degree of axonal degeneration and recovery, the mechanisms that guide neurons back to their correct tonotopic location, the time course of neural recovery, and the importance of innervation in inducing hair cell regeneration are far from well understood. Indeed, it has been a vexing question to clearly trace the degeneration and/or recovery of afferent and efferent neurons after ototrauma.

Transmission electron microscopy of serial sections, antibody labeling of neurofilament proteins in axons, and the application of synapsin or syntaxin-specific immunohistochemistry, have been used to gain some idea of the loss and recovery of afferent and efferent elements in the cochlear nerve. Four scenarios for the fate of cochlear neurons have been proposed (Cotanche et al. 1994): (1) The axons remain connected to the synaptic bouton, and the bouton itself separates from the dying hair cell. In this situation the neural elements remain in place, and simply await the reemergence of new hair cells with which to make contact. (2) The afferent and efferent terminals remain attached to the hair cell with the more central axon projection breaking off. Despite the loss of the synaptic terminal, as the hair cell degenerates, the nerve fibers remain intact, healthy, and in position. In this case the sprouting of new terminals would be required for reinnervation with new hair cells. (3) The synaptic endings of the neurons are lost, which leads to degeneration of afferent and efferent fibers. As these fibers degenerate they retract toward the cochlear ganglion cell body. Repair, in this situation, would require axonal regrowth, and the formation of a new synaptic bouton before reinnervation could occur. (4) Axonal degeneration might be so severe that it leads to the loss of ganglion cell bodies, and in this scenerio repair might be impossible.

Synapses have been observed on newly regenerated hair cells 3–4 days after acoustic trauma, indicating that reinnervation does indeed take place (Ryals and Dooling 1996). Moreover, neurofilament labeling of the cochlear nerve, in moderately damaged ears, revealed that the synaptic endings, while separated from the axon, appeared to remain attached to dying hair cells. The axons, even though separated from the bouton, remain in the immediate vicinity of the hair cells to which they were originally connected (Ofsie and Cotanche 1996; Ofsie et al. 1997). In addition, there was no apparent loss in the number of fibers or their pattern of distribution on the papilla. The tonotopic specificity of the papilla remained intact, and this was supported by fiber tracing, which revealed the papilla frequency map unchanged after hair cell regeneration (Chen et al. 1996). It has also been reported that efferent fibers remained in contact with surviving hair cells in the lesion, implying that the afferent connections on these cells were also intact.

Immunolabeling of the papilla with synapsin, a marker for small synaptic vesicles in neurons, indicated that new hair cells were not initially associated with the presence of efferent fiber terminals after acoustic overstimulation (Ofsie et al. 1997). However, within 7 days of the emergence of new hair cells the first efferent terminals were identified, thus providing additional evidence that the presence of nerve endings was not a requisite for the emergence of new hair cells. With severe acoustic damage to the papilla (e.g., damage in which both hair cells and supporting cells are destroyed), there appeared to be a loss of both the synaptic ending and the nerve fiber (Cotanche 1999). The recovery of the cochlear nerve in this situation has yet to be described, but may well take longer and be less effective.

The restoration of efferent nerve terminals after gentamicin destruction of hair cells has also been traced (Hennig and Cotanche 1998). The ototoxic recovery of neural connections occurred in several stages and takes longer than after acoustic trauma, which may account for the extended functional recovery reported in these ears (see Section 2.4). There also appeared to be differences in the regeneration of efferent terminals after ototoxic injuries to the papilla, again reflecting the more extensive damage to the sensory surface after aminoglycoside trauma.

2.3 Cochlear Nerve Coding of the Acoustic Stimulus

Access to the cochlear nerve may be achieved via an opening in the bony skull overlying the recussus scala tympani, which reveals the cochlear nerve lying beneath the distal wall of this chamber. Passing a microelectrode through the perilymph and penetrating the wall gains access to cochlear ganglion cells and nerve fibers. Remarkably, opening the scala tympani does not alter cochlear mechanics.

Parameters of the cochlear nerve response can assess various functions of the peripheral ear. These can be organized into categories that include spontaneous activity, threshold sensitivity, frequency selectivity, intensity coding, synchronization or phase locking, transient detection and adaptation, and neuronal suppression. The findings below are largely from the young chick, but they are nearly identical to those reported in the adult animal.

2.3.1 Tuning Curves

Frequency selectivity is most commonly determined through the threshold tuning curve or the response area curve. These curves describe either iso-response or iso-stimulus contours of unit activity at many frequencies. Threshold tuning curves use an algorithm that determines the sound pressure level (SPL) needed to achieve a sound-driven criterion response level of, for example, 2 spikes/s (S/s) above the level of spontaneous activity. When measured at many frequencies, it forms an iso-response contour. The response area curve is an iso-stimulus plot describing discharge activity to a constant intensity tonal stimulus of varying frequency. When the discharge activity produced by a random matrix of tone bursts at different frequency and intensity combinations is reconstructed into a tuning curve, it is referred to as a spectral response plot. Tuning curves reflect not only the frequency selective properties of the inner ear but also the tonotopic distribution of frequency along the sensory epithelium. Auditory nerve fibers have been mapped to their hair cell of origin in the chick and pigeon (Chen et al. 1994; Smolders et al. 1995).

Examples of spectral response plot tuning curves in the chick are seen in Figure 3.3. Tuning curves arise from frequency selective mechanisms acting on or within hair cells. One of these is a traveling wave on the basilar membrane. In birds, the locations of the maximum membrane deflection change with frequency, exhibit tonotopic organization along the length of the membrane, and exhibit

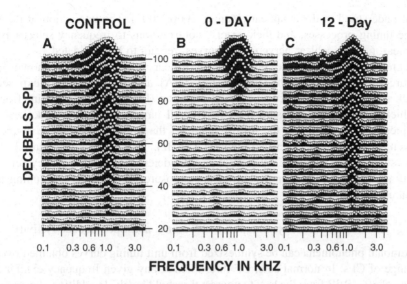

CONTROL   0 - DAY   12 - Day

FIGURE 3.3. (A–C) Three spectral response tuning curves are seen for each of three units with a similar CF. The tuning curves from exposed chicks were obtained at 0 days of recovery (**B**) or after 12 days of postexposure recovery (**C**), and should be compared against the age-matched control tuning curve at the equivalent of 0 days of recovery (**A**). Immediately after removal from the exposure, the CF threshold is elevated and the tuing curve is less frequency selective. The unit recorded after 12 days of recovery is identical to the control unit indicating complete recovery. (Data from Saunders et al. 1996a, with permission.)

relatively poor fequency selectivity (von Békésy et al. 1960; Gummer et al. 1987). If the level of frequency selectivity seen in the basilar membrane response were communicated to the hair cell stereocilia, with no other tuning influence, then the resulting neural tuning curve seen in Figure 3.3A would be much wider. The fact that the cochlear nerve tuning curves are sharper than what would be predicted by the mechanical response of the basilar membrane indicates additional sources of tuning.

A second frequency selective mechanism is found within the hair cell itself, and is associated with electrical tuning of the membrane (Fuchs et al. 1988). Electrical tuning arises from the interaction between voltage-gated Ca^{2+} and calcium-gated K^+ channels and their splice variants. When the movements of the sensory hairs are at the same frequency as the resonance of the hair cell membrane, the magnitude of membrane depolarization is greatest. As a consequence, activity in the cochlear nerves attached to the hair cell is greatest relative to the activity generated by any other frequency at the same level.

Third, a compressive nonlinearity may be due to motor activity that changes the mechanical properties of the sensory hair bundles of the hair cell. This activity might alter the stiffness of the bundle and thus influence hair cell frequency selectivity. Such mechanisms have been identified in anuran ears, but are less well understood in the avian ear (Manley 2000, 2001). The interaction of these three tuning processes, and their precise contributions to frequency selectivity, as measured in cochlear nerve tuning curves, remain to be elucidated.

Three spectral response plots from cochlear nerve units are shown in Figure 3.3: one from a normal control (Fig. 3.3A), another from a noise-exposed chick immediately after acoustic overstimulation (Fig. 3.3B), and a third from a chick 12 days postexposure (Fig. 3.3C). All have a common characteristic frequency (CF), which is the lowest point on the tuning curve where sound-driven activity can be identified (about 1.15 kHz). The example in Figure 3.3B shows the typical loss of CF threshold and the deterioration of frequency selectivity seen shortly after removal from the exposure. The recovery of tuning is apparent in Figure 3.3C.

2.3.2 Threshold Sensitivity, Sharpness of Tuning, and Spontaneous Activity

Additional phenomena can be synthesized from unit tuning curves obtained over a range of CFs. In normal ears, CF thresholds at any given frequency exhibit a range of 30–40 dB from the best to poorest threshold levels. In addition, the sensitivity of the lowest threshold at CFs, across frequency approximate the behavioral threshold curve of the chicken (Gray and Rubel 1985). In young and adult chickens, the absolute threshold of hearing is U shaped with thresholds of approximately 0.1–0.2 kHz and 2.8–4.0 kHz showing the poorest, and those between 0.9 and 1.3 kHz, the best sensitivity. Figure 3.4A shows the CF thresholds in a large sample of units recorded from a group of 3-day-old nonexposed control chicks. Figure 3.4B shows CF thresholds in age-matched units recorded shortly after removal from an intense pure-tone exposure. A comparison of Figures 3.4A and B reveals that units with CFs above 0.4 kHz show threshold shifts corresponding to frequencies near and above the 120 dB SPL exposure tone of 0.9 kHz. The CF thresholds in Figure 3.4B are from units in exposed animals allowed 12 days to recover, and demonstrate completely recovery. The rate of threshold recovery over time is too difficult to trace with single-unit studies, but has been effectively followed using evoked responses directly recorded from the chick brainstem (McFadden and Saunders 1989).

Frequency selectivity is quantified by measuring the Q of the tuning curve. This is an index of filter sharpness defined by the ratio of the characteristic frequency (CF in Hz), divided by the bandwidth of the tuning curve in Hz at a criterion level above the CF (typically 10 dB). Since Q is a ratio, the sharpness of tuning can be directly compared across units with different CFs. The larger the Q value is, the sharper the filter. Figure 3.4D shows Q_{10dB} ratios in cochlear nerve units recorded from 3-day-old control chicks. The value of Q_{10dB} increases (as does the variability) with increasing CF, meaning that frequency selectivity becomes sharper in units originating from higher (more basal) frequency regions of the basilar papilla. Interestingly, the most selective units in birds exhibit sharper tuning than seen at equivalent frequencies in mammals (Manley 2001). Figures 3.4E and F show what happens to frequency selectivity just after removal from an intense sound exposure, and after 12 days of recovery. Tuning curves

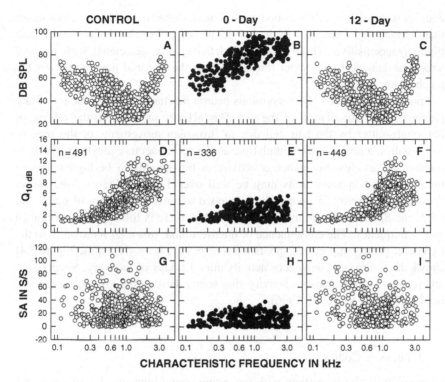

FIGURE 3.4. (**A**) Open circles show CF threshold from 491 units in control chicks. (**B**) The closed circles reveal the same thresholds in 336 units in exposed but 0-day recovered chicks. The threshold shift of units in the 0-day recovery group with CFs between 0.5 and 3.0 kHz is apparent. (**C**) CF thresholds are plotted for exposed chicks allowed 12 days of postexposure recovery. (**D–F**) Similarly organized data for frequency selectivity, as measured by the Q_{10dB} metric. (**G–I**) Data organized for spontaneous activity (SA). A comparison between the control data (**A**, **D**, **G**) with results in exposed units after 12 days of recovery reveals a complete return to normal. (Data from Saunders et al. 1996a, with permission.)

become much broader (have a lower Q) after the exposure, but within 12 days return to normal.

Threshold and tuning were also measured in adult chickens after exposures (48 h, 0.53 kHz, 120 dB SPL) that destroyed hair cells and the tectorial membrane. The CF thresholds were similarly elevated, and iso-response tuning curves were more broadly tuned immediately after the exposure. Threshold and tuning were partially recovered by 5 days postexposure; however, peculiar W-shaped tuning curves with multiple tips and hypersensitive tails, similar to those seen in the developing ear, were also reported (Chen et al. 1996). Single-fiber labeling showed a normal cochlear frequency-place map despite the elevated CF thresholds and broad tuning. After 28 days of recovery, tuning and thresholds were almost normal. Nevertheless, the CF thresholds of the most sensitive neurons were still slightly elevated, tuning curve slopes below CF were shallower than normal, and thresholds in the low-frequency tail of the tuning curves were often hypersensitive. These functional deficits were associated with residual structural damage to the upper fibrous layer of the tectorial membrane over the patch lesion.

Spontaneous activity (SA) represents neuron discharges in the absence of any planned sound stimulation of the ear. This activity arises from the release of neurotransmitter by the hair cell due to Brownian movements of the sensory hair bundles, ambient noise stimulation, and/or spontaneous exocytosis of neurotransmitter vesicles. Spontaneous activity in birds tends to be higher than in mammals, and in some units may be well over 110 S/s. Figure 3.4G shows spontaneous activity in young chicks plotted against CF in control units, and while the SA from unit to unit is quite variable, there is little systematic change across frequency. The consequence of intense sound, just after removal from the exposure, greatly suppresses spontaneous activity (Fig. 3.4H), while Figure 3.4I shows the return of spontaneous activity after 12 days of recovery. Suppression and recovery of spontaneous activity after sound damage has also been observed in adult birds (Chen et al. 1996).

2.3.3 Intensity Coding

Changes in discharge activity with increasing sound intensity describe the rate-intensity function. These functions have been reported for the pigeon, owl, emu, starling, and chick. In the chicken, four types of functions have been identified: saturating, sloping upward, straight, and sloping downward (Saunders et al. 2002). Rate-level functions arise from several processes. Changes in stimulus intensity result in different levels of basilar membrane displacement and hence hair bundle displacements. This in turn varies the level of hair cell membrane depolarization, neurotransmitter release, and neuron excitation. While basilar membrane movements with changes in intensity have been described as linear in the pigeon (Gummer et al. 1987), the rate-intensity function shows a compressive nonlinearity at higher stimulus levels. In mammals, OHC contractility is thought to be the principal mechanism of cochlear nonlinearity. There is no evidence of somatic motility in the avian hair cells (He et al. 2003; Köppl et al. 2004). However, this compression may arise from a nonlinearity thought to be associated with the response properties of individual hair cells.

Figures 3.5A–C show rate-level functions for low-, mid-, or high-frequency units (each curve is the average of a number of sloping-up units with a common CF) from control (open circles) and exposed (closed circles) ears shortly after removal from the exposure. At the lowest intensities, only spontaneous activity is seen. Sound-driven responses emerge after a threshold stimulus level is achieved, and discharge activity then increases with sound level (over an approximate 30–40 dB range in control units). Above that range, the slope of the growth

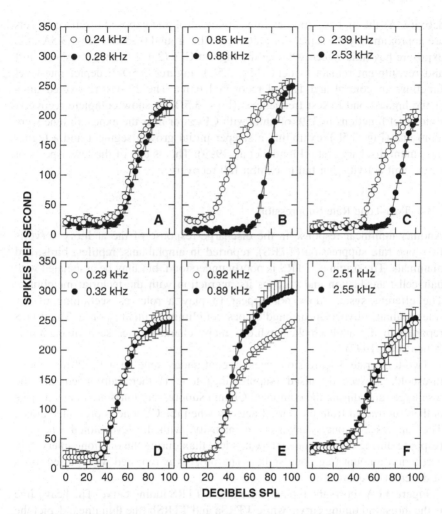

FIGURE 3.5. (**A–C**) Rate-level functions for units with the indicated CFs are shown from exposed chicks at 0 days of recovery (black circles) and age-matched nonexposed controls (white circles). The most severe change in the exposed functions occurs near the 0.9-kHz exposure frequency (**B**) with a pronounced threshold shift and steeper growth of activity. (**D–F**) Similar data in exposed units 12 days postexposure (black circles) and age-matched control units (white circles) appear. The functions in **D** and **F** show complete recovery. Exposed units with CFs near the exposure frequency (**E**) show a steeper growth segment and a larger discharge rate at the higher intensities. These plots represent the average of 6–9 units, each with the same approximate CF. (Data from Saunders et al., 1996a and Plontke et al. 1999, with permission.)

function becomes shallower. The threshold shift for exposed functions with CFs near the exposure frequency (Fig. 3.5B) is obvious, and the initial growth segment is steeper than in the control, suggesting some form of "recruitment" in discharge activity. At the highest stimulus intensity, control and exposed discharge levels are approximately the same. For units with CFs around 0.25 kHz (Fig. 3.5A), the exposure has little effect, while at the highest CFs (2.4–2.6 kHz) threshold shift and recruitment remain evident (Fig. 3.5C). Figures 3.5D–F depict rate-level functions in control and 12-day recovered units. The 12-day recovered units at the highest and lowest frequencies (Figs. 3.5D, F) show complete recovery. Rate-level functions in 35% of units with CFs at or near the exposure tone were abnormal (Fig. 3.5E), exhibiting a steeper initial growth segment and a higher maximum discharge rate (Plontke et al. 1999). This is one of the few aspects of single-unit activity that fail to exhibit full recovery.

2.3.4 Two-Tone Rate Suppression

Another nonlinear response in the discharge response of the auditory nerve is two-tone rate suppression (TTRS), reported in amphibians, reptiles, birds, and mammals. The origin of TTRS is poorly understood, but may be associated with hair cell transduction and perhaps its interaction with the tectorial membrane. The efferent system does not appear to play a role, as sectioning of the (mammalian) olivocochlear bundle does not eliminate this response. The TTRS reported in the adult chicken exhibits many characteristics seen in mammals (Chen et al. 1997).

Two-tone rate suppression uses a continuous tone at CF, 20 dB above threshold. A second pulsed (suppressing) tone is then introduced and the discharge rates during the combined CF and suppressing tones are counted. This is then subtracted from the rate of activity when the CF tone is presented alone. The suppressing tone is adjusted in intensity, with the level noted when the response during the two-tone interval is less than during the one-tone interval by a criterion amount (e.g., one spike). The process is repeated with pulsed tones at different frequencies.

Figure 3.6A shows the typical pattern of a TTRS tuning curve. The heavy line is the threshold tuning curve, while TTRSa and TTRSb (the thin lines) depict the suppression threshold above and below the CF of the tuning curve. Figure 3.6B shows what happens shortly after removal from an intense sound exposure. As would be expected, the single tone tuning curve exhibits a threshold shift and broadening of tuning. The suppression slopes are shallower, the best suppression thresholds are elevated, and TTRSa and TTRSb are first identified at a frequency more distant from the CF than in the control (Fig. 3.6A). In addition, acoustic trauma reduced the number of neurons exhibiting these boundaries from 53% to 8% for TTRS below CF, and from 88% to 47% for boundaries above CF. The change in number of neurons showing TTRS boundaries in control and exposed animals, after different recovery intervals, appears in Figure 3.6C. The boundaries below CF have yet to fully recover after 28 days postexposure, while above CF, they returned to normal within 14 days. The incomplete recovery of the lower TTRS boundary has been related to the incompletely healed tectorial membrane in the overstimulated chicken ear.

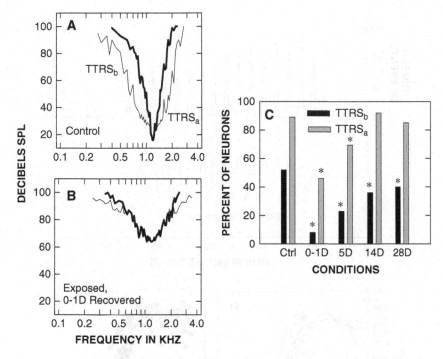

FIGURE 3.6. (**A**) A tuning curve from a nonexposed adult chicken cochlear nerve unit showing two-tone rate suppression boundaries above (TTRSa) and below (TTRSb) the unit CF. (**B**) A unit with a similar CF recorded shortly after removal from an intense sound exposure. The tuning curve is wider and CF threshold shows reduced sensitivity. The TTRS boundaries emerge at a much higher level along the skirts of the tuning curve. (**C**) The proportion of neurons in which TTRS revealed an upper or lower sideband in controls (Ctrl) or chickens allowed to recover 0–1, 5, 14, or 28 days. The suppression boundary above CF exhibits a normal incidence of occurrence 14 and 28 days postexposure, while suppression boundaries below CF never recover to normal. The asterisks indicate a significant difference from the control condition. (Data from Chen et al. 1997, with permission.)

2.3.5 Synchronization and Phase Locking

The ability of a cochlear nerve unit to discharge in concert with the cycles of a sinusoidal signal is referred to as synchronization. When synchronization occurs at a fixed phase angle, it is said to be "phase locked." Figure 3.7A shows the phasic discharges in a peristimulus time (PST) histogram for a unit (CF = 0.42 kHz) stimulated repeatedly with a 0.42-kHz tone burst. The waveform of the tone burst is also plotted. The degree to which each cycle of the stimulus produces a phase-locked discharge can be used to calculate vector strength (VS) and the value of VS would equal one if every cycle of the stimulus produced a discharge. Inherent jitter at the hair cell synapse, limitations in the resistance/capacitance of the hair cell membrane, and limitiations in the influx

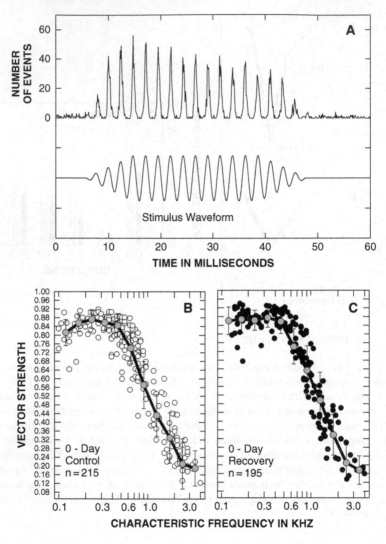

FIGURE 3.7. (**A**) Upper portion shows a phase-locked PST histogram in a chick cochlear nerve unit with a CF of 0.43 kHz. The PSTH was compiled from 376 tone burst presentations. (**B,C**) Vector strength of units in age-matched control and exposed chicks plotted against the CF of the unit. Statisitcal analysis of the control and exposed vector strength reveals that they are identical. (Data from Furman et al. 2006, with permission.)

and diffusion of calcium at neurotransmitter release sites in the hair cell preclude perfect synchrony and restrict the frequency range over which synchronization can occur (Furman et al. 2006).

Figure 3.7B shows VS in a sample of control units plotted against the unit CF. The VS is calculated for phase locked tone bursts 20 dB above the CF threshold. As shown, synchronization is best in neurons with low CF, then deteriorates as frequency increases, and between 3.0 and 4.0 kHz is lost in the noise. Figure 3.7C shows the VS in a sample of units just removed from the 0.9-kHz, 120-dB sound exposure, and these results reveal that VS in exposed and control units are the same. This observation is interesting because these exposed units exhibited the same changes in CF threshold, tuning sharpness, spontaneous activity, and rate-level activity shown in Figures 3.3, 3.4, and 3.5. The fact that synchronization was unaffected, even though overlain on substantial changes in discharge activity, suggests that the hair cell membrane and synaptic properties that determine the degree of synchronization were unaffected by the exposure.

2.3.6 Adaptation

Adaptation in the cochlear nerve response represents a progressive decline in discharge activity with the passage of stimulus time. Adaptation is identified in poststimulus time histograms (PSTH), which describe the accumulation of spike discharges in successive time intervals (bins) that occur with repeated tone-burst presentations. The PSTH exhibits several stages of adaptation; a phasic portion at stimulus onset where the number of discharges per bin are high, followed by a decline in activity to a tonic level where the number of discharges per bin are lower and relatively constant. The decline in discharge activity is orderly, and characterized by fitting the histogram data to an exponential decay function. Figure 3.8A shows a PST histogram for a tone burst with discharge activity collected at a 0.1-ms resolution. Figure 3.8B reduces the data to 1.0-ms bins, and fits the results with an exponential decay (solid line), revealing a time constant of 16.8 ms. Adaptation in chick occurs in three different time compartments called rapid, short term, and long term with time constants of approximately 2–3 ms, 15–17 ms, and 20–40 s. An example of long-term adaptation appears in Figure 3.8C, where discharge rates to a continuous tone were measured in 2-s intervals, every 4 s, for 43 min. The time constant was 30.9 s, but what is remarkable about this curve is that tonic activity (approximately 65 S/s) was maintained for the next 2530 s.

The presynaptic membrane of the hair cell appears to control adaptation through the number and release kinetics of neurotransmitter vesicles (Spassova et al. 2004; Crumling and Saunders 2006). One possible scenario is that vesicles lying beneath the dense body (the synaptic ribbon in chick) in immediate proximity to the hair cell plasma membrane, undergo exocytosis first (this is the immediately releasable pool). This is then followed by the exocytosis of those vesicles tethered to the dense body (the readily releasable pool). Finally, a steady level of vesicle mobilization, probably from the hair cell cytoplasm, is transported to the dense body and hence the release sites (this is the mobilizable pool). The three stages may correspond to the initial burst of activity, followed by the rapid decay and then tonic levels of activity, as vesicles in the various releasable pools are exhausted. Mobilization of cytoplasmic vesicles to the release site permits sustained discharge activity over long time intervals.

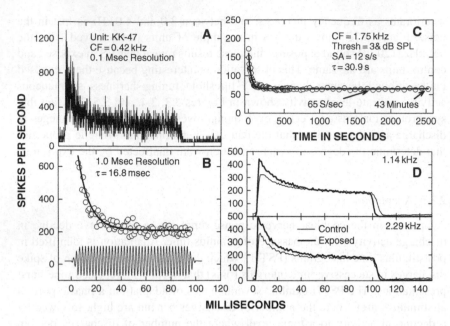

FIGURE 3.8. (**A**) Peristimulus time histogram plotted at a resolution of 0.1 ms. (**B**) Activity is binned to a resolution of 1.0 ms and fitted with a single component exponential decay function. The waveform of the non-phase-locked tone-burst stimulus is seen below. (**C**) Adaptation in a unit recorded with a 4-s resolution over a 2550-s interval. The value of τ indicates the time constant. The upper and lower sections of (**D**) show PST histograms of neural activity to tone burst stimuli, averaged across a sample of units (14–30 units per curve) with approximately the same indicated CF. The data were obtained from age-matched 0-day exposed and control animals. The early phasic component of the PSTH is elevated in the exposed adaptation curves while adapted rates exhibit the same discharge rate in both groups. (Data from Crumling 2006, with permission.)

Figure 3.8D shows changes in adaptation for units measured shortly after removal from the 120-dB, 0.9-kHz pure-tone exposure. The curves in each panel depict an average PST histogram (14–30 units per curve) in control or exposed units, all with approximately the same CF. Exposed units with CFs at or above the exposure frequency show an increase in the phasic discharge rate, but the same level of activity for the tonic region of the curve, compared to the control PSTH. The time constant of these exposed and control PSTHs were much the same (Crumling and Saunders 2006). These results indicated that the kinetics of discharge decay were unaffected by the exposure (as revealed by the similar time constants); however, the phasic portion of the response was increased in exposed units. Adaptation functions for control and exposed units with CFs below the exposure frequency (0.9 kHz) were nearly identical.

The mechanisms controlling vesicle release rate were unaffected by the exposure; however, the number of vesicles released (and the volume of neurotransmitter available to influence the postsynaptic membrane) was larger, but only for the vesicle pools associated with the initial level of exocitic activity. How this can be is not clear at present because the dense body normally has a 60% vesicle packing density. It could be that the exposure causes more vesicles to cluster at the membrane surface, filling the remaining 40% of the dense body, or the dense body gets larger in exposed hair cells. Contributions from efferent feedback or changes in the postsynaptic membrane have yet to be defined, but they could also play a role in this observation. It is interesting to speculate further that the enhanced phasic response is some form of compensatory mechanism by the hair cell to maintain a measure of discharge sensitivity in the face of destructive overstimulation.

2.3.7 Spatial Tuning Curves

The issue of the incompletely healed tectorial membrane and its effect on the chick peripheral auditory system has been noted above and revisited by examining spatial excitation for CFs distributed across the basilar papilla surface (Lifshitz et al. 2004). Spectral response tuning curves were examined in age-matched control chicks, and in chicks exposed to intense sound, but allowed to recover for 12 days. The discharge activity at criterion frequencies and intensities were noted for each unit and then plotted against the CF of the unit. When the CF of the unit was the same as the criterion frequency, unit discharge rate was expectedly high. As units were examined with CFs above or below the criterion frequency, activity diminished. Finally, when the unit CF was sufficiently distant from the criterion frequency, only spontaneous activity was observed. These spatial tuning curves were constructed for a number of criterion frequencies, and the excitation patterns provided a proxy of the mechanical activity on the papilla surface. Spatial tuning curves for control chicks at 10 criterion frequencies appear in Figure 3.9 at one criterion intensity (40 dB SPL). Using an equation that described the distribution of unit CFs along the tonotopic axis of the basilar papilla (Chen et al. 1994), the spatial tuning curve data were plotted against axes for the percent distance and tonotopic frequency from the apical end of the papilla.

Figure 3.9B illustrates the excitation patterns in chicks exposed to a 120-dB SPL, 48-hour, 0.9-kHz pure tone, and then allowed 12 days to recover. Above the spatial tuning curves is a cartoon of the papilla depicting the location and extent of the partially healed, patch lesion. The accuracy of the equation transforming unit CF to papilla location is seen in the fact that the spatial tuning curve in Figure 3.9B with a criterion frequency of 0.95 kHz is centered almost in the middle of the patch created by a 0.9-kHz exposure tone.

The spatial tuning curves with criterion frequencies within the patch lesion are the same as those in the age-matched control group in Figure 3.9A. Thus, the damage to the tectroial membrane does not appear to play a role in the processes that determine the spatial tuning of units across the sensory surface of the basilar papilla (Lifshitz et al. 2004).

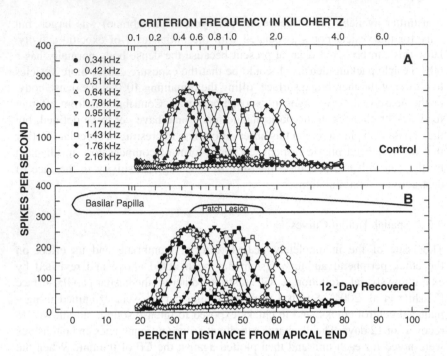

FIGURE 3.9. (A) Spatial tuning curves (see text for description) synthesized from chick cochlear nerve unit activity that are plotted against the spatial distance of the papilla (lower axis) and the tonotopic frequency distribution along the papilla (upper axis). (B) Similar data obtained in age-matched chicks exposed to an intense pure tone, but allowed 12 days of recovery. The surface of the basilar papilla is represented against these data with the location of the incompletely healed patch lesion indicated. The spatial distribution of neural activity and the sharpness of tuning are essentially the same in both groups. (Data from Lifshitz et al. 2004, with permission.)

2.3.8 Summary

The story of acoustic trauma in the young and adult chicken ear can be summarized quite succinctly. Exposures that produce moderate to severe functional loss (e.g., 60–70 dB threshold shifts) will show nearly complete recovery of function. Nevertheless, some residual loss after 12–28 days of recovery is seen in rate-level functions and in TTRS activity. The units exhibiting this abnormal behavior were typically those with CFs at or near the exposure frequency. It may be that with additional recovery time these abnormal responses will return to normal. With the exception of the abnormal appearance of the tectorial membrane, almost all inner ear damage is reversed (Cotanche 1999). Whether or not the tectorial membrane lesion is responsible for the rate-level or TTRS dysfunction remains to be seen.

2.4 Changes in Cochlear Nerve Activity After Ototoxic Trauma

2.4.1 Ototoxic Damage and Recovery to the Papilla

The application of ototoxic agents such as gentamicin or kanamycin is potentially lethal due to nephrotoxic side effects, and animal survival depends on the number of daily injections and drug dose level. These treatments are very destructive to the receptor cells and cause the loss of all neural and abneural hair cells beginning in the high-frequency region of the papilla. As dosing progresses, the lesion extends into the low frequencies, but even with long treatment regimens, rarely goes beyond the basal 50% of the papilla. Despite the hair cell losses, the injury to the overall papilla is less traumatic than with acoustic overstimulation. The tectorial membrane, for example, appears normal, although it is disconnected from the reticular surface of the papilla.

Single, high-dose injections have been used to more precisely time the course of injury and recovery, and these also cause hair cell loss to the approximate basal half of the papilla. A refinement in procedure applies gentamicin-loaded collagen sponges directly to the round window membrane. Nephrotoxicity is greatly reduced and complete loss of all hair cells now occurs over as much as 70% of the papilla (Husmann et al. 1998; Müller and Smolders 1998, 1999).

The time of emergence of new hair cells is difficult to determine with multiple injections, but with a single-dose treatment new hair cells are evident after 3–5 days in young chicks. In general, within 4 weeks posttreatment the sensory epithelium is completely repopulated with new hair cells, although hair bundle orientation may remain abnormal. Also, when a local application of gentamicin is used, it results in fewer regenerated hair cells than found in the original population. Reinnervation of the regenerated hair cells also occurs. Full details of the development and recovery of ototoxic damage to the papilla can be found in Cotanche (1999).

2.4.2 Functional Changes

Figure 3.10A illustrates auditory brainstem response (ABR) threshold shifts in the adult pigeon after application of a gentamicin sponge to the round window (Müller and Smolders 1999). The parameter is recovery duration. At 5 days after the sponge application, the threshold shift is between 63 and 69 dB and between 1.41 and 5.66 kHz. Over the next 37 days considerable recovery occured, but after 70 days there was little further improvement. Similar ABR results have been reported in the chick after aminoglycoside injections that began just after hatching. Thresholds recovered with time, but after durations as long as 14–20 weeks a high-frequency (>1.5 kHz) permanent shift of 20–30 dB was noted (Tucci and Rubel 1990; Girod et al. 1991; Duckert and Rubel 1993).

Cochlear nerve unit activity has been reported after either systemic kanamycin injections in the adult chicken or localized gentamicin treatment in the pigeon (Salvi et al. 1994; Müller and Smolders 1998). The results of both studies were largely

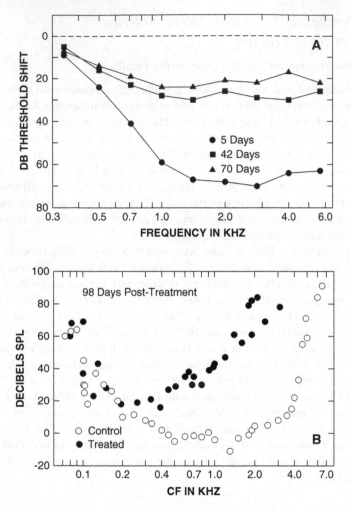

FIGURE 3.10. (**A**) Auditory brainstem thresholds in pigeons exposed to a gentamicin-impregnated sponge (applied to the round window membrane) were subtracted from similar thresholds obtained from control animals to obtain an indication of threshold shift at different frequencies. The parameter is the post-application recovery duration. Thresholds recover over 70 days, but a permanent threshold shift of approximately 20 dB remains. (Data from Müller and Smolders 1999.) (**B**) Cochlear nerve CF thresholds are plotted for the most sensitive units recorded from adult pigeons. The treated animals received a localized application of gentamicin (round window) and the data were obtained 98 days post-application. Control and treated best CFs are much the same up to about 0.2 kHz. The thresholds in the treated units show poorer CF thresholds as frequency increases. This is a permanent change in CF threshold. (Data from Müller and Smolders 1998, with permission.)

consistent with each other. Shortly after the end of treatment, lower frequency CFs showed abnormal thresholds, tuning curves, lower spontaneous activity, and changes in the proportion of units showing preferred intervals of spontaneous activity. It was difficult if not impossible to record units with higher frequency CFs because the basal region of the papilla was completely stripped of hair cells. After long recovery intervals (approximately 10–20 weeks), activity in low-frequency units exhibited complete recovery. Units with higher CFs could be identified as the basal papilla hair cells regenerated and matured; however, they often failed to exhibit normal levels of activity. These effects are illustrated in Figure 3.10B by plotting only the cochlear nerve units that showed the most sensitive CF thresholds in control animals and in gentamicin-treated pigeons (Müller and Smolders 1998). The units in the treated animals were recorded 14 weeks after the aminoglycoside application. The "best" CF thresholds are approximately the same in both groups up to about 0.25 kHz, and then differ by increasing amounts as CF rises. The differences are startling given that hair cell complements in the high-frequency region of the papilla have returned to nearly normal levels. Measures of Q_{10dB}, spontaneous activity, and the dynamic range of rate-intensity functions were also diminished in units with CFs above 0.25 kHz.

2.4.3 Differences Between Acoustic and Ototoxic Trauma

The effects of acoustic and ototoxic trauma to the papilla offer interesting and contrasting models of damage. Acoustic lesions are often confined to the abneural hair cells at a restricted location on the sensory surface that corresponds to the location stimulated by the exposure frequency. Overstimulation produces lesions that result in hair cell loss, cellular disruption of the sensory epithelium, and severe damage to the tectorial membrane. The latter damage does not appear to recover (Cotanche 1999). An ototoxic lesion destroys all neural and abneural hair cells over a large area of the sensory sheet, extending perhaps from the mid-papilla to the basal end. With ototoxic damage, the tectorial membrane remains intact, though disconnected from the reticular papilla surface in the area where hair cells disappear. Hair cell regeneration and neuronal reinnervation occur after acoustic and amino-glycoside trauma. The patterns of functional recovery accompanying these lesions differ with nearly complete return of function seen after acoustic damage and only partial functional recovery after ototoxic damage. These differences offer important comparative possibilities for understanding the consequences of papilla repair. Unfortunately, the utility of these comparisons has yet to be fully exploited.

3. Otoacoustic Emissions in Normal and Damaged Bird Ears

3.1 Acoustically Evoked Otoacoustic Emissions

As noted in the preceding text, single-neuron activity that describes rate-level functions and two-tone rate suppression reflect nonlinear activity within the cochlea. Another peripheral auditory response originating from nonlinear cochlear activity is the otoacoustic emission. These emissions have served as an indicator of active cochlear processes in both mammals and birds (Manley 2001).

Mammalian otoacoustic emissions are closely linked to somatic motility of OHCs (Brownell 1990). This motility is an active process driven by the motor protein prestin, which is heavily expressed along the lateral wall of OHCs (Dallos and Fakler 2002; Zheng et al. 2002). A sound-evoked change in the OHC membrane potential elicits a contractile response that causes feedback of mechanical energy into the cochlear partition. This in turn enhances the sensitivity and frequency selectivity of the inner ear, and as a by product gives, rise to otoacoustic emissions. These emissions are propagated out of the cochlea via the middle ear, and can be identified as acoustic signals in the ear canal. Support for this mechanism of producing emissions is found in lesion studies, which show that selective destruction of OHCs abolishes the response (Brown et al. 1989; Hofstetter et al. 1997; Wang et al. 1997).

Synchronously evoked otoacoustic emissions were first detected from the ear of the starling (*Sternus vulgaris*; Manley et al. 1987), and sound-evoked distortion product otoacoustic emissions (DPOAEs) have been observed in the chicken (Norton and Rubel 1990; Burkard et al. 1996). The DPOAE consists of acoustic components not identified in the stimuli used to generate them (Probst et al. 1991). The cubic difference tone is considered the most prominent DPOAE. It is defined mathematically as $2f_1 - f_2$, where f_1 and f_2 are the frequencies of two equally intense primary stimulus tones that share the relation $f_1 < f_2$.

The cubic difference tone component (which we now refer to, by convention, as the DPOAE) was examined in adult chickens after intense sound exposures (120 dB SPL) at either 1.5 or 0.53 kHz (Froymovich et al. 1995; Trautwein et al. 1996). Shortly after removal from the exposure, DPOAE amplitude exhibited a substantial reduction at or above the exposure frequency. After 8–16 weeks of recovery, DPOAE frequencies at or near the exposure tone showed little recovery whereas the frequencies lying further away from the exposure returned to normal levels. The failure of the DPOAE to exhibit recovery was attributed to the persistent tectorial membrane lesion, and suggested that this membrane was somehow involved in the production of the emission signal.

Changes in the DPOAE have also been examined in young chicks after intense tone exposure (Ipakchi et al. 2005), and Figures 3.11A and B show DP-grams and response-level functions from two groups of control animals (3 and 15 days old). The DP-grams (Fig. 3.11A) were obtained by averaging the DPOAE levels at each test frequency between 0.2 and 2.54 kHz across primary levels between 75 and 90 dB SPL. The emission level declines at a rate of approximately 2 dB/octave as frequency increases. The response-level functions (Fig. 3.11B) averaged emission levels across frequency (0.4–2.54 kHz), at each primary tone level, from 60 to 100 dB SPL. This input/output curve shows systematic increases in the DPOAE response as the primary tone levels increase. A slight but reliable difference between the two age groups is also seen.

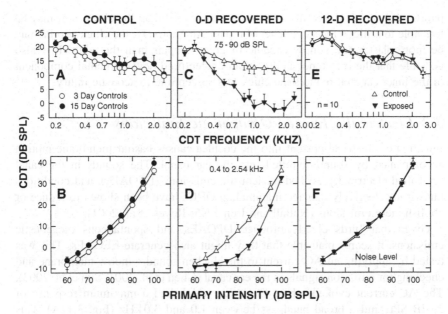

FIGURE 3.11. (**A,B**) Distortion product otoacoustic emission (DPOAE)-grams and response-level functions are plotted for 3-day and 15-day controls. (**C,E**) The DPOAE-grams in 0-day recovered and 12-day recovered chicks are compared to age-matched controls. DP-grams show a severe loss in the DPOAE response immediately (**C**) after removal from a pure-tone exposure and show complete recovery 12 days later (**E**). (**D,F**) The same pattern of response loss and recovery is seen in the response-level functions. (Data from Ipacki et al. 2005, with permission.)

Chicks were exposed to a 0.9-kHz, 120-dB tone beginning at 1 day of age for 48 hours, and thus were 3 days old when removed from the sound. Figures 3.11C and 3.11D compare DP-grams and response-level functions just after removal from the exposure (0–day recovered). A significant shift in the emission response is apparent, with the greatest effects between 1.0 and 2.0 kHz for primaries between 75 and 90 dB SPL.

Figures 3.11E and F show similar data, but for exposed animals allowed 12 days of recovery (e.g., 15 days old). Both the DP-gram and response-level functions exhibit complete recovery. This was different from that seen in the adult chicken (see earlier), where recovery was incomplete after substantially longer recovery durations. The inability of the adult to exhibit recovery was attributed to the incompletely healed tectorial membrane, but this conclusion may need refinement because the same lesion persisted in the chick, but with complete recovery of the DPOAE.

Although the same exposure conditions were used with adults and chicks, the effect of intense sound in the adults may be greater because middle-ear conduction is more efficient. Thus, the inner ear of adults may be more vigorously overstimulated than that of youngsters, and the consequence might be greater papilla damage. If the adult ear is more traumatized than the chick is, it may be less able to achieve the same level of repair. Alternatively, the chick papilla may be more plastic and simply capable of greater repair than the adult. It is also possible that the difference is due to an unidentified developmental component in the inner ear that renders the chick less susceptible to acoustic injury.

3.2 Electrically Evoked Otoacoustic Emissions

Injection of electrical current into the cochlea causes basilar membrane motion accompanied by sounds emitted into the ear canal. The sounds in the canal are called electrically evoked otoacoustic emissions (EEOAEs), and conditions known to selectively damage mammalian OHCs have been shown to reduce or abolish these emissions (Nuttall and Ren 1995; Reyes et al. 2001).

Given that birds clearly produce DPOAEs and spontaneous otoacoustic emissions, it seems plausible that they might also generate EEOAEs. This was tested by applying an AC current to the chicken round window membrane and checking for emission signals in the ear canal (Chen et al. 2001; Sun et al. 2002). The AC current evoked robust EEOAEs, which had a maximum response of 27 dB SPL and a broad bandpass between 1.0 and 3.0 kHz (Fig. 3.12A). This band encompassed the region of maximum threshold sensitivity in the chicken audibility curve.

Transient EEOAEs were also found in chicken ears by applying brief current pulses to the round window. These transient stimuli-evoked emissions were characterized by a damped oscillation with an instantaneous frequency near

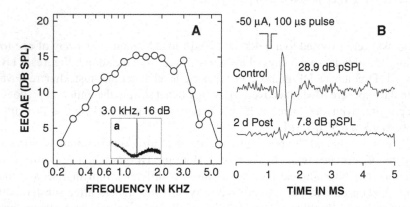

FIGURE 3.12. (**A**) Amplitude (dB SPL) of electrically evoked otoacoustic emission (EEOAE) obtained with a 50-mA sinusoidal current presented as a function of frequency. The inset (**a**) shows spectrum of EEOAE obtained at 3.0 kHz; EEOAE amplitude is 16 dB SPL. (**B**) Top trace shows a transient EEOAE from a normal control chicken (peak EEOAE amplitude was 28.9 dB SPL) evoked by a 50-μA pulse of 100 μs duration. Bottom trace shows an EEOAE obtained from a chicken treated with kanamycin (400 mg/kg perday for 10 days) with an EEOAE of 7.8 dB SPL that is barely detectable above the noise floor. (Data from Chen et al. 2001, with permission.)

2.0 kHz (Fig. 3.12B). The transient EEOAEs showed phase reverse when current polarity was reversed, and were abolished with anoxia or perfusion of paraformaldehyde into the inner ear. These later observations suggest that the EEOAE depends on normal metabolic function, and is not an artifact of the electrical stimulation itself.

Kanamycin was used to explore the role of hair cells in EEOAE generation. The dose employed spared hair cells on the apical papilla below approximately 500 Hz. As expected, the DPOAE and cochlear microphonic responses were considerably reduced after hair cell destruction. In addition, the EEOAE amplitude evoked by a 3.0-kHz electrical stimulus decreased from 15 dB SPL in normal ears to 3 dB at 1–4 days posttreatment. A similar reduction in the EEOAE response was also seen for pulse stimuli (see "2d Post Trace" in Fig. 3.12B). These results suggested that hair cells are the source of avian EEOAEs. Surprisingly, EEOAE amplitude and high-frequency DPOAEs 11–14 weeks after the kanamycin treatment were still greatly reduced, despite regeneration of almost all the hair cells lost to the ototoxic treatment. The cochlear microphonic amplitude showed substantial but incomplete recovery after 11–14 weeks. The major improvement in CM suggested that the mechanisms responsible for the hair cell transduction current had recovered substantially. In contrast, the minimal recovery of EEOAEs and DPOAEs implied that the mechanisms responsible for avian otoacoustic emissions failed to recover. An intriguing question that remains to be addressed is whether the EEOAE and DPOAE ever fully recover in adult birds after aminoglycoside damage.

Avian neural and abneural hair cells respectively are often considered homologous to mammalian outer and inner hair cells, largely because of similarities in the afferent and efferent innervation patterns. These similarities led to speculation that the avian hair cells might possess electromotile properties (Brix and Manley 1994; He et al. 2003). The somatic electromotility of isolated chick hair cells were compared to those of mammalian OHCs via the microchamber (Fig. 3.13B) and whole-cell voltage-clamp techniques (He et al. 2003). In response to sinusoidal voltage, gerbil OHCs changed length on a cycle by cycle basis and exhibited nonlinear capacitance, the hallmarks of electromotility (Fig. 3.13B). The OHC movements were also asymmetric in that the cells shortened more than they elongated. In contrast, neural (tall) and abneural (short) hair cells from the chick failed to exhibit somatic motility and nonlinear capacitance.

The absence of somatic motility in chick hair cells suggested that EEOAEs must originate from some other voltage-sensitive process. This "other" process may be associated with small stereocilia bundle movements, which change direction with current polarity. These electrically driven bundle movements have been reported in amphibian and reptilian hair bundles and suggested for avian bundles (Crawford and Fettiplace 1985; Manley et al. 2001). The movements disappear when the tip-links are disrupted or the hair cell transducer channel is blocked (Crawford and Fettiplace 1985; Ricci et al. 2000). If these stereocilia bundle movements are responsible for generating EEOAEs then their movements

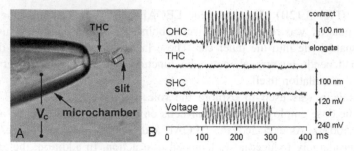

FIGURE 3.13. (**A**) Photomicrograph showing a microchamber with the base of a tall hair cell (THC) in the lumen. The cuticular plate is positioned over a slit that records axial movements during stimulation with sinusoidal current. (**B**) Electromotile responses showing elongation and contraction of outer hair cell (OHC), tall hair cell (THC), and short hair cell (SHC) to sinusoidal stimulation. Note robust movements from OHC and absence of response from THC and SHC. (Data from He et al. 2003, with permission.)

would need to be synchronized across a number of hair cells to generate sufficient pressure in the surrounding fluids to produce the EEOAEs.

4. Cochlear Nerve Reversible and Irreversible Excitotoxic Damage

A difference between the damage caused by aminoglycoside antibiotics and acoustic overstimulation is that intense sound stimulation causes a glutamate-induced excitotoxic injury to the afferent terminals beneath the hair cells (Henry and Mulroy 1995; Pujol and Puel 1999). Kainic acid (KA), a potent glutamate analog that preferentially activates kainate-sensitive ionotropic glutamate receptors, has been used to identify regions of excitotoxic damage in the avian inner ear (Shero et al. 1998). Within hours of applying KA to the round window membrane of adult chickens, a massive swelling occurred in the afferent dendrites beneath the neural hair cells of the papilla, but with no apparent trauma to the hair cells themselves (Fig. 3.14A, B). Efferent terminals appeared normal. The afferent terminal swelling extended over 80% of the papilla from the apical end (Fig. 3.14C). These results indicate that glutamate is the major neurotransmitter between neural hair cells and afferent nerve fibers. The absence of afferent terminal swelling beneath abneural hair cells suggested further that another neurotransmitter may be active in this papilla region.

Light and electron microscopic observations revealed that the application of both low and high KA doses produced rapid and severe swelling of the chicken afferent terminals immediately after application (Sun et al., 2001). The swelling of the afferent terminals, however, followed different patterns with the passage of time in the two groups. While there was no evidence of cochlear ganglion neuron loss in the low-dose group, there was, nevertheless, a substantial percentage of cell bodies with vacant spaces between the myelin sheath and soma, as well as

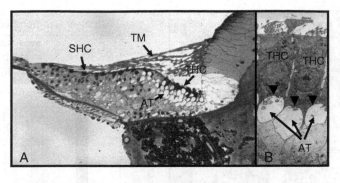

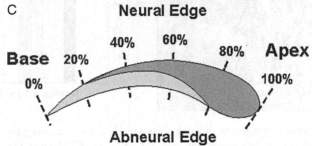

FIGURE 3.14. (A) Cross section of adult chicken basilar papilla showing tectorial membrane (TM), the short hair cell (SHC) and tall hair cell (THC) region, and swollen afferent terminals (AT) beneath the THC. (B) High-magnification view showing swollen afferent terminals (AT) beneath the tall hair cells (THC). (C) Papilla cartoon with the dark gray area revealing the location of swollen afferent terminals. (Data from Sun et al. 2001, with permission.)

changes in the distribution of nuclear chromatin. These initial effects of the KA treatment rapidly disappeared, and the postsynaptic terminals appeared normal in the low-dose group within 1 day.

At 30 min after KA treatment, cochlear ganglion neurons appeared normal (Fig. 3.15A). However, 4 weeks later many cochlear ganglion neurons were surrounded by vacuoles (Fig. 3.15B, arrows) and had a shrunken appearance (arrowheads), a morphological feature characteristic of apoptosis. The vacant spaces beneath the tall hair cells, in the high-dose group, persisted for months indicating a lack of repair. In addition, ears treated with the high dose had a considerable number of missing cochlear ganglion neurons 4 weeks after treatment (Fig. 3.15C).

The DPOAE and chronic CM measurements were used to assess the functional status of hair cells after the application of low- or high-dose KA injections into the scala tympani of adult chickens (Sun et al. 2000). Emission and microphonic responses remained normal after treatment, indicating functioning hair cells. In contrast, both doses caused a large and almost immediate reduction in the compound action potential (CAP, Fig. 3.16A) and the slow positive neural

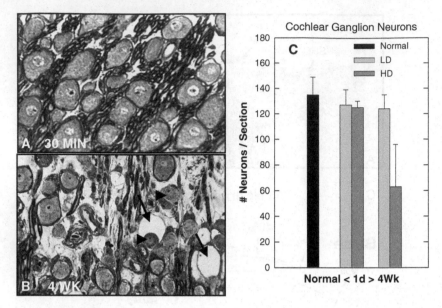

FIGURE 3.15. Photomicrographs showing condition of chicken cochlear ganglion neurons (**A**) 30 min and (**B**) 4 weeks after a high dose of kainic acid. (**C**) Mean number of cochlear ganglion neurons per section from normal chickens versus chickens treated with a low dose (LD) or high dose (HD) of kainic acid; measurements obtained after less than 1 day or greater than 4 weeks. (Data from Sun et al. 2001, with permission). Vacuole (arrow) surrounding shrunken cochlear ganglion some (arrowhead).

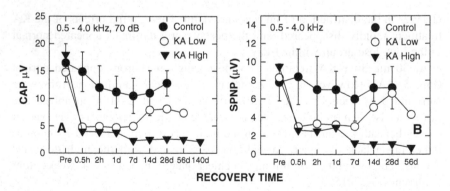

FIGURE 3.16. (**A**) Recovery of compound action potential (CAP) amplitude for adult chicken treated with a low dose or high dose of kainic acid versus a control group. Average amplitude of CAP for 0,5-, 1.0-, 2.0-, and 4.0-kHz tone bursts presented at 70 dB SPL. (**B**) Similar results as in (**A**) for the slow positive neural potential (SPNP) response following low or high dose of kainic acid treatment. (Data from Sun et al. 2000, with permission.)

potential (SPNP; Fig. 3.16B). The CAP amplitudes showed a 50–70% recovery within 2–4 weeks following the low-dose treatment, but no recovery with the high–dose treatment.

The total lack of functional recovery in the high-dose group is consistent with the persistent damage to the afferent terminals and the loss of cochlear ganglion neurons. Despite the loss of neurons in the high-dose group, the hair cells retained a normal appearance, indicating that avian hair cell survival depends on factors other than the synaptic contacts or trophic influences provided by the presence of ganglion axons. A similar conclusion was reached in mammals (Ernfors et al. 1995). The partial recovery of CAP and SPNP amplitude in the lowdose group is consistent with the morphological recovery of the afferent terminals. There remained, however, a major discrepancy between the rate and degree of recovery in the low-dose group (Fig. 3.16A, B). The afferent terminals largely regenerated within 1 day of the low-dose injection, whereas the CAP and SPNP responses showed little recovery until 14 days later. Moreover, these responses were still depressed 1–2 months after treatment, even though the afferent terminals and number of ganglion neurons returned to normal. The incomplete recovery in the low-dose group might be related to the subtle morpho-logical pathologies seen in approximately 35% of the neurons.

Afferent terminal damage with recovery patterns similar to that seen in the chicken were observed in pigeons after infusion of the glutamate agonist α-amino-3-hydroxy-5-methylisoxazole-4- propionic acid (AMPA) into scala tympani (Reng et al. 2001). The partial but incomplete recovery of CAP responses after excitotoxic damage is also seen in pigeon cochlear ganglion unit responses 3–4 months after AMPA infusions. Overall, the mean spontaneous discharge rate, for units between 0.1 and 1.0 kHz, increased substantially over that seen in control units. Similarly, the maximum sound-evoked discharge rate was higher in nerve fibers from AMPA-treated pigeons than in controls. Finally, as CFs increased, AMPA-treated pigeons showed tuning curves with poorer CF thresholds and broader tuning.

Collectively, the results from chicken and pigeon indicated that regeneration of the afferent synapse after mild excitotoxic damage caused incomplete recovery of postsynaptic function. In cases of more severe excitotoxic damage, significant numbers of cochlear ganglion neurons degenerate, and, more importantly, fail to regenerate. Cochlear ganglion cell degeneration in the quail was reported after acoustic overstimulation beginning around 30 days postexposure and continued out to 90 days (Ryals et al. 1989), and may be due to a loss of trophic input from the hair cells or delayed cell death from acoustically induced excitotoxic damage. The loss of cochlear ganglion neurons after severe excitotoxic damage or acoustic overstimulation suggests that this region of the peripheral pathway does not contain a pool of resident progenitor cells that can divide and differentiate into neurons that reestablish peripheral and central synaptic contacts.

5. Higher Brain Centers

5.1 Cellular Changes

The anatomical effects of acoustic or ototoxic trauma at the level of the chicken receptor organ have been investigated in some detail at the next auditory relay, the cochlear nuclei, and the results clearly indicate that inner-ear structural damage can extend beyond the receptor epithelium and affect neurons in the central nervous system (CNS). The physiological consequences of these changes in cochlear nuclei neurons, however, are poorly understood. The chick cochlear nuclei consist of the nucleus magnocellularis and nucleus angularis and their cellular structure and innervation have been described in detail (Ryugo and Parks 2003). Recordings of normal discharge activity in both nuclei have also been reported (Warchol and Dallos 1990).

It is well known that removal of afferent input causes the death or atrophy of postsynaptic cells throughout the CNS (Rubel et al. 1990). This may be due to disruption of the cell signaling pathways controlling protein synthesis or the regulation of postsynaptic calcium entry into the cell. The normal maintenance of postsynaptic proteins and calcium is controlled in part by afferent input. Because afferent input can be effectively manipulated by damaging the auditory periphery in chick, the cochlear nucleus has proved an interesting model for studying these signaling pathways. Unilateral cochlear removal, oval window perforation, aminoglycoside damage, and acoustic trauma caused neuron loss or damage to cells. It appears that a tonic level of afferent input is needed to maintain normal cell size and function at this level of the brainstem.

Gentamicin-induced hair cell loss caused a reduction in cochlear nerve spontaneous activity, and in turn a reduction in cochlear nucleus cells and cell shrinkage. With sufficient recovery time cell size returned to normal and unexpectedly the number of cochlear nucleus cells recovered (Durham et al. 2000). Overstimulation in the chick ear also caused a reduction in cochlear nerve units spontaneous activity (see Fig. 3.4H) and shrinkage of cells in the nucleus magnocellularis. Within 43 days of recovery, cell size returned to normal. This acoustic trauma, however, yielded no evidence of neuronal cell loss in nucleus magnocellularis (Saunders et al. 1998).

Discharge activity in nucleus magnocellularis cells has been examined after intense sound exposure. A deterioration in CF thresholds, a reduction in the sharpness of tuning, changes in rate-intensity functions, and a substantial reduction in spontaneous activity were reported (Cohen and Saunders 1993). After 12 days of recovery, all these parameters showed full recovery. Despite the shrinkage in nucleus magnocellularis cell size, functional recovery appeared to be a simple reflection of the recovery exhibited in the cochlear nerve. The fact that it took longer for cell size to return to normal than for function to return suggested that shrinkage of these cells did not alter their function.

In summary, reduced afferent input to the chick cochlear nuclei, after peripheral ototrauma, caused cells in this area of the medulla to die or undergo substantial shrinkage. The effect is most likely due to a postsynaptic reduction in protein synthesis or the disruption of intracellular calcium regulation. With sufficient recovery duration cell size returned to normal and there is some evidence of a return to normal cell numbers. The milder damage caused by acoustic trauma resulted in only cell shrinkage, with complete recovery within 6 weeks postexposure. There appeared to be no physiological changes in neuron activity unique to cochlear nucleus cell shrinkage after acoustic trauma, because all the reported changes could be attributed to more peripheral losses in function.

5.2 Central Auditory System Neurogenesis Stimulated by Hair Cell Loss and Regeneration

For many years, a dogma of neurobiology was that most neurons are born during embryogenesis and that neurons that die later in life are not replaced. The discovery of injury-induced hair cell regeneration in the avian ear challenged this long held tenant and begs the question of where the new neurons come from. As noted in the preceding text, destruction of avian cochlear hair cells leads to diverse changes in nucleus magnocellularis. These changes are followed by recovery of cell size and number as the hair cells regenerate (Lippe 1991; Park et al. 1998; Saunders et al. 1998). If neuron number recovers after cochlear injury, are new neurons "born" to replace those lost as a result of cochlear injury? To test this idea, tritiated thymidine, an exogenous nucleic acid incorporated into the DNA of dividing cells, was administered to postnatal chicks for 14 days via osmotic pumps (Park et al. 2002). One day after starting thymidine administration, chicks were injected with gentamicin or saline, and then sacrificed 16 days later for autoradiographic and histologic analysis of the number of newborn, thymidine-positive neurons, or glial cells in nucleus magnocellularis. Newborn neurons and glia were found throughout nucleus magnocellularis in both the saline- and gentamicin-treated chicks. The vast majority of tritiated labeled cells, however, were glia. Nevertheless, the number of newborn neurons in nucleus magnocellularis was significantly greater in gentamicin-treated chicks than in saline controls. However, the total number of tritiated-positive neurons in the aminoglycoside ears was relatively small considering that the gentamicin treatment destroyed all the hair cells in the basal third of the basilar papilla. Although the hair cell lesion was confined to the basal region of the papilla, neurogenesis occurred throughout the nucleus magnocellularis. Many newborn cells were also seen in the cerebellum and brainstem of saline- and gentamicin-treated chicks, a result that suggested the avian CNS retained substantial potential for regeneration under both normal and pathologic conditions. When viewed in this context, hair cell regeneration represented only one small part of a much larger phenomenon of neurogenesis involving brain plasticity and repair in birds (Alvarez-Buylla et al. 1994), and perhaps even mammals (Alvarez-Buylla et al. 2002).

6. Summary

This final section aims to form some coherent conclusions from the large array of information offered in the preceding text. One interesting observation can be offered immediately. We may know more about the consequences of ototoxic and acoustic insults to the inner ear of birds than in any other animal model. This formidable literature is traceable to the discovery of hair cell regeneration on the basilar papilla (Cotanche 1987a,b). While the underlying molecular and cellular mechanisms of this regenerative process remain elusive, the functional consequences of papilla damage and recovery are well understood.

Another important conclusion is the remarkable consistency of the avian ototrauma data. Part of this may be due to the relatively few trauma-inducing conditions appearing in this literature. For example, the extensive number of papers using a pure tone exposure of 120 dB for 48 hours, in both adult and young chickens, has enabled comparisons among reports dealing with different parameters of the peripheral auditory response. This has permitted a measure of synthesis, as seen in this chapter, that has yet to be realized in other animal models. Another important observation is that intra-animal variability, which has frustrated a more systematic description of acoustic injury in the mammalian ear, seems to be remarkably low in the avian ear.

There is little argument that the disruption of tissue along the papilla and the loss of hair cells should have an impact on threshold sensitivity. Moreover, if the structure completely heals itself, there is the logical assumption that the functional loss should reverse itself. Figures 3.1, 3.3, 3.4, and 3.5 give credence to these assumptions, but at best the relationship remains a correlation. Despite the available data, analytic models that attempt to predict a more formalistic relationship between structural injury and repair, and functional loss or recovery of CF threshold, have yet to enter the avian literature. The extensive data base suggests that the time is ripe for introducing such modeling efforts. This would permit the manipulation of parameters whose model results could be tested against observation, and, in a synergistic fashion, further observation could be used to modify the entry parameters of the model. This admonition could be extended to any of the functional parameters discussed earlier, because the specific basis of threshold shift, loss of frequency selectivity, reductions in spontaneous activity, a reduced dynamic range in rate-level functions, changes in two-tone rate suppression, or PSTH changes in adaptation functions, or any other measure of function following any form of avian ototrauma, are poorly appreciated. Nevertheless, the field is ready for an overreaching theory about avian ototrauma to guide thinking toward a grander scheme of the phenomenon, and there is optimism that this will happen because of the extensive and systematic data base that has been accumulated in the last 15–20 years.

Fortunately, many mysteries remain to be solved. For instance, why is it that there is nearly complete functional recovery with acoustic damage to the papilla, yet permanent functional loss with ototoxic trauma? Papilla affected by both sources of ototrauma show complete hair cell regeneration and reinnervation.

So why is there a persistent functional difference that results from the chemical treatment of the ear? In addition, it would be important to exploit the differences arising from acoustic and ototoxic trauma, as they may reveal some, as yet unappreciated underlying principles, not only of inner ear function, but also of the consequences of trauma. There are also differences between the effects of acoustic injury between adult chickens and juveniles in the EP and DPOAE. The basis for these differences are elusive, and a simple explanation utilizing age and the level of maturity of inner structure reveals relatively little.

Finally, there is the enigma of the persistent tectorial membrane damage seen on the basilar papilla after acoustic trauma. The inner ear is an organ exquisitely evolved to detect miniscule mechanical movements of tissue. In the mammalian ear, these movements are discussed at the level of Ångstroms, and there is every reason to assume similar magnitudes are involved in the avian ear. The tectorial membrane injury constitutes a relatively huge defect on this exquisitely sensitive organ. The available evidence has yet to determine definitively that this injury has any substantial impact on auditory function. The fascinating unresolved issue is to determine what this observation is telling us about the role of the tectorial membrane.

The intense and fruitful odyssey that has been the study of hair cell regeneration may now be winding down as the phenomenon turns its focus on the mammalian inner ear. Nevertheless, ototrauma in the avian ear remains an important comparative animal model for studying the relationship between structural damage and repair and functional loss and recovery. Understanding the mechanisms that give rise to this remarkable recovery of function may provide important insights for what needs to be accomplished in mammalian inner ears for hearing to be restored. Future investigations hold the promise of understanding the relationship between these two phenomena in greater detail, and that will yield important insight into the operation of the inner ear.

Acknowledgment. This work was supported by NIH grants R01 DC06 630-01 and P01 DC03600 to R.J.S., and R01 DC 000710-12, and years of funding from the Pennsylvania Lions Hearing Research Foundation to J.C.S. The initial preparation of this chapter occurred while the first author was a Visiting Professor at the University de Bordeaux 2, Inserm Laboratoire de Biologie Cellulaire et Moleculaire de L'Audition, Hopital Pellegrin, Bordeaux, France.

References

Alvarez-Buylla A, Ling CY, Yu WS (1994) Contribution of neurons born during embryonic, juvenile, and adult life to the brain of adult canaries: regional specificity and delayed birth of neurons in the song-control nuclei. J Comp Neurol 347:233–248.

Alvarez-Buylla A, Seri B, Doetsch F (2002) Identification of neural stem cells in the adult vertebrate brain. Brain Res Bull 57:751–758.

Askew CH, Bateman K, Saunders JC, Gratton MA (2006) Ultrastuctural analysis of the tegmentum vasculosum after intense sound. Abs Assoc Res Otolaryngol 29:14.

Brix J, Manley GA (1994) Mechanical and electromechanical properties of the stereovillar bundles of isolated and cultured hair cells of the chicken. Hear Res 76:147–157.

Brown AM, McDowell B, Forge A (1989) Acoustic distortion products can be used to monitor the effects of chronic gentamicin treatment. Hear Res 42:143–156.

Brownell WE (1990) Outer hair cell electromotility and otoacoustic emissions. Ear Hear 11:82–92.

Burkard R, Salvi R, Chen L (1996) 2f1–f2 distortion product otoacoustic emissions in White Leghorn chickens (*Gallus domesticus*): effects of frequency ratio and relative level. Audiol Neurootol 1:197–213.

Chen L, Salvi RJ, Hashino E (1993) Recovery of CAP threshold and amplitude in chickens following kanamycin ototoxicity. Hear Res 69:15–24.

Chen L, Salvi R, Shero M (1994) Cochlear frequency-place map in adult chickens: intracellular biocytin labeling. Hear Res 81:130–136.

Chen L, Trautwein PG, Miller K, Salvi RJ (1995) Effects of kanamycin ototoxicity and hair cell regeneration on the DC endocochlear potential in adult chickens. Hear Res 89:28–34.

Chen L, Trautwein PG, Shero M, Salvi RJ (1996) Tuning, spontaneous activity and tonotopic map in chicken cochlear ganglion neurons following sound-induced hair cell loss and regeneration. Hear Res 98:152–164.

Chen L, Trautwein PG, Powers N, Salvi RJ (1997) Two-tone rate suppression boundaries of cochlear ganglion neurons in chickens following acoustic trauma. J Acoust Soc Am 102:2245–2254.

Chen L, Sun W, Salvi RJ (2001) Electrically evoked otoacoustic emissions from the chicken ear. Hear Res 161:54–64.

Cohen YE, Saunders JC (1993) The effects of sound overexposure on the spectral response patterns of nucleus magnocellularis in the neonatal chick. Exp Brain Res 95:202–212.

Cotanche DA (1987a) Regeneration of hair cell stereociliary bundles in the chick cochlea following severe acoustic trauma. Hear Res 30:181–196.

Cotanche DA (1987b) Regeneration of the tectorial membrane in the chick cochlea following severe acoustic trauma. Hear Res 30:197–206.

Cotanche DA (1997) Hair cell regeneration in the avian cochlea. Ann Otol Rhinol Laryngol Suppl 168:9–15.

Cotanche DA (1999) Structural recovery from sound and aminoglycoside damage in the avian cochlea. Audiol Neurootol 4:271–285.

Cotanche DA, Lee KH, Stone JS, Picard DA (1994) Hair cell regeneration in the bird cochlea following noise damage or ototoxic drug damage. Anat Embryol 189:1–18.

Crawford AC, Fettiplace R (1985) The mechanical properties of ciliary bundles of turtle cochlear hair cells. J Physiol 364:359–379.

Crumling MA, Saunders JC (2007) Tonotopic distribution of short-term adaptation properties in the cochlear nerve of normal and acoustically overexposed chicks. J Assoc Res Otolaryngol 8:54–68.

Dallos P, Fakler B (2002) Prestin, a new type of motor protein. Nat Rev Mol Cell Biol 3:104–111.

Duckert LG, Rubel EW (1993) Morphological correlates of functional recovery in the chicken inner ear after gentamycin treatment. J Comp Neurol 331:75–96.

Durham D, Park DL, Girod DA (2000) Central nervous system plasticity during hair cell loss and regeneration. Hear Res 147:145–159.

Ernfors P, Van De Water T, Loring J, Jaenisch R (1995) Complementary roles of BDNF and NT-3 in vestibular and auditory development. Neuron 14:1153–1164.

Fischer FP (1994) Quantitative analysis of the innervation of the chicken basilar papilla. Hear Res 61:167–178.

Fischer FP, Miltz C, Singer I, Manley GA (1992) Morphological gradients in the starling basilar papilla. J Morphol 213:225–240.

Forge A (1996) Sensory cell regeneration and functional recovery: a review. In Axelsson A, Borchgrevink H, Hamernik RP, Hellstrom P-A, Henderson D, Salvi RJ (eds) Scientific Basis of Noise-Induced Hearing Loss. New York: Thieme, pp. 3–32.

Froymovich O, Rebala V, Salvi RJ, Rassael H (1995) Long-term effect of acoustic trauma on distortion product otoacoustic emissions in chickens. J Acoust Soc Am 97:3021–3029.

Fuchs PA, Nagai T, Evans MG (1988) Electrical tuning in hair cells isolated from the chick cochlea. J Neurosci 8:2460–2467.

Furman AC, Avissar M, Saunders JC (2006) Phase locking in cochlear nerve units of the chick (Gallus domesticus) exposed to intense sound. Eur J Neurosci 24:2003–2010.

Girod DA, Tucci DL, Rubel EW (1991) Anatomical correlates of functional recovery in the avian inner ear following aminoglycoside ototoxicity. Laryngoscope 101:1139–1149.

Gray L, Rubel EW (1985) Development of absolute threshold in chickens. J Acoust Soc Am 77:1162–1172.

Gummer AW, Smolders JW, Klinke R (1987) Basilar membrane motion in the pigeon measured with the Mössbauer technique. Hear Res 29:63–92.

Hara J, Plymale DR, Shepard DL, Hara H, Garry RF, Yoshihara T, Zenner HP, Bolton M, Kalkeri R, Fermin CD (2002) Avian dark cells. Eur Arch Otorhinolaryngol 259:121–141.

He DZ, Beisel KW, Chen L, Ding DL, Jia S, Fritzsch B, Salvi R (2003) Chick hair cells do not exhibit voltage-dependent somatic motility. J Physiol 546:511–520.

Hennig AK, Cotanche DA (1998) Regeneration of cochlear efferent nerve terminals after gentamycin damage. J Neurosci 18:3282–3296.

Henry WR, Mulroy MJ (1995) Afferent synaptic changes in auditory hair cells during noise-induced temporary threshold shift. Hear Res. 84, 81–90.

Hofstetter P, Ding D, Powers N, Salvi RJ (1997) Quantitative relationship of carboplatin dose to magnitude of inner and outer hair cell loss and the reduction in distortion product otoacoustic emission amplitude in chinchillas. Hear Res 112:199–215.

Husmann KR, Morgan AS, Girod DA, Durham D (1998) Round window administration of gentamicin: a new method for the study of ototoxicity of cochlear hair cells. Hear Res 125:109–119.

Ipakchi R, Kyin T, Saunders JC (2005) Loss and recovery of sound evoked otoacoustic emissions in young chick following acoustic trauma. Audiol Neurootol 10:209–219.

Kachar B, Parakkal M, Kurc M, Zhao Y, Gillespie PG (2000) High-resolution structure of hair-cell tip links. Proc Natl Acad Sci USA 97:13336–13341.

Köppl C, Wegscheider A, Gleich O, Manley GA (2000) A quantitative study of cochlear afferent axons in birds. Hear Res 139:123–143.

Köppl C, Forge A, Manley GA (2004) Low density of membrane particles in auditory hair cells of lizards and birds suggests an absence of somatic motility. J Comp Neurol 479:149–155.

Lifshitz J, Furman AC, Altman KW, Saunders JC (2004) Spatial tuning curves along the chick basilar papilla in normal and sound-exposed ears. J Assoc Res Otolaryngol 5:171–184.

Lippe WR (1991) Reduction and recovery of neuronal size in the cochlear nucleus of the chicken following aminoglycoside intoxication. Hear Res 51:193–202.

Manley GA (2000) Cochlear mechanisms from a phylogenetic viewpoint. Proc Natl Acad Sci USA 97:11736–11743.

Manley GA (2001) Evidence for an active process and a cochlear amplifier in non mammals. J Neurophysiol 86:541–549.

Manley GA, Schulze M, Oeckinghaus H (1987) Otoacoustic emissions in a song bird. Hear Res 26:257–266.

Manley GA, Gleich O, Kaiser A, Brix J (1989) Functional differentiation of sensory cells in the avian auditory periphery. J Comp Physiol A 164:289–296.

Manley GA, Kirk DL, Köppl C, Yates GK (2001) In vivo evidence for a cochlear amplifier in the hair-cell bundle of lizards. Proc Natl Acad Sci USA 98:2826–2831.

Markin VS, Hudspeth AJ (1995) Gating-spring models of mechanoelectrical transduction by hair cells of the internal ear. Annu Rev Biophys Biomol Struct 24:59–83.

McFadden EA, Saunders JC (1989) Recovery of auditory function following intense sound exposure in the neonatal chick. Hear Res 41:205–216.

Müller M, Smolders JW (1998) Hair cell regeneration after local application of gentamicin at the round window of the cochlea in the pigeon. Hear Res 120:25–36.

Müller M, Smolders JW (1999) Responses of auditory nerve fibers innervating regenerated hair cells after local application of gentamicin at the round window of the cochlea in the pigeon. Hear Res 131:153–169.

Müller M, Smolders JW, Ding-Pfennigdorff D, Klinke R (1996) Regeneration after tall hair cell damage following severe acoustic trauma in adult pigeons: correlation between cochlear morphology, compound action potential responses and single fiber properties in single animals. Hear Res 102:133–154.

Müller M, Smolders J, Ding-Pfennigdorff D, Klink R (1997) Discharge properties of pigeon single auditory nerve fibers after recovery from severe acoustic trauma. Int J Dev Neurosci 15:401–416.

Norton SJ, Rubel EW (1990) Active and passive ADP components in mammalian and avian ears. In Dallos P, Geisler CD, Matthews JW, Ruggero MA, Steele CR (eds) Mechanics and Biophysics of Hearing. New York: Springer-Verlag, pp. 219–226.

Nuttall AL, Ren T (1995) Electromotile hearing: evidence from basilar membrane motion and otoacoustic emissions. Hear Res 92:170–177.

Ofsie MS, Cotanche DA (1996) Distribution of nerve fibers in the basilar papilla of normal and sound-damaged chick cochleae. J Comp Neurol 370:281–294.

Ofsie MS, Hennig AK, Messana EP, Cotanche DA (1997) Sound damage and gentamicin treatment produce different patterns of damage to the efferent innervation of the chick cochlea. Hear Res 113:207–223.

Park DL, Girod DA, Durham D (1998) Evidence for loss and recovery of chick brainstem auditory neurons during gentamicin-induced cochlear damage and regeneration. Hear Res 126:84–98.

Park DL, Girod DA, Durham D (2002) Avian brainstem neurogenesis is stimulated during cochlear hair cell regeneration. Brain Res 949:1–10.

Plontke SK, Lifshitz J, Saunders J.C (1999) Distribution of rate-intensity function types in chick cochlear nerve after exposure to intense sound. Brain Res 842:262–274.

Poje CP, Sewell DA, Saunders JC (1995) The effects of exposure to intense sounds on the DC endocochlear potential in the chick. Hear Res 82:197–204.

Probst R, Lonsbury-Martin BL, Martin GK (1991) A review of otoacoustic emissions. J Acoust Soc Am 89:2027–2067.

Pujol R, Puel JL (1999) Excitotoxicity, synaptic repair, and functional recovery in the mammalian cochlea: a review of recent findings. Ann NY Acad Sci 884:249–254.

Ramakrishna R, Kurian R, Saunders JC, Gratton MA (2004) Recovery of the tegmentum vasculosum in the noise exposed chick. Abstr Assoc Res Otolaryngol 27:65.

Reng D, Müller M, Smolders JW (2001) Functional recovery of hearing following ampa-induced reversible disruption of hair cell afferent synapses in the avian inner ear. Audiol Neurootol 6:66–78.

Reyes S, Ding D, Sun W, Salvi R (2001) Effect of inner and outer hair cell lesions on electrically evoked otoacoustic emissions. Hear Res 158:139–150.

Ricci AJ, Crawford AC, Fettiplace R (2000) Active hair bundle motion linked to fast transducer adaptation in auditory hair cells. J Neurosci 20:7131–7142.

Rubel EW, Hyson RL, Durham D (1990) Afferent regulation of neurons in the brain stem auditory system. J Neurobiol 21:169–196.

Ryals BM, Dooling RJ (1996) Changes in innervation and auditory sensitivity following acoustic trauma and hair cell regeneration in birds. In Salvi RJ, Henderson D, Fiorino F, Colletti V (eds) Auditory Plasticity and Regeneration: Basic Science and Clinical Implications. New York: Thieme, pp. 84–99.

Ryals BM, Ten Eyck B, Westbrook EW (1989) Ganglion cell loss continues during hair cell regeneration. Hear Res 43:81–90.

Ryals BM, Stalford MD, Lambert PR, Westbrook EW (1995) Recovery of noise-induced changes in the dark cells of the quail tegmentum vasculosum. Hear Res 83:51–61.

Ryugo DK, Parks TN (2003) Primary innervation of the avian and mammalian cochlear nucleus. Brain Res Bull 60:435–456.

Salt AN, Melichar I, Thalmann R (1987) Mechanisms of endocochlear potential generation by stria vascularis. Laryngoscope 97:984–991.

Salvi RJ, Saunders SS, Hashino E, Chen L (1994) Discharge patterns of chicken cochlear ganglion neurons following kanamycin-induced hair cell loss and regeneration. J Comp Physiol A 174:351–369.

Saunders JC, Doan DE, Poje CP, Fisher, KA (1996a) Cochlear nerve activity after intense sound exposure in neonatal chicks. J Neurophysiol 76:770–787.

Saunders JC, Doan DE, Cohen YE, Adler HJ, Poje CP (1996b) Recent observations on the recovery of structure and function in the sound damaged chick ear. In Salvi RJ, Henderson D, Fiorino F, Colletti V (eds) Auditory Plasticity and Regeneration: Basic Science and Clinical Implications. New York: Thieme, pp. 62–83.

Saunders JC, Adler HJ, Cohen YE, Smullen S, Kazahaya K (1998) Morphometric changes in the chick nucleus magnocellularis following acoustic overstimulation. J Comp Neurol 390:412–426.

Saunders JC, Ventetuolo CE, Plontke SK, Weiss BA (2002) Coding of sound intensity in the chick cochlear nerve. J Neurophysiol 88:2887–2898.

Shero M, Salvi RJ, Chen L, Hashino E (1998) Excitotoxic effect of kainic acid on chicken cochlear afferent neurons. Neurosci Lett 257:81–84.

Smith CA (1985) Inner ear. In King A, MacLeland J (eds) Form and Function in Birds. New York: Academic Press, pp. 273–310.

Smolders JWT (1999) Functional recovery in the avian ear after hair cell regeneration. Audiol Neurootol 4:286–302.

Smolders JWT, Ding-Pfenningdorff D, Klinke R (1995) A functional map of the pigeon basilar papilla: correlation of the properties of single auditory nerve fibers and their peripheral origin. Hear Res 92:151–169.

Spassova MA, Avissar M, Furman AC, Crumling MA, Saunders JC, Parsons TD (2004) Evidence that rapid vesicle replenishment of the synaptic ribbon mediates recovery from short-term adaptation at the hair cell afferent synapse. J Assoc Res Otolaryngol 5:376–90.

Sterkers O, Ferrary E, Amiel C (1988) Production of inner ear fluids. Physiol Rev 68:1083–1128.

Stone JS, Rubel EW (2000) Cellular studies of auditory hair cell regeneration in birds. Proc Natl Acad Sci USA 97:11714–11721.

Sun H, Salvi RJ, Ding, DL, Hashino DE, Shero M, Zheng, XY (2000) Excitotoxic effect of kainic acid on chicken otoacoustic emissions and cochlear potentials. J Acoust Soc Amer 107:36–2142.

Sun H, Hashino E, Ding DL, Salvi RJ (2001) Reversible and irreversible damage to cochlear afferent neurons by kainic acid excitotoxicity. J Comp Neurol 430:172–181.

Sun W, Chen L, Salvi RJ (2002) Acoustic modulation of electrically evoked otoacoustic emission in chickens. Audiol Neurootol 7:206–213.

Trautwein P, Salvi RJ, Miller K, Shero M, Hashino E (1996) Incomplete recovery of chicken distortion product otoacoustic emissions following acoustic overstimulation. Audiol Neurootol 1:86–103.

Trautwein PG, Chen L, Salvi RJ (1997) Steady state EP is not responsible for hearing loss in adult chickens following acoustic trauma. Hear Res 110:266–270.

Tucci DL, Rubel EW (1990) Physiologic status of regenerated hair cells in the avian inner ear following aminoglycoside ototoxicity. Otolaryngol Head Neck Surg 103:443–50.

von Békésy G (1952) Gross localization of the place of origin of the cochlear microphonics. J Acoust Soc Am 24:399–409.

von Békésy G (1960) Experiments in Hearing. New York: John Wiley & Sons.

Vossieck T, Schermuly L, Klinke R (1991) The influence of DC-polarization of the endocochlear potential on single fibre activity in the pigeon cochlear nerve. Hear Res 56:93–100.

Wang J, Powers NL, Hofstetter P, Trautwein P, Ding D, Salvi R (1997) Effects of selective inner hair cell loss on auditory nerve fiber threshold, tuning and spontaneous and driven discharge rate. Hear Res 107:67–82.

Warchol ME, Dallos P (1990) Neural coding in the chick cochlear nucleus. J Comp Physiol A 166:721–734.

Zheng J, Madison LD, Oliver D, Fakler B, Dallos P (2002) Prestin, the motor protein of outer hair cells. Audiol Neurootol 7:9–12.

4
Functional Recovery After Hair Cell Regeneration in Birds

ROBERT J. DOOLING, MICHEAL L. DENT, AMANDA M. LAUER, AND BRENDA M. RYALS

1. Introduction

In response to either acoustic trauma or insult from ototoxic drugs, both young and adult birds show a temporary period of hair cell loss and regeneration, usually culminating in considerable anatomical, physiological, and even behavioral recovery within several weeks (Corwin and Cotanche 1988; Ryals and Rubel 1988; Tucci and Rubel 1990; Girod et al. 1991; Hashino et al. 1991; Lippe et al. 1991; Saunders et al. 1992, 1996; Ryals et al. 1999b). Recent reviews of the recovery of auditory function after hair cell regeneration have focused on physiological measures of the auditory nerve and brainstem (compound action potential [CAP], auditory brainstem response [ABR], or changes in hair cell responses using distortion product emissions (e.g., Smolders 1999; see Saunders and Salvi, Chapter 3). All of these measures are highly correlated with the return of hearing, but behavioral measures of hearing address, most directly, the actual recovery of auditory perception. This chapter emphasizes studies that address the behavioral recovery of hearing after hair cell loss and regeneration in birds.

As far as we know, birds provide the only animal model in which it is possible to restore hearing through renewed sensory cell input and then examine the effect of this hearing recovery on the learning and production of vocalizations. We review several studies that have addressed the effects of hair cell loss and regeneration on complex vocal production. As one can imagine, the issue of whether a "new" auditory periphery results in sufficient functional recovery so that an adult bird can perceive, learn, and produce complex acoustic communication signals has considerable health relevance, as current research efforts are focused on triggering hair cell regeneration in the mammalian auditory system (e.g., Izumikawa et al. 2005). Understanding the fine detail of hearing recovery in birds may tell us something about how bird ears function, add to our knowledge of plasticity in both peripheral and central auditory nervous system structures, and expose the common features of sensorimotor interfaces across vertebrates.

2. Changes in Absolute Sensitivity

There have been principally two ways to damage auditory sensory cells with resulting changes in sensory function: acoustic trauma and administration of ototoxic drugs. The two approaches typically lead to different patterns of hair cell damage and loss and different patterns of hair cell regeneration and functional recovery (Saunders et al. 1995; Salvi et al. 1998; Cotanche 1999; Smolders 1999). Here, we discuss the extents and time courses of hearing loss and functional recovery to these two types of peripheral auditory system trauma in birds.

2.1 Acoustic Overexposure

Morphological assessment of hair cell loss and regeneration after acoustic trauma has shown that acoustic overstimulation generally results in the loss of some, but not all, hair cells in a specific location on the basilar papilla, depending on the type, intensity, and duration of the acoustic trauma (reviewed in Cotanche 1999).

Temporary hearing loss in birds after acoustic overexposure, first described in budgerigars (*Melopsittacus undulatus*), revealed some differences between birds and mammals (Saunders and Dooling 1974; Dooling 1980). Budgerigars exposed to 1/3 octave bands of noise centered at 2 kHz for 72 hours at levels of 76–106 dB sound pressure level (SPL) showed maximum hearing losses at 2 kHz, and the threshold shift ranged from 10 to 40 dB depending on the level of the exposure. A permanent threshold shift was observed only with the 106-dB exposure, suggesting that birds, compared to mammals, are more resistant to damage from noise (Dooling 1980). Temporary threshold shifts in these birds were also of a shorter duration than typically seen in mammals and were also restricted to a narrower range of frequencies (e.g., Luz and Hodge 1971; Price 1979; Dooling 1980; Henderson and Hamernik 1986). In addition to showing little spread across frequencies, in budgerigars the maximum threshold shift occurred at higher frequencies than the exposure frequency. Hashino and colleagues (1988) extended these bird–mammal differences to impulse noise exposures. Two 169-dB SPL impulse noises produced by pistols caused more low-frequency than high-frequency hearing loss with the return to preexposure hearing levels occurring at a faster rate for high than for low frequencies. These results are unique and intriguing, and their confirmation could provide insight into the functioning of the avian ear.

Japanese quail (*Coturnix coturnix*) exposed to a 1.5-kHz octave band noise at 116 dB SPL for 4 hours showed elevated thresholds for pure tones up to 50 dB immediately after exposure (Niemiec et al. 1994). Thresholds were most severely affected at frequencies of 1.0 kHz and above, although there was considerable variation among subjects. Thresholds improved rapidly within the first week after exposure, recovering to preexposure levels by 8–10 days. Damaged hair cells were still observed up to 2 weeks postexposure but not by 5 weeks postexposure. Similar patterns of threshold shifts and recoveries were seen after repeated exposures to noise, although recovery times increased with increasing numbers of exposures. Interestingly, Niemiec et al. (1994) found that structural abnormalities (elongated stereocilia, supporting cell expansion, and stress links between hair cells) remained for at least 4 weeks after pure tone sensitivity had recovered. They suggested that while these abnormalities may not influence absolute threshold sensitivity, they may be involved in other aspects of functional hearing. Other investigators have suggested that lingering structural abnormalities within the tectorial membrane (lack of upper fibrous layer) may be related to incomplete recovery of other aspects of auditory function such as frequency resolution (Salvi et al. 1998; Lifshitz et al. 2004). While neural correlates of frequency resolution tend to corroborate poorer frequency resolving capacity after acoustic trauma and hair cell regeneration, corresponding behavioral studies are still lacking. Behavioral studies of frequency resolution after hair cell regeneration following ototoxic insult have been performed and are described later in this chapter.

Ryals and colleagues (1999b) found that the amount of hearing loss and the time course of recovery varied considerably among different species of adult (sexually mature) birds even when exposure conditions and test conditions were identical. In their study, quail and budgerigars were exposed to intense pure tones centered in their region of best hearing at 112–118 dB SPL for 12 hours. Quail showed much greater susceptibility to acoustic trauma than did budgerigars, with significantly larger threshold shifts and hair cell loss. Quail showed a threshold shift of 70 dB at 2.86 kHz 1 day after overexposure. Thresholds for the quail remained virtually unchanged for 8–9 days postexposure, and then began to improve by about 2 dB/day until day 30. Quail experienced a permanent threshold shift of approximately 20 dB, which remained even when tested 1 year after exposure. Budgerigars exhibited a threshold shift of about 35–40 dB at 0.5 days after exposure, but showed a much faster recovery than quail. By 3 days postexposure, budgerigars' thresholds had improved to within 10 dB of normal. Chickens exposed to a 120-dB pure tone at 525 Hz for 48 hours (Saunders et al. 1995) showed similar initial threshold shifts and rates of recovery as the budgerigars. CAP measurements in pigeons exposed to a 142-dB pure tone at 700 Hz for 1 hour (Ding-Pfennigdorff et al. 1998) showed intermediate threshold shifts between the quail and the budgerigars and chickens.

In a more comprehensive study, budgerigars, canaries (*Serinus canaria*), and zebra finches (*Taeniopygia guttata*) were exposed to a bandpass noise (2–6 kHz) at 120 dB SPL for 24 hours (Ryals et al. 1999b). These birds showed thresholds at 1.0 kHz that were elevated by 10–30 dB and that improved to within normal limits by about 10 days postexposure in all three species. At 2.86 kHz, the center of the exposure band, budgerigars, canaries, and zebra finches showed a 50-dB threshold shift. Recovery began immediately afterward for canaries and finches, and threshold improved to within 10 dB of normal by about 30 days postexposure. In budgerigars, threshold recovery did not begin until 10 days postexposure. By 50 days postexposure, thresholds recovered to about 20 dB above normal, and no further improvement occurred by 70 days, at which point the loss was assumed to be permanent. Overall, results showed a significantly more rapid recovery in canaries and zebra finches than in budgerigars. Histological analysis in all of these birds quantified hair cell loss and recovery in the region of damage before, during, and after threshold recovery. In general, the more severe the initial degree of hair cell loss (width of damage and decrease in hair cell number) was, the more severe the initial threshold shift.

Ryals et al. (1995), Salvi et al. (1998), and Cotanche (1999) have shown that structures such as the tegmentum vasculosum, which provides the endolymphatic potential in birds, the tectorial membrane, and neural synapses may also be damaged immediately after acoustic trauma. There are two important conclusions from these studies. One is that even when exposure and test conditions are identical, the amount of damage and the time course of loss and recovery from acoustic trauma are quite different among species. The second conclusion is that determination of the direct role of regenerated hair cells in the recovery of hearing after acoustic overstimulation is confounded by the continuing presence of nonregenerated hair cells on the papilla after initial acoustic trauma, the initial and continuing damage to other structures within the inner ear such as the tectorial membrane, and the fact that a considerable amount of hearing can return before a full complement of hair cells is replaced through regeneration.

These behavioral results are paralleled by a wealth of physiological data. Measures of CAP and evoked potential (EP) thresholds, for instance, have also shown a recovery from acoustic trauma in birds. In pigeons, CAP thresholds increased immediately after exposure to a 0.7-kHz tone at 136–142 dB SPL for 1 hour, but showed some recovery in most subjects (Müller et al. 1996, 1997). The time course of recovery varied somewhat among individual subjects, and some animals showed no recovery. A residual threshold shift of 26.3 dB remained at 2.0 kHz for some of the animals that recovered, while others showed normal thresholds within 3 weeks after exposure. Newborn chicks and adult chickens also show increased CAP and EP thresholds immediately after acoustic trauma, but animals with longer survival times showed near-normal thresholds (McFadden and Saunders 1989; Adler et al. 1992, 1993; Pugliano et al. 1993). Interestingly, less recovery occurred when chicks were exposed a second time (Adler et al. 1993).

Threshold shifts to a pure tone overexposure (Fig. 4.1A) and to narrowband noise overexposure (Fig. 4.1B) are shown across several species of birds. Although the exposure durations, intensities, and frequencies differed across species, these recovery curves represent virtually all of the existing data available on experiments that tested the same subjects repeatedly. Given the large differences in susceptibility to acoustic trauma both within and across species, these types of experiments are important for understanding the nature of the time course of recovery in individual subjects. These figures highlight the similarities in recovery times across species, even when the extent of the initial hearing loss is quite different.

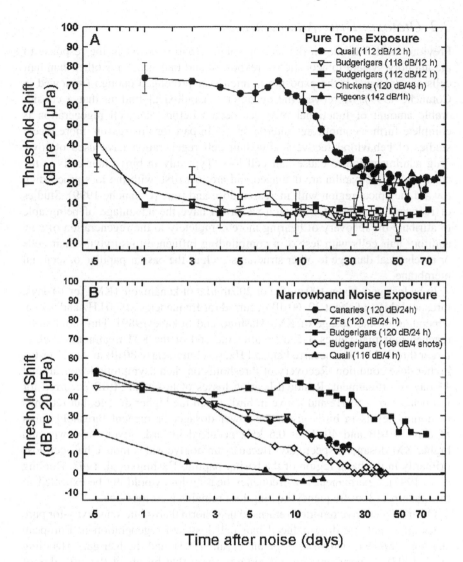

FIGURE 4.1. (**A**) Behaviorally measured threshold shifts after acoustic overexposure to a 2.86-kHz pure tone for Japanese quail ($n = 3$) and budgerigars ($n = 2$ at 112 dB and $n = 3$ at 118 dB) (replotted from Ryals et al. 1999b) and to a 525-Hz pure tone for chickens ($n = 4$) (replotted from Saunders et al. 1995). Shown for comparison are threshold shifts after a 700-Hz pure tone exposure as measured by the compound action potential (CAP) in pigeons (replotted from Ding-Pfennigdorff et al. 1998). (**B**) Threshold shifts after continuous narrowband noise overexposures for canaries ($n = 2$), budgerigars ($n = 5$), zebra finches (ZF, $n = 3$), and Japanese quail ($n = 3$; canaries, budgerigars, zebra finches replotted from Ryals et al. 1999b and Japanese quail replotted from Niemiec et al. 1994). For comparison with these continuous exposures, results from budgerigars exposed to 4 gunshot impulses ($n = 2$) are also shown (replotted from Hashino et al. 1988).

2.2 Ototoxic Drug Administration

Previous work has shown that considerable variation exists in the response of the auditory system to acoustic overexposure and that structures other than hair cells are damaged if the exposure is intense or prolonged enough (reviewed in Cotanche 1999). Moreover, the apparently paradoxical finding that a considerable amount of functional recovery occurs before hair cell regeneration is complete further complicates interpretation. In part for this reason, more recent studies of behavioral recovery after hair cell regeneration have used ototoxic drug administration to cause hair cell loss. Typically in birds, hair cells in the basal end of the papilla are damaged and are lost first, with the loss proceeding toward the apical region with increased dose and time (Cotanche 1999). Studies of recovery from insult with ototoxic drugs have the advantage of being able to attribute the recovery of hearing more completely to the regeneration of new auditory hair cells with less of a confounding influence of surviving hair cells or mechanical damage to other structures such as the basilar papillae or tectorial membrane.

In budgerigars given 100 mg/kg or 200 mg/kg of kanamycin (KM) for 10 days, threshold shifts of about 20–60 dB occurred for frequencies of 2.0 kHz and above, depending on the dosage of KM (Hashino and Sokabe 1989). Threshold shifts reached maximal values 3–5 days after the end of the KM treatment. Slightly larger threshold shifts occurred at 0.5 kHz, reaching nearly 80 dB at day 3 for the higher dose condition. Recovery of thresholds reached asymptotic levels around 15 days posttreatment. Residual hearing losses of less than 20 dB occurred for frequencies of 2.0 kHz and above in birds given the higher dosage, but returned to normal levels in birds given the lower dosage. Permanent threshold shifts of about 10 dB and 40 dB at 0.5 kHz occurred in birds given the lower and higher KM dosage, respectively. Because aminoglycosides induce hair cell loss primarily in the basal region of the basilar papilla (Hashino et al. 1992; Dooling et al. 1997), a permanent low-frequency hearing loss would not be expected so this pattern of low-frequency hearing loss remains a curious one.

Other studies have reported returns to near-normal absolute sensitivity for pure tones after ototoxic drug-induced hair cell loss and regeneration in European starlings (*Sturnus vulgaris*; Marean et al. 1993) and budgerigars (Dooling et al. 1997). Experiments on budgerigars show that by about the fifth day of injections, absolute thresholds began to increase dramatically, and the threshold shift at 2.86 kHz by the time injections were completed (10 days) was greater than 60 dB. Immediately after KM treatment, the greatest shift occurred for frequencies above 2.0 kHz (Dooling et al. 1997). Recovery of hearing began immediately after cessation of injections. By 8 weeks postinjection, absolute thresholds improved to within 5–15 dB of normal for frequencies below 2.0 kHz and to within 20–30 dB of normal above 2.0 kHz. Recovery of threshold eventually reached an asymptote with a permanent threshold shift on average of 23 dB.

In a more recent study on budgerigars, Dooling et al. (2006) measured thresholds at six frequencies and found them to be differentially affected by the

KM injections. Initial threshold shift was greater at high frequencies (50–60 dB above 2 kHz) than it was at low frequencies (10–30 dB below 2 kHz). Thresholds gradually recovered to within about 15–25 dB of preinjection thresholds by 8 weeks after injections, with the most rapid recovery occurring at the lowest frequencies. Absolute thresholds for all frequencies reached asymptote by 8 weeks after cessation of KM injections with a return to within 5–15 dB of normal for frequencies below 2 kHz and within 20–30 dB of normal above 2 kHz (Dooling et al. 2006).

Marean and colleagues (1995) compared behavioral detection thresholds from starlings before, during, and after 11 days of subcutaneous injections of KM. The starlings were given 100 mg/kg injections for 2 days and then 200 mg/kg injections for 9 days. The birds showed large threshold shifts initially at high frequencies and then at progressively lower frequencies. These shifts were in excess of 60 dB at 4–7 kHz but none of the birds showed any change in auditory thresholds for frequencies below 3 kHz. The recovery of hearing began within a few days of the end of the KM injections, and lasted for approximately 50 days. After a second course of KM injections, the starlings showed less of a threshold shift than after the first injections. Subsequent studies showed that the smaller threshold shift after the second course of injections was likely due to metabolic changes sustained after the first course of antibiotics and not to a resistance to antibiotic damage by regenerated hair cells (Marean et al. 1995). The time course of recovery and the amount of permanent threshold shift in starlings, compared to budgerigars, suggests that they may be less sensitive to damage from aminoglycoside antibiotics, similar to Bengalese finches (*Lonchura domestica*; Marean et al. 1993, 1998; Woolley et al. 2001).

As is the case with acoustic overexposure, there are a number of physiological estimates of threshold recovery after ototoxic drug administration that in general parallel behavioral findings. Tucci and Rubel (1990) used frequency-specific auditory evoked potentials and tested 16- to 20-week-old chicks (*Gallus gallus domesticus*) after gentamicin injections. EPs were increased immediately after injections ceased, and continued to increase for up to 5 weeks afterward. This is in contrast to behavioral thresholds that generally show a decrease in thresholds following termination of injections. They suggested that the increase in EP threshold during the early stage of recovery was due to a lack of neural synchrony. Partial recovery of thresholds was seen by 16–20 weeks, predominantly at low and mid frequencies. Residual hearing loss was greatest at the highest frequencies. Interestingly, recovery of thresholds lagged hair cell regeneration by 14–18 weeks. Administration of KM also results in temporarily increased CAP and EP thresholds, primarily at higher frequencies in chickens and quail (Chen et al. 1993; Lou et al. 1994; Trautwein et al. 1998). Some residual loss persists at the highest frequencies tested, however.

In a more recent study, Woolley et al. (2001) measured hair cell regeneration and the recovery of auditory thresholds in Bengalese finches using the aminoglycoside amikacin (alternating doses of 150 mg/kg and 300 mg/kg daily for 7 days). ABR thresholds were measured for frequencies ranging from 0.25

to 6 kHz. Heavy hair cell losses on the basal end of the basilar papilla correlated with hearing losses above 2 kHz. Recovery of auditory thresholds began at about 1 week after treatments and ceased approximately 4 weeks later. Thresholds at high frequencies remained elevated at 8–12 weeks. One conclusion is that the temporal course of hearing recovery is more similar among closely related species because the pattern in finches is more similar to that of the starling (Marean et al. 1993) than that of either the budgerigar (Dooling et al. 1997) or the chick (Tucci and Rubel 1990).

In aggregate, these comparative results are similar in some respects to the comparative results on recovery of hearing after noise exposures, where canaries and zebra finches were similar to each other but different from budgerigars and quail (Ryals et al. 1999b). It is also important to realize the difficulties inherent in comparisons between behavioral and EP measures of hearing. Hearing is a behavior whereas EPs are a measure of neural synchrony due to stimulus onset, and usually correlate well with stimulus intensity. However, the minimum stimulus intensity required to produce a measurable evoked potential is typically 10–30 dB higher than behavioral detection thresholds. Thus, behavioral measures of hearing remain the gold standard for determining the effect of ototoxicity or acoustic overstimulation on hearing.

Figure 4.2 shows behavioral audiograms from several species of birds measured behaviorally (A-budgerigars and B-starlings) compared with ABR (C-pigeons and D-Bengalese finches), and CAP (E-chickens) threshold curves. These functions show the preinjection thresholds for each species, along with data from several time periods after the ototoxic drug administration. In general, the findings are quite similar across species. Hearing losses are greatest at the high frequencies, minimal at the lowest frequencies, and recovery generally proceeds from the lowest frequencies first to the higher frequencies at later time periods.

These findings from several species of birds show that hair cell damage from ototoxic drug administration proceeds from the base to the apex of the basilar papilla, that hearing loss is greatest at high frequencies, and that functional recovery of hearing follows the hair cell replacement that one might expect from a place representation of frequency on the basilar membrane. The only exception from this general pattern is an unusual report that budgerigars experienced a low-frequency hearing loss to KM-induced hair cell destruction (Hashino and Sokabe 1989). The exact reasons for the anomalous results are still unknown.

An unusual bird, the Belgian *Waterslager* (BW) canary, also shows an interesting response to hair cell loss and regeneration. These birds have been bred for a distinctive low-frequency song and have an inherited auditory pathology resulting in, on average, 30% fewer hair cells on their basilar papilla, a 12% reduction in the number of auditory nerve fibers, and abnormal cochlear nuclei cell volumes compared to non-BW canaries (Gleich et al. 1994, 2001; Kubke et al. 2002). Hair cells on the abneural edge of the papilla, which are primarily innervated by efferent fibers, appear to be most affected, and this pathology develops after hatching and extends the length of the papilla (Gleich et al. 1994;

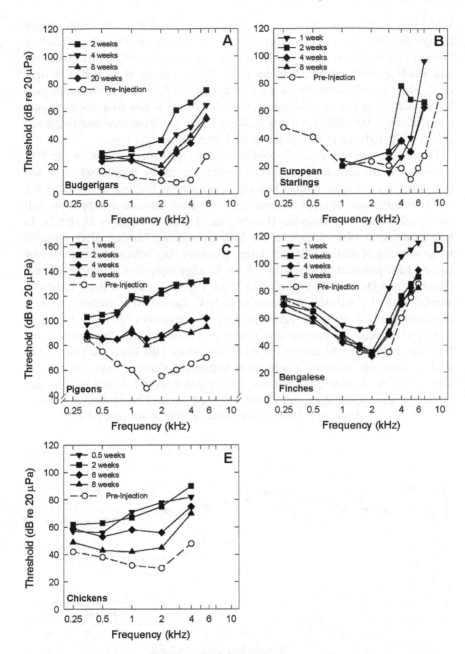

FIGURE 4.2. Behavioral data are shown for (**A**) budgerigars (Dooling et al. 2006) and (**B**) European starlings (Marean et al. 1993). ABRs are shown for (**C**) pigeons (Muller and Smolders 1998, 1999) and (**D**) Bengalese finches (Woolley et al. 2001), and (**E**) CAPs are shown for chickens (Chen et al. 1993).

Weisleder et al. 1996; Ryals et al. 2001; Ryals and Dooling 2002). Even though the loss of hair cells is evident throughout the papilla, the hereditary hearing loss in BW is clearly greatest at frequencies above 2.0 kHz (Okanoya and Dooling 1985, 1987; Okanoya et al. 1990; Gleich et al. 1995; Wright et al. 2004). Absolute thresholds are elevated by 20–40 dB and critical ratios are larger than normal at high frequencies, while normal at low frequencies (Lauer and Dooling 2002), indicating that auditory filters are broadened and frequency selectivity is reduced in BW in the area of greatest hearing loss.

Despite an ongoing rate of spontaneous hair cell regeneration in the BW canary, new hair cells never completely repair the basilar papilla, which would potentially lead to more normal hearing. Interestingly, both noise overexposure and aminoglycoside-induced damage cause a further increase in supporting cell proliferation and differentiation (Dooling et al. 1997; Gleich et al. 1997). In BW canaries, absolute thresholds for high frequencies increase after systemic administration of KM in BW, but during recovery they return to slightly below (better than) preinjection levels by 13 weeks after injections of the drug cease (Dooling and Dent 2001; unpublished data). Figure 4.3 shows these results in greater detail for 4.0 kHz. Although both BW and non-BW canaries show a threshold shift following the KM injections, the amounts of both the threshold shifts and the recovery patterns differ between the two groups. Interestingly, the final thresholds for BW canaries at 4 kHz are about 5 dB lower than the preinjection thresholds, showing a permanent improvement rather than a permanent loss as in normal canaries. There was no improvement in hair cell number and morphology in these BW canaries with improved thresholds (Ryals et al. 1999a). Thus, the cause of this threshold improvement in BW is still not clear.

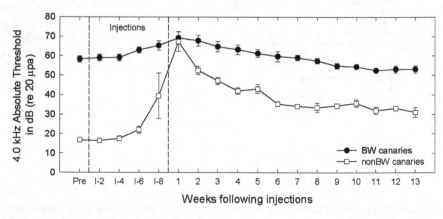

FIGURE 4.3. Absolute thresholds at 4.0 kHz before, during, and after KM injections in BW (black circles) and non-BW (white squares) canaries. Normal canaries show the typical pattern of a small residual permanent threshold shift after KM treatment. BW canaries also show a temporary threshold shift but then show a permanent improvement in hearing sensitivity at 4 kHz as a result of KM treatment.

3. Effects of Hair Cell Loss and Regeneration on Auditory Discrimination

Not surprisingly, almost all of the work on the effect of hair cell loss and regeneration on hearing has focused on the audiogram. However, a subject's ability to discriminate complex features of sounds after hair cell regeneration may be equally or more important than simply assessing threshold. Several recent studies have addressed this issue.

3.1 Discrimination Tests: Changes in Spectral, Intensity, and Temporal Resolution

Behavioral studies of auditory masking and discrimination have now been conducted on several species of birds including starlings, budgerigars, and chickens (Saunders and Salvi 1995; Dooling et al. 1997; Marean et al. 1998).

3.1.1 Spectral and Intensive Measures

Saunders and Salvi (1995) examined pure tone masking patterns in chickens using a tone-on-tone paradigm before and after exposure to a 525-Hz pure tone at 120 dB SPL for 48 hours. Two months after the exposure, masking patterns were reassessed in these birds and the postexposure masking patterns were virtually identical to the preexposure masking patterns. Frequency selectivity has also been shown to decrease in birds immediately after aminoglycoside treatment. Critical ratios, a crude estimate of auditory filter bandwidth, have been shown to increase immediately after KM treatment in budgerigars (Hashino and Sokabe 1989), indicating a broadening of the auditory filters. Marean et al. (1998) measured auditory filter widths in starlings using notched noise maskers before, during, and after KM injections. Auditory filters at 5 kHz were significantly wider after hair cell regeneration while spectral resolution was virtually unchanged at the other frequencies. In budgerigars, intensity (IDLs) and frequency difference limens (FDLs) were measured before, during, and after 10 days of KM administration for pure tones at 1.0 and 2.86 kHz. In spite of mild but permanently elevated absolute thresholds 4 weeks after KM treatment, IDLs were not significantly affected by the KM injections and FDLs were only slightly elevated at both frequencies as shown in Figure 4.4 (Dooling et al. 2006).

3.1.2 Temporal Measures

Marean et al. (1998) also measured minimum temporal resolution in starlings over the time course of hair cell loss and regeneration using temporal modulation transfer functions (TMTFs). TMTFs were mostly unaffected after KM administration. Some decrease in temporal resolution was evident for band-limited noise centered at 5.0 kHz, but these observed effects may have been due to poor

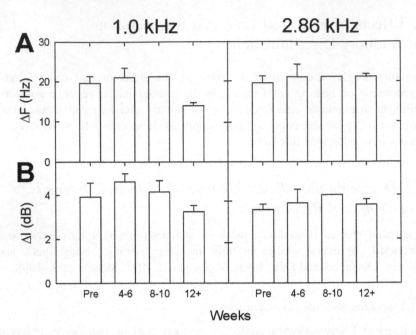

FIGURE 4.4. (**A**) FDLs and (**B**) IDLs for 1.0- and 2.86-kHz pure tones before and at three time periods after injections for three subjects. Error bars represent standard errors. (Replotted from Dooling et al. 2006.)

audibility of the signal as a result of threshold shifts, rather than to disruption of temporal resolution per se.

Maximum temporal integration (increasing threshold with decreasing stimulus duration) in birds, as in most vertebrates, is typically around 200–300 ms (Dooling et al. 2000). Saunders et al. (1995) showed that maximum temporal integration decreases immediately after acoustic trauma in chickens; it then returned to normal approximately 10–20 days after exposure. The slopes of threshold-duration functions decreased immediately after exposure, but recovered to near-normal levels after hair cell regeneration. Figure 4.5 shows the maximum temporal integration functions for the chicken at 1 kHz after acoustic overexposure to a 525-Hz pure tone (Saunders and Salvi, 1993; Saunders et al. 1995).

3.2 Discrimination Tests: Vocalizations

Hearing in humans and animals is typically assessed with simple sounds. But we know from work with both normal hearing and hearing impaired listeners that auditory tests with pure tones are less than perfect predictors of how well a human listener can detect, discriminate, or understand speech. Recent tests on budgerigars recovering from KM treatment were designed with parallels to human speech perception in mind. Budgerigars have been particularly useful

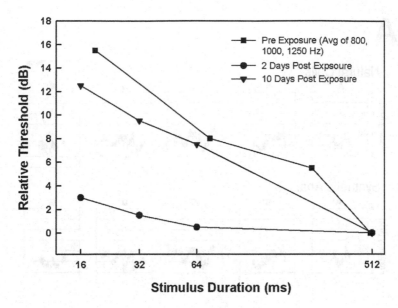

FIGURE 4.5. (**A**) Maximum temporal integration functions at 1 kHz for chickens from Saunders et al. (1995) and Saunders and Salvi (1993) for two subjects. Thresholds are shown relative to the subjects' thresholds for the 512-ms stimulus condition.

avian models for behavioral studies of auditory perception as well as complex vocal production. These birds learn new vocalizations throughout life, especially in response to changes in their social milieu (Dooling 1986; Dooling et al. 1987; Farabaugh et al. 1994; Hile et al. 2000) and they require hearing and auditory feedback for the maintenance of their adult vocal repertoire. Deafening, by cochlear removal during development or in adulthood, results in an impoverished vocal repertoire (Dooling et al. 1987; Heaton et al. 1999). Moreover, budgerigars trained to produce particular vocalizations under controlled experimental conditions show strong evidence of the Lombard effect (increases in vocal intensity when placed in a noisy environment), suggesting a real time monitoring of vocal output (Manabe et al. 1998).

In a series of tests on the abilities of budgerigars to discriminate among complex vocal signals before and after KM treatment, birds were tested on their ability to discriminate between pairs of calls in a test set of five different species-specific budgerigar contact calls and their synthetic analogs for a total of 10 calls in the test set (see Fig. 4.6A). Each bird was tested on the entire 10-call set both before and approximately every 4–6 weeks after injections up to about 24 weeks after cessation of KM injections. Discriminations between contact call types were relatively easy while discrimination between a natural contact call and its synthetic analog was difficult (Dooling et al. 2006).

The relatively easy discriminations between contact call types as measured by percent correct response returned to pretreatment levels by only 4 weeks after cessation of KM injections while the more difficult discriminations between a

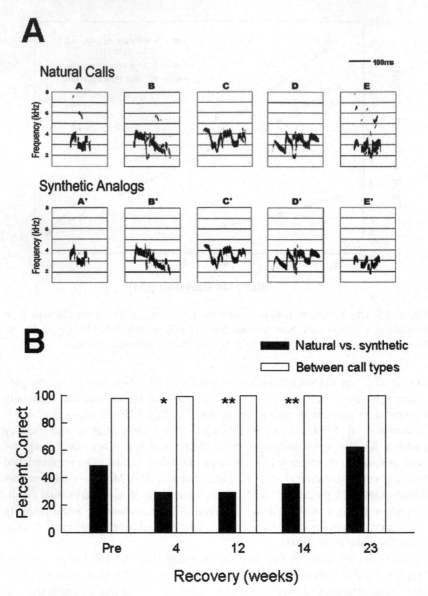

FIGURE 4.6. (**A**) Sonograms of contact calls from five different birds and their synthetic analogs plotted as frequency by time (taken from Dooling and Okanoya 1995). (**B**) The average percentage correct for three budgerigars discriminating among the five natural contact calls and their synthetic analogs before treatment with KM and at 4, 12, 14, and 23 weeks after injections. Asterisks represent significantly different results from the preinjection condition (*$p = 0.05$, **$p < 0.05$).

natural call and its synthetic analog showed prolonged disruption (Fig. 4.6B). Additional analyses of response latencies also show that the birds' perceptual space for contact calls was still somewhat distorted four weeks into recovery. Some calls that were perceptually distinct before KM administration were being perceived as similar many weeks into the recovery process (Dooling et al. 2006). These data on the budgerigar suggest a more complicated picture of hearing recovery in these birds when the task involves the discrimination of complex vocalizations.

3.3 Identification Tests: Recognition of Contact Calls

The previous results focused on the recovery in the detection and discrimination of simple and complex sounds after hair cell regeneration. However, a higher order task that needs to be addressed after hair cell regeneration is the ability to understand what is being said. This reminds us that it is one thing to hear speech, and even to discriminate among speech sounds but it is quite another thing to understand what is being said. A recent experiment addressed the effect of hair cell loss and regeneration on this broader question by testing whether budgerigars recognized contact calls that were previously familiar. This task is inherently more difficult than the discrimination task described in the preceding text in that the birds had to remember from trial to trial which call was the "go" call and which call was the "no go" call.

Figure 4.7A shows the average percent correct for the six budgerigars on the classification task before, during, and after 10 days of KM injections. Before injections of KM, the birds' performance was well above the 85% criterion level. However, performance fell to chance levels (50%) after several days of KM injections and remained there when assessed in a single 100-trial test session 24 and 38 days later. At 38 days (4 weeks after cessation of KM injections), the birds were retrained and tested daily on the same task in 100-trial daily sessions. The birds returned to preinjection performance levels in 4 days.

Response latencies for "go" stimuli remained below 1 s and relatively unchanged throughout the experiment, attesting to the birds' excellent health and behavioral responsiveness (Fig. 4.7B). Response latencies to the "no-go" stimuli averaged near 5 s before and during the first few days of treatment with KM. But, as testing continued and the birds began to respond by pecking the report key to both "go" and "no-go" stimuli, response latencies to the "no-go" approached the levels recorded to the "go" stimuli.

The six KM-treated birds tested after 4 weeks of recovery required an average of more than four 100-trial sessions to reach criterion instead of the average of less than two 100-trial sessions to reach criterion after a 4-week pause in testing but with no KM treatment. The results of important control experiments necessary for interpretation are summarized in Figure 4.7C. The first two control conditions are for four birds that were not injected with KM but were nonetheless given either a 2-week or a 4-week pause in testing. When testing resumed, the number of trials to reach criterion was much less than the time required to learn a

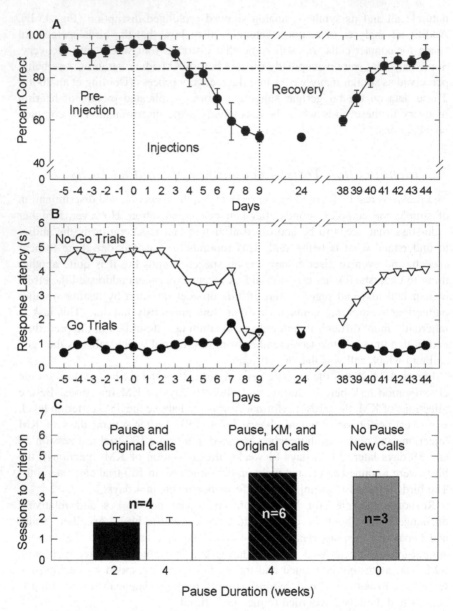

FIGURE 4.7. (**A**) Average percentage correct responses of six budgerigars on a Go/No-Go recognition task involving two contact calls before, during, and after treatment with KM. Twenty-four days after the end of treatment, classification performance had not yet returned to preinjection levels. At retesting on day 38, however, performance improved to preinjection levels within 4 days of testing. Error bars represent standard errors. (**B**) The average response latencies to both "go" and "no-go" stimuli for the six budgerigars before, during, and after KM treatment. Response latencies for the "go" stimuli remained low throughout the experiment, attesting to the birds' health and attentiveness.

new classification because the calls were familiar to the birds. Birds performing above criterion on one pair of contact calls (original calls) and switched to another pair (new calls) also required slightly over four 100-trial sessions to reach criterion. These results suggest that previously familiar contact calls do not sound the same to budgerigars that have been treated with KM and subsequently regenerated new hair cells. Instead, they sound like unfamiliar calls based on the time required to relearn a classification task involving these calls.

4. Vocal Production

One of the most devastating consequences of severe hearing impairment in humans is the loss of acoustic communication and the inevitable decline in the precision of speech production (Binnie et al. 1982; Waldstein 1990). With their dependence on hearing to develop and maintain a normal, species-specific vocal repertoire, birds provide a model for testing the effects of hearing loss and recovery on vocal production. To test this, operant conditioning with food reward was used to train birds to reliably and consistently produce species-specific contact calls with a high degree of precision (see, e.g., Manabe et al. 1998). Contact calls produced by the bird in the operant test chamber were compared online to a digitally stored template of the call. Contact calls that matched a stored template resulted in food reinforcement, while calls that did not match the template were not reinforced with food. These contingencies ensured that the birds were highly motivated to produce contact calls matching the template with the utmost precision. This precision with which birds produced contact calls under these controlled conditions was assessed during and after treatment with high doses of KM.

Figure 4.8A shows template matching performance for three budgerigars, each producing two different call types under operant control before, during, and after an 8-day course of KM. All three birds showed some loss in vocal precision during KM treatment; however, the loss was variable across the three birds. All three birds recovered to pretreatment levels, even those showing the greatest loss, within 10–15 days after the injections. On average, the ability to produce a precise vocal match to a stored contact call recovered to preinjection levels long before auditory recovery reached asymptote at about 8 weeks.

◄───

FIGURE 4.7. (C) Summary of the number of trials to reach criterion in the KM experiment and various control experiments indicating that KM treatment renders previously familiar calls unrecognizable. Control animals (no drug condition) take less than two sessions after 2 (black bar) or 4 (white bar) weeks of not running in any experiments to again reach criterion. Birds given a 4-week pause in running plus KM (striped bar) took an average of four sessions to reach criterion, a similar number of sessions as no-drug animals given a new set of calls to learn (gray bar). Error bars represent standard errors.

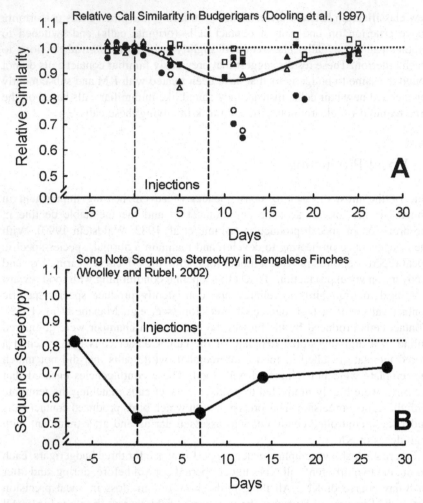

FIGURE 4.8. (**A**) The relative similarity of contact calls produced by three different budgerigars to their respective templates before, during, and up to 25 days after 8 days of KM injections. Closed symbols (square, circle, and triangle) represent template 1 and open symbols represent template 2 for each bird. The line represents the best fit to the mean data from the three birds (two calls from each bird for a total of six calls). Individual data from two of the birds were published previously (Dooling et al. 1997). (**B**) The mean song note sequence stereotypy for eight birds that sang a degraded song before, during, and after Amikacin injections. (Data are replotted from Woolley and Rubel 2002.)

In another set of experiments on vocal recovery, Woolley and Rubel (2002) examined vocal memory and vocal learning in adult Bengalese finches after hair cell loss and regeneration. They used a combination of noise and the ototoxic drug amikacin to induce the extensive hair cell loss to maximize song degradation (see Woolley and Rubel 1999). The results were interesting but complicated.

Bengalese finches normally sing a multinote song with energy ranging from 1 to 10 kHz. After 1 week of this treatment, the stereotypy of syllable sequences was reduced and syllable structure changed in 8 of the 15 birds tested. The remaining birds maintained normal song despite treatment. Scanning electron microscopy analysis of the auditory papillae and ABR responses in these birds revealed that some had hair cells remaining in the apical region of the papillae and also had normal low-frequency hearing. Over the course of hair cell regeneration, song gradually recovered. By 4 weeks posttreatment, the stereotypy of syllable sequences returned to normal. Figure 4.8B shows this degradation and then recovery in syllable order of songs in Bengalese finches from Woolley and Rubel (2002).

Though most syllables returned to their original structures by 8 weeks posttreatment, some syllables showed changes in note placement and changes in acoustic structure. The time courses of recovery are quite similar to those seen in the budgerigars, despite differences in the situation where the birds produced these songs. Three of the eight birds whose song was disrupted by hair cell loss showed additional song modifications between 4 and 8 weeks posttreatment. Approximately half of these birds' syllables changed after the song had recovered to near normal. Changes included breaking the notes of original syllables apart, dropping some notes from the repertoire, modifying the acoustic structure of remaining notes, recombining notes to form new syllables, and creating new sequences of syllables. These changes were linked to changes in the song of the birds' cagemates and were not spontaneous.

5. Summary

These studies address recovery of hearing and auditory perception of vocalizations after hair cell damage and regeneration and the recovery in the ability to produce precise vocalizations under experimental control. Results show that both hearing and vocal production are affected when many hair cells are damaged or lost, but both behaviors return to normal or near normal over time.

In humans, hearing loss as measured by pure tone detection or discrimination tasks involving simple sounds often does not predict the effect of hearing loss on the perception, understanding, or production of speech. Similarly, difficult discriminations between contact calls in budgerigars were affected for up to 20 weeks after KM treatment. At 4 weeks of recovery, perceptual maps for contact calls were distorted compared to pretreatment maps; however, the maps recovered to normal after 6 months posttreatment. So, in budgerigars, at least, results show a large recovery in vocal discrimination within a few weeks but some residual perceptual problems persist for up to 5–6 months.

A common refrain of hearing impaired humans is that speech can be heard as well as before but not understood (Newby 1964). This is an intriguing phenomenon and one that is even more interesting in an organism that has the capability of auditory hair cell regeneration. When trained to recognize or "label" two different contact calls, budgerigars completely lose the ability to label correctly when their hair cells have been lost to KM ototoxicity. Four weeks into recovery, these birds quickly relearn the classification of previously familiar contact calls to a high level, but they do so with a time course that suggests that the previously familiar calls now sound unfamiliar. In other words, though the ability to detect, discriminate, and classify complex acoustic sounds approaches pretreatment levels even as soon as 4 weeks into recovery, the perceptual world of complex vocalizations does not sound the same as it did before hair cells were lost. These results bear some similarity to the difficulties postlingually deafened humans have with new cochlear implants or even reprogrammed cochlear implants (e.g., Tyler et al. 1997; Hamzavi et al. 2003). If recent studies on hair cell regeneration in mammals prove someday applicable to humans, one can anticipate from these studies that a considerable period of adjustment may be necessary before a human can successfully interpret a new acoustic world.

Another consequence of profound hearing loss in humans is disruption of normal auditory feedback that serves to guide the precision of vocal production leading to degraded speech quality. The broad outlines of a similar phenomenon have been known to exist in birds for some time. Budgerigars deafened as young develop extremely abnormal contact calls and songs. Birds deafened in adulthood also eventually come to produce extremely abnormal vocalizations. These results, and those from many other species, establish the importance of auditory feedback in vocal learning in birds.

Budgerigars trained to produce specific contact calls under operant control show a loss of vocal precision and considerable variability at the peak of their hearing loss from ototoxicity. The effect of hearing loss on the quality of speech production in humans is also characterized by acute variability. In birds, this effect is transient, lasting no more than 2 weeks, and recovers well before absolute auditory sensitivity recovers substantially. These results suggest that even a little hearing goes a long way in guiding vocal production in these birds and that proprioceptive feedback may provide a temporary substitute for auditory feedback in this case of hearing loss.

Interestingly, while the precision of vocal production is initially affected by hearing loss from hair cell damage, this precision recovers long before the papilla is repopulated with new, functional hair cells. This suggests that, even in the absence of veridical auditory feedback, budgerigars, like humans, can also rely on long-term memory combined perhaps with feedback from other sensory systems to guide vocal production. Because both young and adult budgerigars deafened by cochlear removal show permanent changes in vocal output, the questions now should focus on how long the sensorimotor interfaces can do without appropriate sensory feedback before functional recovery in vocal production is no longer possible. The answer to this question, even in a bird, should have immediate relevance for treatment of hearing loss in humans, the timing of auditory prosthetic devices such as cochlear implants, and the ultimate hope for restoration of human hearing with regenerated hair cells.

Acknowledgments. This work was supported by NIDCD grants DC-01372 to RJD and BMR, DC-000198 and DC-004664 to RJD; DC-005450 to AML; and T32-DC-00046 to MLD.

References

Adler HJ, Kenealy JF, Dedio RM, Saunders JC (1992) Threshold shift, hair cell loss, and hair bundle stiffness following exposure to 120 and 125 dB pure tones in the neonatal chick. Acta Otolaryngol 112:444–454.

Adler HJ, Poje CP, Saunders JC (1993) Recovery of auditory function and structure in the chick after two intense pure tone exposures. Hear Res 71:214–224.

Binnie CA, Daniloff RG, Buckingham HW (1982) Phonetic disintegration in a five year old following sudden hearing loss. J Speech Hear Dis 47:181–189.

Chen L, Salvi RJ, Hashino E (1993) Recovery of CAP threshold and amplitude in chickens following kanamycin ototoxicity. Hear Res 69:15–24.

Corwin JT, Cotanche DA (1988) Regeneration of sensory hair cells after acoustic trauma. Science 240:1772–1774.

Cotanche DA (1999) Structural recovery from sound and aminoglycoside damage in the avian cochlea. Audiol Neurootol 4:271–285.

Ding-Pfennigdorff D, Smolders JWTh., Muller M, Klinke R (1998) Hair cell loss and regeneration after severe acoustic overstimulation in the adult pigeon. Hear Res 120:109–120.

Dooling RJ (1980) Behavior and psychophysics of hearing in birds. In Popper AN, Fay RR (eds) Comparative Studies of Hearing in Vertebrates. New York: Springer-Verlag, pp. 261–288.

Dooling RJ (1986) Perception of vocal signals by budgerigars. Exp Biol 45:195–218.

Dooling RJ, Dent ML (2001) New studies on hair cell regeneration in birds. Acoust Sci Tech 22:93–100.

Dooling RJ, Okanoya K (1995) The method of constant stimuli in testing auditory sensitivity in small birds. In Klump GM, Dooling RJ, Fay RR, Stebbins WC (eds) Methods in Comparative Psychoacoustics. Basel, Switzerland: Birkhauser-Verlag, pp. 161–169.

Dooling RJ, Gephart BF, Price PH, McHale C, Brauth SE (1987) Effects of deafening on the contact call of the budgerigar, *Melopsittacus undulatus.* Anim Behav 35:1264–1266.

Dooling RJ, Ryals BM, Manabe K (1997) Recovery of hearing and vocal behavior after hair cell regeneration. Proc Natl Acad Sci USA 94:14206–14210.

Dooling RJ, Lohr B, Dent ML (2000) Hearing in birds and reptiles. In Dooling RJ, Fay RR, Popper AN (eds) Comparative Hearing: Birds and Reptiles. New York: Springer-Verlag, pp. 308–359.

Dooling RJ, Ryals BM, Dent ML, Reid TL (2006) Perception of complex sounds in budgerigars (*Melopsittacus undulatus*) with temporary hearing loss. J Acoust Soc Am 119:2524–2532.

Farabaugh SM, Linzenbold A, Dooling RJ (1994) Vocal plasticity in budgerigars (*Melopsittacus undulatus*): evidence for social factors in the learning of contact calls. J Comp Psychol 108:81–92.

Girod DA, Tucci DL, Rubel EW (1991) Anatomical correlates of functional recovery in the avian inner ear following aminoglycoside ototoxicity. Laryngoscope 101:1139–1149.

Gleich O, Dooling RJ, Manley GA (1994) Inner-ear abnormalities and their functional consequences in Belgian Waterslager canaries (*Serinus canarius*). Hear Res 79:123–136.

138   R.J. Dooling et al.

Gleich O, Klump GM, Dooling RJ (1995) Peripheral basis for the auditory deficit in Belgian Waterslager canaries (*Serinus canarius*). Hear Res 82:100–108.

Gleich O, Dooling RJ, Presson JC (1997) Evidence for supporting cell proliferation and hair cell differentiation in the basilar papilla of adult Belgian Waterslager canaries (*Serinus canarius*). J Comp Neurol 377:5–14.

Gleich O, Dooling RJ, Ryals BM (2001) A quantitative analysis of the nerve fibers in the VIIIth nerve of Belgian Waterslager canaries with a hereditary sensorineural hearing loss. Hear Res 151:141–148.

Hamzavi J, Baumgartner WD, Pok SM, Franz P, Gstoettner W (2003) Variables affecting speech perception in postlingually deaf adults following cochlear implantation. Acta Otolaryngol 123:493–498.

Hashino E, Sokabe M (1989) Kanamycin induced low-frequency hearing loss in the budgerigar (*Melopsittacus undulatus*). J Acoust Soc Am 85:289–294.

Hashino E, Sokabe M, Miyamoto K (1988) Frequency susceptibility to acoustic trauma in the budgerigar (*Melopsittacus undulatus*). J Acoust Soc Am 83:2450–2453.

Hashino E, Sokabe M, Tanaka R (1991) Function-structure correlation during recovery from aminoglycoside ototoxicity in the avian auditory system. In Dancer AL, Henderson D, Salvi RJ, Hamernik RP (eds) Noise-Induced Hearing Loss. St. Louis: Mosby Year Book, pp. 228–236.

Hashino E, Tanaka R, Salvi RJ, Sokabe M (1992) Hair cell regeneration in the adult budgerigar after kanamycin ototoxicity. Hear Res 59:46–58.

Heaton JT, Dooling RJ, Farabaugh SM (1999) Effects of deafening on the calls and warble song of adult budgerigars (*Melopsittacus undulatus*). J Acoust Soc Am 105:2010–2019.

Henderson D, Hamernik RP (1986) Impulse noise: critical review. J Acoust Soc Am 80:569–584.

Hile AG, Plummer TK, Streidter GF (2000) Male vocal imitation produces call convergence during pair bonding in budgerigars (*Melopsittacus undulatus*). Anim Behav 59:1209–1218.

Izumikawa M, Minoda R, Kawamoto K, Abrashkin KA, Swiderski DL, Dolan DF, Brough DE, Raphael Y (2005) Auditory hair cell replacement and hearing improvement by *Atoh1* gene therapy in deaf mammals. Nat Med 11:271–276.

Kubke MF, Dent ML, Hodos W, Carr CE, Dooling RJ (2002) Nucleus magnocellularis and nucleus laminaris in Belgian Waterslager and normal strain canaries. Hear Res 164:19–28.

Lauer AM, Dooling RJ (2002) Frequency selectivity in canaries with a hereditary hearing loss. J Acoust Soc Am 111:2392.

Lifshitz J, Furman AC, Altman KW Saunders JC (2004) Spatial tuning curves along the chick basilar papilla in normal and sound-exposed ears. J Res Otolaryngol 5:171–184.

Lippe WR, Westbrook EW, Ryals BM (1991) Hair cell regeneration in the chick cochlea following aminoglycoside ototoxicity. Hear Res 56:203–210.

Lou W, Dong M, Dong M (1994) Hair cell damage and regeneration in the quail cochlea following kanamycin ototoxicity. Zhonghua Er Bi Yan Hou Ke Za Zhi 29:85–88.

Luz GA, Hodge DC (1971) Recovery from impulse-noise induced TTS in monkeys and men: a descriptive model. J Acoust Soc Am 49:1770–1777.

Manabe K, Sadr EI, Dooling RJ (1998) Control of vocal intensity in budgerigars (*Melopsittacus undulatus*): differential reinforcement of vocal intensity and the Lombard effect. J Acoust Soc Am 103:1190–1198.

Marean GC, Burt JM, Beecher MD, Rubel EW (1993) Hair cell regeneration in the European starling (*Sturnus vulgaris*): recovery of pure-tone detection thresholds. Hear Res 71:125–136.

Marean GC, Cunningham D, Burt JM, Beecher MD, Rubel EW (1995) Regenerated hair cells in the European starling: Are they more resistant to kanamycin ototoxicity than original hair cells? Hear Res 82:267–276.

Marean GC, Burt JM, Beecher MD, Rubel EW (1998) Auditory perception following hair cell regeneration in European starling (*Sturnus vulgaris*): frequency and temporal resolution. J Acoust Soc Am 103:3567–3580.

McFadden EA, Saunders JC (1989) Recovery of auditory function following intense sound exposure in the neonatal chick. Hear Res 41:205–215.

Müller M, Smolders JW, Ding-Pfennigdorff D, Klinke R (1996) Regeneration after tall hair cell damage following severe acoustic trauma in adult pigeons: correlation between cochlear morphology, compound action potential responses, and single fiber properties in single animals. Hear Res 102:133–154.

Müller M, Smolders JW, Ding-Pfennigdorff D, Klinke R (1997) Discharge properties of pigeon single auditory nerve fibers after recovery from severe acoustic trauma. Int J Dev Neurosci 15:401–416.

Newby HA (1964) Disorders of Hearing. In Newby HA (ed) Audiology. New York: Appleton-Century-Crofts, pp. 45–53.

Niemiec AJ, Raphael Y, Moody DB (1994) Return of auditory function following structural regeneration after acoustic trauma: Behavioral measures from quail. Hear Res 79:1–16.

Okanoya K, Dooling RJ (1985) Colony differences in auditory thresholds in the canary (*Serinus canarius*). J Acoust Soc Am 78:1170–1176.

Okanoya K, Dooling RJ (1987) Hearing in passerine and psittacine birds: a comparative study of masked and absolute auditory thresholds. J Comp Psych 101:7–15.

Okanoya K, Dooling RJ, Downing JD (1990) Hearing and vocalizations in hybrid Waterslager-Roller canaries (*Serinus canarius*). Hear Res 46:271–275.

Price GR (1979) Loss of auditory sensitivity following exposure to spectrally narrow impulses. J Acoust Soc Am 66:456–465.

Pugliano FA, Wilcox TO, Rossiter J, Saunders JC (1993) Recovery of auditory structure and function in neonatal chicks exposed to intense sound for 8 days. Neurosci Lett 151:214–218.

Ryals BM, Rubel EW (1988) Hair cell regeneration after acoustic trauma in adult coturnix quail. Science 240:1774–1776.

Ryals BM, Dooling RJ (2002) Development of hair cell stereovilli bundle abnormalities in Belgian Waterslager canary. Assoc for Res Otol Abs 25:942.

Ryals BM, Stalford MD, Lambert PR, Westbrook EW (1995) Recovery of noise-induced changes in the dark cells of the quail tegmentum vasculosum. Hear Res 83:51–61.

Ryals BM, Young J, Dooling RJ, Gleich O (1999a) Morphological correlates of improved hearing after kanamycin in Belgian Waterslager Canaries. Assoc Res Otol Abstr 22:522.

Ryals BM, Dooling RJ, Westbrook E, Dent ML, MacKenzie A, Larsen ON (1999b) Avian species differences in susceptibility to noise exposure. Hear Res 131:71–88.

Ryals BM, Gleich O, Tucker T (2001) Inner ear abnormalities in Belgian Waterslager Canary are not congenital. Assoc Res Otol Abstr 24:777.

Salvi RJ, Chen L, Trautwein P, Powers N, Shero M (1998) Hair cell regeneration and recovery of function in the avian auditory system. Scand Audiol Suppl 48:7–14.

Saunders J, Dooling RJ (1974) Noise-induced threshold shift in the parakeet (*Melopsittacus undulatus*). Proc Natl Acad Sci USA 71:1962–1965.

Saunders SS, Salvi RJ (1993) Psychoacoustics of normal adult chickens: thresholds and temporal integration. J Acoust Soc Am 94:83–90.

Saunders SS, Salvi RJ (1995) Pure tone masking patterns in adult chickens before and after recovery from acoustic trauma. J Acoust Soc Am 98:365–371.

Saunders JC, Adler H, Pugliano F (1992) The structural and functional aspects of hair cell regeneration in the chick as a result of exposure to intense sound. Exp Neurol 115:13–17.

Saunders SS, Salvi RJ, Miller KM (1995) Recovery of thresholds and temporal integration in adult chickens after high-level 525–Hz pure-tone exposure. J Acoust Soc Am 97:1150–1164.

Saunders JC, Doan DE, Cohen YE, Adler HJ, Poje CP (1996) Recent observations on the recovery of structure and function in the sound damaged chick ear. In Salvi RJ, Henderson D, Fiorino F, Colletti V (eds) Auditory System Plasticity and Regeneration. New York: Thieme, pp. 62–84.

Smolders JW (1999) Functional recovery in the avian ear after hair cell regeneration. Audiol Neurootol 4:286–302.

Trautwein PG, Hashino E, Salvi RJ (1998) Regenerated hair cells become functional during continuous administration of kanamycin. Audiol Neurotol 3:229–239.

Tucci DL, Rubel EW (1990) Physiological status of regenerated hair cells in the avian inner ear following aminoglycoside ototoxicity. Otolaryngol Head Neck Surg 103:443–445.

Tyler R, Parkinson AJ, Fryauf-Bertchy H, Lowder MW, Parkinson WS, Gantz BJ, Kelsay DM (1997) Speech perception by prelingually deaf children and postlingually deaf adults with cochlear implant. Scand Audiol Suppl 46:65–71.

Waldstein RS (1990) Effects of postlingual deafness on speech production: implications for the role of auditory feedback. J Acoust Soc Am 88:2099–2114.

Weisleder P, Liu Y, Park TJ (1996) Anatomical basis of a congenital hearing impairment: basilar papilla dysplasia in the Belgian Waterslager canary. J Comp Neurol 369:292–301.

Woolley SM, Rubel EW (1999) High-frequency auditory feedback is not required for adult song maintenance in Bengalese finches. J Neurosci 19:358–371.

Woolley SM, Rubel EW (2002) Vocal memory and learning in adult Bengalese Finches with regenerated hair cells. J Neurosci 22:7774–7787.

Woolley SM, Wissman AM, Rubel EW (2001) Hair cell regeneration and recovery of auditory thresholds following aminoglycoside ototoxicity in Bengalese finches. Hear Res 153:181–195.

Wright T, Brittan-Powell EF, Dooling RJ, Mundinger- PC (2004) Sex-linkage of deafness and song frequency spectrum in the Waterslager strain of domestic canary. Proc R Soc Lond B (Suppl) 271:s409–s412.

5
Hair Cell Regeneration: Mechanisms Guiding Cellular Proliferation and Differentiation

ELIZABETH C. OESTERLE AND JENNIFER S. STONE

1. Introduction

Hair cells (HCs) are vertebrate sensory mechanoreceptors that detect signals from the environment associated with hearing, balance, body orientation, and external water movement. They are derived from special cranial placodes (e.g., otic placode, lateral line placodes) that are transformed into the membranous labyrinth of the inner ear or the lateral line neuromasts during development. HCs are located along with nonsensory supporting cells in specific sensory epithelia in the inner ear of terrestrial vertebrates and also in neuromasts of the lateral line in aquatic animals. Loss of HCs or HC function typically leads to severe sensory deficits. In humans, many deafness-causing mutations that affect HC structure and organization have been identified (Cryns and Van Camp 2004). Epigenetic factors, such as intense or prolonged noise, ototoxic drugs, and certain microbial infections, also lead to HC damage and loss. The effects of HC loss are most pronounced in the elderly human population, of whom as many as 65% experience sensorineural hearing loss and 30% experience vestibular problems such as vertigo. Because HCs are not spontaneously replaced in humans once they are lost, hearing and balance deficits caused by HC loss at any stage of life are irreversible.

1.1 Nonmammals

Spontaneous HC production and replacement occur in many nonmammalian vertebrates after embryogenesis has ceased. The number of HCs increases for some time after birth in the otolithic organs of fish (Platt 1977; Corwin 1981, 1983; Popper and Hoxter 1984), toads (Corwin 1985), turtles (Severinsen et al. 2003), and birds (Goodyear et al. 1999). Further, HCs are added after birth to lateral line organs in axolotls (*Ambystoma mexicanuum*; Jørgensen and Flock 1976) and zebrafish (*Danio rerio*; Williams and Holder 2000; Higgs et al. 2002). In fish and amphibia, this increase in HC production is mirrored

141

by a concomitant growth of each otolithic organ over much of the lives of these animals (Platt 1977; Corwin 1981, 1983; Popper and Hoxter 1984). In contrast, the area of the utricular epithelium of birds ceases to expand shortly after hatching (Jørgensen 1991; Goodyear et al. 1999), although new HCs continue to be produced at a slow rate throughout this tissue and in other vestibular epithelia postembryonically (Jørgensen and Mathiesen 1988; Roberson et al. 1992). Because HCs in avian vestibular organs die spontaneously at a steady rate (Jørgensen 1991; Kil et al. 1997), it is assumed that postembryonic HC production serves to replace dead HCs and to maintain a relatively constant HC number. In vestibular and lateral line sensory organs of mature nonmammalian vertebrates, rates of HC production are upregulated in response to experimentally induced HC loss (Weisleder and Rubel 1992; Baird et al. 1993; Kil et al. 1997; Williams and Holder 2000; Avallone et al. 2003; Harris et al. 2003).

In the auditory sensory epithelium (SE) of nonmammals, normal HC and SC production appears to be largely confined to embryogenesis (Katayama and Corwin 1989, 1993). In the auditory SE (basilar papilla) of mature birds, only negligible cell division is detectable, and no new HCs are added (Corwin and Cotanche 1988; Ryals and Rubel 1988; Oesterle and Rubel 1993). However, experimental induction of HC loss leads to a significant upregulation in cell division and to the production of a full set of replacement HCs (Cotanche 1987a; Cruz et al. 1987; Corwin and Cotanche 1988; Ryals and Rubel 1988), assuming epithelial trauma is not too severe (see Cotanche et al. 1995; Ding-Pfennigdorff et al. 1998).

1.2 Mammals

In contrast to nonmammalian species, in mammals HC production ceases during the late embryonic period (Ruben 1967). Cell division does not occur spontaneously in mature mammalian auditory or vestibular SE (Roberson and Rubel 1994; Kuntz and Oesterle 1998a). No new auditory (organ of Corti) HCs are produced, either under normal conditions or after experimentally induced HC lesions (Sobkowicz et al. 1992; Roberson and Rubel 1994; Chardin and Romand 1995). However, rare mitotic activity is triggered in vestibular SE (Warchol et al. 1993; Lambert 1994; Rubel et al. 1995; Li and Forge 1997; Kuntz and Oesterle 1998a). Recent investigations have begun to explore the potential for inducing mitotic HC replacement in mammalian species via the addition of mitogenic growth factors to cultured mammalian tissue (Lambert 1994; Yamashita and Oesterle 1995) and to intact animals (Kuntz and Oesterle 1998a; Kopke et al. 2001) and by disrupting normal gene function (Chen and Segil 1999; Löwenheim et al. 1999; Mantela et al. 2005; Sage et al. 2005). Investigations have also begun to explore HC production in mammals by a nonmitotic process, through either repair or direct conversion of another cell type (e.g., supporting cells) into HCs (Zheng et al. 2000; Kawamoto et al. 2003; Izumikawa et al. 2005).

At this point, the following critical questions related to mammalian HC regeneration must be explored further: (1) Are cells with the potential to form new

HCs, either mitotically or nonmitotically, retained in the mature mammalian SE? (2) If so, how are these cells prevented from dividing after HC damage? (3) Can such cells be triggered to generate new HCs using a noninvasive, harmless method? In this chapter, mechanisms used by nonmammalian species to regenerate HCs after birth are discussed, with a focus on factors regulating SE proliferation and differentiation. Then, the mammalian response to HC trauma is discussed, and currently promising approaches toward triggering replacement of mammalian HCs are considered.

2. Hair Cell Regeneration in Nonmammalian Species

Because most of the research on regeneration of auditory HCs in nonmammalian species has been conducted in birds, most of the discussion in this section focuses on studies in birds. Research on other animals is discussed as appropriate.

2.1 Cellular Proliferation

As discussed in the preceding text, mitotic activity in the SE gives rise to new HCs in many mature nonmammalian animals. Several studies have addressed the identity of progenitor cells in this regenerative process and have examined the timing, magnitude, and cellular/molecular events associated with progenitor cell division to gain insight into potential regulatory mechanisms used by species capable of mitotic HC regeneration. These studies are the focus of the next section.

2.1.1 Progenitor Cells: Identity, Location, and Behavior

All HC epithelia have two primary cell types: sensory HCs and nonsensory supporting cells (SCs) (Fig. 5.1). Supporting cells are interposed between HCs. While the cell bodies of HCs are confined to the lumenal surface of the SE, the somata of all SCs appear to extend from the lumen to the basal lamina. The nucleus of most SCs is located in the basal (adlumenal) half of the epithelium. For many years, SCs were suspected to serve as progenitors to new HCs during postembryonic HC production in mature nonmammalian vertebrates. The earliest evidence to support this hypothesis emerged from studies that demonstrated new SCs are also formed by mitosis, under normal and damaged conditions (Corwin 1981; Cotanche 1987a; Cruz et al. 1987; Corwin and Cotanche 1988; Ryals and Rubel 1988; Girod et al. 1989; Presson and Popper 1990). However, it was not until a decade ago that SCs were "caught in the act" of dividing, and their progeny were shown to go on to form new HCs and SCs. These observations were first made directly, after laser ablation of HCs in the lateral line end organs in live salamanders (Balak et al. 1990; Jones and Corwin 1993). Microscopic time-lapse imaging allowed investigators to determine that SC nuclei migrate from the basal compartment of the epithelium to the lumenal surface, where they undergo

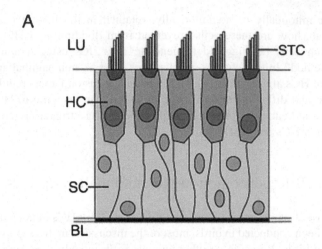

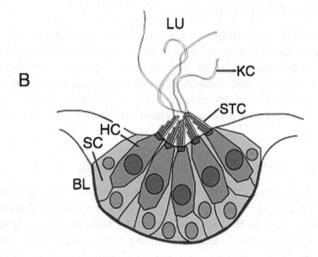

FIGURE 5.1. Cellular organization of hair cell epithelia in nonmammals. This figure depicts the generalized organization of hair cells (HC) and supporting cells (SC) in the inner ear sensory epithelium (**A**) and the lateral line neuromast (**B**). In both diagrams, the cytoplasm of the HCs appears darker than that of the SCs. HCs are distributed along the luminal (LU) surface of the epithelium; their cell bodies do not make contact with the basal lamina (BL). Their bundle of hair-like stereocilia (STC), which are planted in the cuticular plate (dark gray), protrudes apically into the lumen. In the lateral line neuromast (**B**), a single long cilium called the kinocilium (KC) is present in the mature bundle. The cell bodies of SCs extend the entire depth of the epithelium, from the lumen to the basal lamina. SC nuclei reside primarily within the basal half of the epithelium.

mitotic division. Analysis of SE from birds and fish also showed that SCs are the first cells to undergo DNA synthesis after HC damage, with some of the resulting progeny differentiating into HCs (Presson and Popper 1990; Raphael 1992; Hashino and Salvi 1993; Stone and Cotanche 1994; Tsue et al. 1994a; Warchol and Corwin 1996; Fig. 5.2A). Epithelial cells lining the SE called hyaline cells

A   Mitotic regeneration

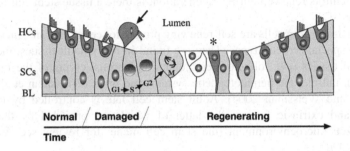

B   Direct transdifferentiation

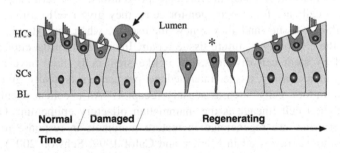

FIGURE 5.2. Temporal progression of cellular events associated with hair cell regeneration in the chicken auditory epithelium (basilar papilla). This schematic depicts the temporal progression of cellular events involved in the two known modes of nonmammalian HC regeneration in the chicken basilar papilla: mitotic regeneration (**A**) and direct transdifferentiation (**B**). Mature and regenerating hair cells (HCs) are depicted with darker cytoplasm than the supporting cells (SCs). In both diagrams, events progress temporally from left to right, with the normal epithelium shown on the left, the damaged and regenerating epithelium in the middle, and the fully repaired epithelium on the right. Arrows point to damaged HCs in the process of being extruded into the lumenal fluid. Asterisks indicate regenerated HCs with elongated morphology. For mitotic regeneration (**A**), the stages of the cell cycle are depicted as G1 (gap1), S (DNA synthesis), G2 (gap2), and M (mitosis). BL, Basal lamina. (Reprinted from Bermingham-McDonogh O, Rubel EW. Hair cell regeneration: winging our way towards a sound future. Curr Opin Neurobiol 13:119–126, Copyright 2003, with permission from Elsevier.)

are induced by severe noise trauma to undergo cell division and to migrate into the damaged SE. However, unlike SCs, hyaline cells do not appear to be capable of giving rise to new HCs (Cotanche et al. 1995).

Over the nearly two decades after the identification of SCs as progenitor cells, substantial work was done to characterize progenitor cell behavior after HC damage, primarily in avian species (reviewed in Stone and Rubel 2000b). Nonetheless, the following questions remain unanswered: (1) Are distinct subsets of SCs specialized to serve as HC progenitors? (2) Do all dividing cells give rise to HCs, or are there other cell lineages present in inner ear SE? (3) How are HC progenitors renewed during regeneration; is there a tissue stem cell found in inner ear SE?

Adult tissue stem cells are self-renewing progenitor cells that can generate one or more specialized cell types. They are found in many adult tissues, including epithelial tissues (e.g., olfactory epithelium, epidermis, crypts of the intestinal mucosa), muscle, the hematotopoetic system, and brain (Weissman et al. 2001; Wagers and Weissman 2004). Adult stem cell fate is controlled by intrinsic factors and extrinsic factors, the latter of which are provided by the cell's specialized microenvironment (the stem cell niche; for review see Watt and Hogan 2000).

In other sensory systems capable of regeneration, nonsensory cells are highly divergent, and distinct cell lineages appear to be present (e.g., retina: Fischer and Reh 2000, 2001; olfactory epithelium: reviewed in Schwob 2002). These lineages include stem cells, intermediate progenitors, and terminally mitotic cells. Stem cells are founder progenitor cells; they give rise to a range of cell types within a tissue, and they renew themselves when they divide, thereby ensuring the capability of future tissue repair. Intermediate progenitor cells are generated by stem cell divisions, but unlike stem cells, they have a limited potential. In a sensory tissue, intermediate progenitors typically give rise to terminally mitotic cells, such as sensory receptors. An excellent example of a neuroepithelial cell lineage is the mammalian olfactory epithelium. Olfactory receptor neurons are regenerated in mature mammals in response to several types of trauma (reviewed in Murray and Calof 1999; Schwob 2002). Careful analysis has demonstrated that, in addition to olfactory neurons, at least five types of nonsensory cells exist in the olfactory epithelium, each with distinct morphological and molecular properties. Sustentacular cells, microvillar cells, and Bowman's gland/duct cells all span the width of the olfactory epithelium. In contrast, the cell bodies of horizontal basal cells and globose basal cells reside near the basal lamina. When olfactory neurons are selectively killed, only globose basal cells show a large increase in mitotic activity (Schwartz Levey et al. 1991). Transplantation of globose basal cells into an olfactory epithelium where more than one cell type has been injured shows that these cells can give rise to a large range of cell types, including sustentacular cells, globose basal cells, and neurons (Goldstein et al. 1998). This finding demonstrates that some globose basal cells have properties of multipotent stem cells. Analyses of cell fate during regeneration show that several cell lineages exist in the olfactory epithelium, and some are more restricted than others. For example, progenitors that give rise to Bowman's gland/duct cells do not appear to generate neurons (Huard et al. 1998). In the neuronal lineage, globose basal cells with stem cell properties are few in number (approximately 1/3600; Mumm et al. 1996) and appear to divide at a slow rate. In the process, they renew themselves and form intermediate progenitors (Calof and Chikaraishi 1989), which are more predominant and divide at a faster rate. The progeny of intermediate progenitors eventually undergo terminal differentiation, as neurons or nonsensory cells.

Knowledge of subsets of nonsensory cells in the olfactory epithelium has enabled relatively rapid progress in identifying signals that regulate the behavior of distinct cell types (reviewed in Beites et al. 2005). In contrast, little is known about SC specialization or cell lineages in mature inner ear epithelia. The presence of a self-renewing progenitor (stem) cell among the SC population in the avian basilar papilla is supported by several studies (reviewed in Parker and Cotanche 2004). The basilar papilla of the Belgian *Waterslager* canary undergoes spontaneous HC death after birth and appears to continually regenerate new HCs and SCs (Gleich et al. 1994, 1997). In starlings and quail, auditory HCs are regenerated after two separate HC injuries, suggesting progenitor cells are renewed following the first insult (Niemiec et al. 1994; Adler and Saunders 1995; Marean et al. 1995). In the damaged chicken basilar papilla, between 1% and 4% of SCs divide twice during the course of recovery from drug-induced HC loss (Stone et al. 1999), suggesting a capacity for self-renewal in at least some SCs. Some progenitor recycling also appears to occur in the chick basilar papilla after noise damage (Stone and Cotanche 1994). Further, some mitotic events in the damaged basilar papilla are asymmetric in nature, producing a HC and a SC (Stone and Rubel 2000a). Asymmetric division leading to the replenishment of a cell with proliferative capacity is one hallmark of stem cells.

Data from nonmammalian vestibular epithelia support the presence of specialized stem cells in that tissue as well. For example, while growth of the chicken utricle slows considerably over time after hatching (Jørgensen 1991; Goodyear et al. 1999), progenitor cell division proceeds at a substantial rate into adulthood (Roberson et al. 1992; Kil et al. 1997). The presence of a stem cell is assumed, as the progenitor population must be replenished in order to maintain life-long mitotic activity. Interestingly, though, it has been difficult to document progenitor cell recycling in the undamaged chicken utricle, perhaps due to a slow cycling rate (Stone et al. 1999; Wilkins et al. 1999). In contrast, in the undamaged fish saccule, as many as 70% of the progenitor cells reenter the cell cycle within days of completing it (Wilkins et al. 1999). Removal of rapidly dividing cells has been achieved by treatment of organs in vitro with cytosine arabinoside (Ara-C), an antineoplastic drug that causes death in cells entering S phase in many tissues, including the brain (Jung et al. 2004). Ara-C treatment in the fish saccule causes the loss of dividing cells and the recruitment of another group of SCs into the cell cycle (Presson et al. 1995). Based on studies of cell lineages in other systems, this finding implies the presence of a rapidly dividing intermediate progenitor (susceptible to Ara-C) in the fish saccule along with a slowly dividing or quiescent stem cell (resistant to Ara-C; reviewed in Hall and Watt 1989).

While several features distinguish HCs from SCs in the nonmammalian SE (e.g., Tanaka and Smith 1978; Ginzberg and Gilula 1979), very few morphological, molecular, or behavioral features clearly delineate subsets of SCs. This is in contrast to the mammalian organ of Corti, which contains several subtypes of SCs that are structurally and chemically distinct. In the late embryonic chicken basilar papilla, two different SC morphologies have been documented: cells with thick apical and basal processes and a centrally located nucleus, and cells with thin, delicate processes and an adlumenally located nucleus (Fekete et al. 1998). It is not known if these subtypes are functionally distinct or if they are retained in mature chickens. Electron microscopic studies of SCs in fish and chickens have revealed no marked systematic differences among SCs at the ultrastructural level (Presson et al. 1996; L.G. Duckert, unpublished observations). However, small morphological and ultrastructural differences have been reported for the several rows of SCs that line the abneural (inferior) edge of the papilla in comparison to SCs that are situated between HCs (Oesterle et al. 1992). Furthermore, nuclei of S-phase SCs in the fish vestibular epithelium display a different shape and position than quiescent SCs (Presson et al. 1996). While some molecular markers of SCs have been characterized (e.g., cytokeratins: Stone et al. 1996; receptor-like protein tyrosine phosphatase [RPTP]: Kruger et al. 1999; tectorins: Goodyear and Richardson 2002), only a handful of markers show systematic variability across the quiescent SC population. For example, mRNA for fibroblast growth factor receptor-3 (Bermingham-McDonogh et al. 2001) is present in high levels in most SCs, but it is absent from those located along the neural (superior) edge of the basilar papilla. In addition, proliferating cell nuclear antigen (PCNA) is immunologically detectable in 3% of SC nuclei, although no overt pattern of spatial distribution is evident (Bhave et al. 1995). Further, expression of the homeobox transcription factor, *cProx1*, reveals molecular diversity among quiescent SCs and among dividing progenitor cells in the basilar papilla (Stone et al. 2004). The relevance of this variation in expression of these different genes among SCs has not been determined. However, all three genes have been implicated in cell cycle regulation in other tissues/species (Wigle et al. 1999; Li and Vaessin 2000; Dyer et al. 2003; Hong and Chakravarti 2003; Inglis-Broadgate et al. 2005). Assuming these proteins play similar roles in the chicken basilar papilla, these findings support the existence of progenitor populations with different potentials for cell division. Alternatively, these differences may reflect stochastic variability in gene expression across the SC population.

For the remainder of this chapter, HC progenitors are presumed to reside among the SC population in nonmammals, so the terms SC and progenitor cell are used interchangeably, with some distinctions made in context. As discussed in Section 3.1.1, several lines of evidence suggest that tissue stem cells are intrinsic to mammalian vestibular epithelia, albeit in limited numbers and in a quiescent or restricted state. Further characterization of nonmammalian stem/progenitor cells would provide information for comparative studies across nonmammals and mammals that may shed light on why mitotic activity and HC regeneration are so highly attenuated in mammals.

2.1.2 Proliferation Signals

In nonmammals, mitotic activity in progenitor cells is robust in response to damage. For example, in the chicken basilar papilla, hundreds of SCs incorporate the nucleotide analog 5-bromo-2-deoxyuridine (BrdU) after a single bolus of gentamicin is delivered intraperitoneally; over the course of recovery, thousands of SCs appear to divide (Stone et al. 1999). The section that follows discusses studies that have laid the groundwork for our current understanding of the cellular processes leading to the entry of HC progenitors into the cell cycle and of the molecules associated with these processes.

2.1.2.1 Mitotic Activity Is Dependent on Location, Timing, and Magnitude of Hair Cell Death

Mitotic activity among HC progenitors is spatially, temporally, and quantitatively correlated with HC death. This is best illustrated by experimental studies in which location, timing, and amount of HC death are known entities, such as after noise-, ototoxic drug-, or laser-induced traumas to the auditory epithelium (discussed in Forge and Van De Water, Chapter 6). Exposure to pure-tone or narrow-band noise stimuli leads to HC lesions in the tonotopically corresponding region of the auditory epithelium. The proportion of HCs (and SCs in extreme cases) that are killed within this region depends on the intensity and duration of acoustic overstimulation. Administration of ototoxic drugs, such as aminoglycoside antibiotics (e.g., gentamicin, neomycin), causes substantial HC death to the proximal, high-frequency portion of the auditory epithelium (reviewed in Cotanche 1999; Forge and Schacht 2000). Prolonged drug exposure, or increased drug dosages, leads to expansion of the damaged region toward the distal, low-frequency end of the organ. Single-dose regimens lead to smaller lesions that are generated in a shorter amount of time, and these have proven effective for studying the timing of cellular events that are presumably interdependent, such as HC death and cell division, discussed later (Bhave et al. 1995; Janas et al. 1995; Roberson et al. 1996). Combinations of noise exposure and drug treatment have profound effects on HCs, causing damage and/or death across nearly the entire auditory organ (e.g., Bone and Ryan 1978; Woolley and Rubel 1999). Another method of killing HCs, laser ablation, is directed at single HCs or groups of HCs in vivo (Jones and Corwin 1993) or in vitro (Warchol and Corwin 1996).

After each method of HC injury, SC proliferation is limited to regions of severe HC trauma and/or HC death (Corwin and Cotanche 1988; Ryals and Rubel 1988; Hashino and Salvi 1993, Hashino et al. 1995; Stone and Cotanche 1994; Warchol and Corwin 1996; Fig. 5.2A). Interestingly, two studies in the chicken basilar papilla show that progenitors residing outside the area of overt HC damage show a limited change in cell cycle status, even though HCs surrounding them do not display clear signs of damage (Bhave et al. 1995; Sliwinska-Kowalska et al. 2000). As discussed in the preceding text, cell division is very rare in the undamaged basilar papilla. This feature is reflected by the fact, that in the undamaged papilla, most SCs (62%) express an antigen characteristic of growth arrest (the G0 marker, statin), whereas 0–3% of SCs express proliferating cell marker (PCNA), an antigen characteristic of G1 and later stages of the cell cycle (Fig. 5.3). Short treatments with the ototoxin gentamicin led to complete HC loss in the proximal (high-frequency) end of the basilar papilla. In response to gentamicin treatment, statin and PCNA expression are significantly altered, with statin expression decreasing and PCNA expression increasing. Remarkably, these changes occur in SC nuclei throughout the basilar papilla, including the distal undamaged area, demonstrating widespread SC transit from growth arrest (G0) to the gap 1 phase of the cell cycle (G1). A similar response was seen in the basilar papilla after exposure to wide-band noise; PCNA immunoreactivity increased in both damaged and undamaged regions (Sliwinska-Kowalska et al. 2000). Importantly, though, progression of SCs to DNA synthesis (S phase) and later stages of the cell cycle only occurred in areas of complete HC loss. These studies demonstrate that, like many other tissues including the central nervous system (Galderisi et al. 2003), cell cycle progression in avian SCs depends on multiple highly controlled signals. They suggest that, after HC damage, signals that trigger SC transit from G0 to G1 are altered throughout the tissue, while signals directing SCs to enter S phase are limited to areas of marked HC damage. Future studies should strive to identify molecules that act regionally to promote each step in cell cycle progression.

The time course of cell division after experimental HC injury has been examined most extensively in chickens (Raphael 1992; Hashino and Salvi 1993; Stone and Cotanche 1994; Bhave et al. 1995; Hashino et al. 1995; Warchol and Corwin 1996; Stone et al. 1999; Roberson and Cotanche 2000). In the basilar papilla, entry of progenitor cells into S phase is first seen between18 and 24 hours after the onset of acoustic overstimulation, drug treatment, or laser ablation. The reason for this delay is not known, but it is undoubtedly influenced by the timing of the signaling cascades directing progenitor cells to transit from growth arrest, their normal state (Oesterle and Rubel 1993), to S phase (discussed in more detail in Section 2.1.2.3). The trigger for this cascade appears to be the onset of HC damage because the latency period for SCs to reach S phase is similar in all damaging paradigms, regardless of the nature and duration of the damaging treatment (e.g., see Stone and Cotanche 1994; Warchol and Corwin 1996; Bhave et al. 1998). Perhaps the most dramatic demonstration of the dependence of progenitor cell cycle entry on the onset of HC damage is provided in a study by Hashino et al. (1995). Chronic treatment of chickens with the aminoglycoside kanamycin creates a HC lesion in the basilar papilla that spreads from proximal to distal over time (Hashino et al. 1991). Labeling of dividing cells with BrdU for short intervals at different periods relative to the start of drug exposure demonstrates that SCs in the proximal region, which are damaged initially, are the first to divide, and SC entry into the cell cycle progresses up the length of

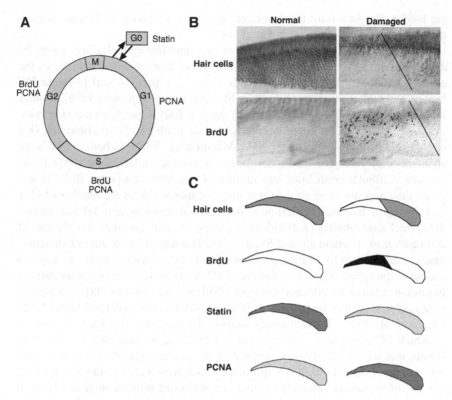

FIGURE 5.3. Cell cycle progression in the chicken basilar papilla after gentamicin treatment. This figure depicts changes that occur with respect to cell cycle status among SCs in different regions of the damaged basilar papilla after gentamicin treatment. Data were derived from Bhave et al. (1995). Cell cycle status was assessed by immunolabeling for different cell cycle-related markers: statin, PCNA, and bromodeoxyuridine (BrdU). (A) Stage of the cell cycle where labeling for each marker is elevated. Stages of the cell cycle are depicted as G1 (gap1), S (DNA synthesis), G2 (gap2), and M (mitosis); G0 is growth arrest. Statin is elevated in cells in G0, but not in cycling cells. PCNA is elevated in cells in G1, S, and M phase, but not in cells in G0. BrdU labels cells in S phase and onward (G2, M, post-M). (B) Immunolabeling for hair cells and BrdU in the basal half of the epithelium of normal and damaged (3 days post-gentamicin) birds. In all panels, the basal tip of the epithelium is to the left, and the middle portion of the epithelium is to the right. The lines in the right-hand panels delineate the border of the HC lesion in the damaged basilar papilla. Note that no BrdU incorporation is seen in the normal epithelium, and that after damage, BrdU incorporation is confined to the region of complete hair cell loss. (C) Spatial patterns of hair cell loss and immunolabeling for BrdU, statin, and PCNA are shown for the epithelium at 3–5 days post-gentamicin. Each image is a full basilar papilla, with the proximal tip to the left and the distal tip to the right. Relative levels of hair cell loss and immunolabeling are indicated by the intensity of the gray label. Note the spatial correlation between hair cell loss and BrdU incorporation, but the lack of correlation between hair cell loss and statin/PCNA labeling.

the basilar papilla toward the distal end over time, mirroring with some latency the timing of HC death (Hashino et al. 1995).

The magnitude of cell proliferation in regenerating epithelia appears to be directly related to the amount of HC death that occurs. For example, in the chicken auditory epithelium, longer noise exposure periods lead to larger HC lesions and higher numbers of dividing SCs (Stone and Cotanche 1994). Studies in the chicken vestibular epithelium and the fish lateral line neuromasts provide evidence that there is a direct link between HC death and SC proliferation (Kil et al. 1997; Williams and Holder 2000; Matsui et al. 2002). In both SE, HCs die spontaneously via a presumed apoptotic mechanism at a relatively low rate. In the chicken vestibular epithelium, the number of SCs that incorporate BrdU at any given time appears to be on the same order of magnitude as the number of HCs undergoing apoptosis, as determined by terminal deoxynucleotidyl transferase-dUTP-nick-end-labeling (Kil et al. 1997). Experimental modulation of the rate of cell death leads to attenuation of SC division. The rate of HC death in fish neuro-masts is diminished by treatment with inhibitors of caspases, which are enzymes central to the apoptotic process. Levels of SC division are decreased after caspase inhibition relative to untreated controls (Williams and Holder 2000). A similar relationship is seen in the chicken vestibular epithelia (Weisleder and Rubel 1995; Matsui et al. 2002). These findings support the presence of a feedback system in which SC division is directly regulated by HCs. Normal HCs may provide signals that inhibit SC division and/or they may release signals that promote SC division upon their death. It is important to note, however, that in some species, the size of the sensory epithelium (SE) grows significantly throughout life, with new cells being added to the organ's periphery (e.g., the saccule of oscar fish, Astronotus ocellatus; Popper and Hoxter 1984). While the location and degree of HC death occurring in species with extensive postembryonic growth are unclear (Jørgensen 1991), the degree and pattern of cell proliferation seen suggest that mitotic activity is not primarily reliant on HC death as a stimulus, and therefore, is not regenerative in nature.

2.1.2.2 Overshooting the Mark

The production of the correct number and types of new cells is not an exact process during HC regeneration. In avian inner ear epithelia, new HCs and SCs are overproduced shortly after damage, and their numbers are subse-quently refined over time to more closely resemble those in controls (Girod et al. 1991; Weisleder and Rubel 1993; Janas et al. 1995; Dye et al. 1999; Kevetter et al. 2000). Hair cell overproduction after damage has been documented in the avian basilar papilla, utricle, and crista, suggesting it may be a general phenomenon in these tissues during regeneration. For example, in the chicken basilar papilla after multiple gentamicin injections, examination of the epithelial surface with SEM revealed that an extra 12–16% of HCs (relative to controls) are detected in the distal third of the epithelium at 20 weeks after drug treatment (Girod et al. 1991). A similar overproduction of HCs was observed by Janas et al. (1995) in the basilar papilla 2 weeks after a single gentamicin injection, although HC numbers returned to control levels by 5 weeks post-gentamicin. Analysis of the pigeon utricle after a single streptomycin injection demonstrated a small, but significant, overproduction of HCs in the striolar region at 180 days following drug treatment as compared to age-matched controls (Dye et al. 1999). Further, a study of the pigeon crista after drug damage demonstrated that both SCs and HCs are present in higher numbers at early phases of regeneration than at later phases (Kevetter et al. 2000). Between 70 days and 5 months, the number of type I and type II HCs increased above control levels by 150%, and SC numbers were increased by120%. Numbers of both HCs and SCs returned to control values after 1 year.

The mechanisms that ultimately lead to regeneration of the correct numbers of HCs and SCs during regeneration have not been characterized. Overproduction of a particular cell type (e.g., HCs) could be achieved via (1) higher rates of cell fate determination of one type over another or (2) selective cell culling. Regulation of cell fate decision-making is discussed in the next section. To date, there have been few detailed analyses of cell death among regenerated cells in the inner ear SE. Wilkins et al. (1999) showed that as many as 50% of mitotically formed cells (type unknown) appear to be extruded from the chicken utricle as early as 7 days post-S phase. Time-lapse studies after laser ablation in lateral line neuromasts of axolotls demonstrate that some newly regenerated HCs and SCs are phagocytosed by macrophages (Jones and Corwin 1993, 1996). However, it is not clear in these studies if the function of apoptosis is to regulate cell numbers and patterning, to remove poorly differentiated cells, or to remove cells that are damaged by lingering effects of the damaging agent. During embryogenesis, developmental cell death occurs in the primordial vestibular SE in frogs (*Xenopus laevis*), zebrafish (Bever and Fekete 1999), and rats (Zheng and Gao 1997). While cell death in the otocyst has been documented in chickens, there are conflicting reports on whether it is instrumental in SE patterning (e.g., Lang et al. 2000; Avallone et al. 2002).

2.1.2.3 Molecular Regulation

Cells in complex organisms constantly receive multiple signals from their environment. Many such signals are derived from diffusible extracellular factors, such as growth factors and hormones, which can act locally or distally. Additional signals are provided by direct interactions with other cells or the extracellular matrix. In both cases, signaling molecules bind to specific receptors and activate intracellular signaling pathways leading to changes in cell behavior. With respect to postembyronic HC growth in the inner ear, perhaps the most critical signals are those controlling whether progenitor cells remain in a quiescent state or are triggered to divide. As discussed in the preceding text, numerous observations provide clues regarding the origin and timing of positive and negative regulators of progenitor cell division in hair-cell epithelia. However, relatively little is known about the specific molecules that confer this regulation. Progress in identifying signaling pathways associated with progenitor cell division during

HC regeneration has relied heavily on the development of tissue culture conditions that retain the SE's ability to undergo HC regeneration while permitting easy manipulation of cell signaling (Oesterle 1993; Warchol et al. 1993; Stone et al. 1996; Warchol 1999).

2.1.2.3.1 Growth Factor Regulation of Proliferation. Early studies in cultured inner ear tissues suggest diffusible molecules play a critical role in regulating progenitor cell proliferation in mature inner ear SE (Tsue et al. 1994b; Warchol and Corwin 1996). Some important signaling molecules appear to be present in the SE itself because isolated vestibular SCs from mature chickens divide and give rise to new HCs when cultured without serum or exogenous growth signals (Oesterle 1993; Warchol et al. 1993; Warchol 1995). At the present time, the best-studied factors regulating stem/progenitor cell behavior across tissue types are growth factors. For example, epidermal growth factor (EGF), transforming growth factor-alpha (TGF-α), insulin-like growth factor-1 (IGF-1), and fibroblast growth factor-2 (FGF-2) are mitogenic for neural stem cells in vivo and in vitro (Richards et al. 1992; Kilpatrick and Bartlett 1993; Gritti et al. 1996, 1999; Aberg et al. 2003). Additional growth factors act to keep the mitotic activity of stem cells in check and/or regulate stem cell differentiation. Dying HCs, VIIIth nerve neurons, roving leukocytes that are attracted to sites of damage, and SCs themselves are all potential sources of growth factors that could regulate inner ear stem/progenitor cell proliferation.

To investigate the specific factors that promote regeneration, several researchers have tested the effects of exogenous growth factors in chicken auditory and vestibular SE in vitro. Two members of the insulin-like growth factor superfamily of growth factors, IGF-1 and insulin, stimulate proliferation in mature avian vestibular, but not auditory, SE in a dose-dependent manner (Oesterle et al. 1997) (Fig. 5.4; Table 5.1). IGF-1 is also mitogenic for developing chicken otic epithelium (León et al. 1995, 1998, 1999) and neonatal mammalian vestibular SE (Zheng et al. 1997), but does not have mitogenic effects by itself in mature mammalian vestibular SE (Yamashita and Oesterle 1995). IGF-1 and insulin (at high levels) both bind the IGF-1 receptor, IGF-1R (Rosenfeld and Roberts 1999), which has been shown to be present in mature chicken and mammalian inner ear SE by polymerase chain reaction (PCR) analysis and immunocytochemical labeling (Lee and Cotanche 1996; Saffer et al. 1996; Pickles and van Heumen 1997; Zheng et al. 1997). Sources of IGF-1 in the ear remain to be delineated. IGF-1 is a secreted hormone that is able to reach the inner ear via the general circulation. IGF-1 and insulin are also expressed locally in the cochlear and vestibular ganglia of mice (Camarero et al. 2001; Varela-Nieto et al. 2004).

The biological activity of IGF-1 is substantially modified by IGF-1 binding proteins (IGFBPs) that typically attenuate IGF-1 action by binding IGF-1 and preventing it from activating IGF-1R. Less often, IGFBPs can enhance the effects of IGF-1 by providing a local source of bioavailable IGF-1 or by preventing autocrine downregulation of IGF-1R. IGFBP expression patterns and involvement in avian and other nonmammalian inner ear SE functioning remain

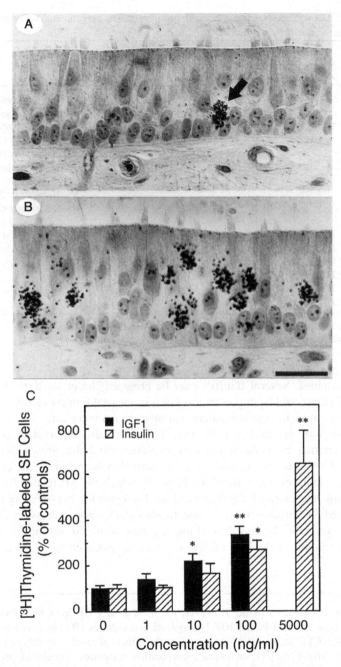

FIGURE 5.4. IGF-1 stimulates proliferation in cultured utricular SE taken from posthatch chickens. (**A, B**) Normal chicken utricles grown in culture for 2 days in control media (**A**) and in media supplemented with IGF-1 at 100 ng/ml (**B**). The cell proliferation marker, [^{3}H] thymidine, was present in media for the entire culture period. Labeled cells are present in control media (**A**, arrow) illustrating the ongoing proliferation seen in this

TABLE 5.1. Growth factor effects on cell proliferation in early neonatal/posthatch and mature inner ear sensory epithelia.

Species	Vestibular SE		Auditory SE		References
	Adult	Neonatal	Adult	Neonatal	
Mammal	EGF and insulin ⇑ TGF-α ⇑	*EGF* ⇑ *FGF-2, 4, 6, 7* ⇑ *GGF-2* ⇑ *HRG* ⇑ *IGF-1, II* ⇑ *NDF-α, β* ⇑ *TGF-α* ⇑ *TGF-β1, 2, 3, 5* ⇓		*EGF* ⇑	Lambert, 1994; Yamashita and Oesterle 1995; Zheng et al. 1997, 1999; Gu et al. 1998, 1999; Montcouquiol and Corwin 2001a,b; Hume et al. 2003; Doetzlhofer et al. 2004;
Bird	FGF-2 ⇓ IGF-1 ⇑ Insulin ⇑ *TGF-α* ⇑ *TGF-β1* ⇓ *TNFα* ⇑		FGF-2 ⇓		Oesterle et al. 1997, 2000; Warchol 1999

Italics indicate isolated sheet or dissociated cell cultures; regular text indicates organotypic cultures. ⇑ = mitogenic effect; ⇓ = inhibitory effect.

to be determined. Several IGFBPs may be present (Cho et al. 2001; Klockars et al. 2002) that could be important modifiers of the proliferative effects of IGF-1 on SE progenitors in nonmammalian and mammalian inner ear.

A member of the EGF superfamily, TGF-α, is also reported to stimulate cell proliferation in isolated chicken vestibular epithelia grown in culture (Warchol 1999), but it does not appear to have significant mitogenic effects in organotypic cultures (Oesterle et al. 1997), in which the SE retains its normal contact with the extracellular matrix. The discrepancies between organotypic and isolated SE cultures may be due to altered receptor expression resulting from the enzymatic and mechanical manipulation of the SE during its isolation for tissue culture, or the loss of contact with the extracellular matrix. TGF-α

◄ ──

FIGURE 5.4. (Continued) tissue. However, the number of [^{3}H] thymidine-labeled cells is significantly elevated in the IGF-1–supplemented cultures (**B**) compared to control cultures (**A**). (**C**) Effects of IGF-1, along with the effects of another member of the IGF superfamily, insulin, on [^{3}H] thymidine incorporation in cultured utricular SE are shown. Each datum value represents the mean value ± SEM. Mean values are expressed as a percentage of concordant control cultures, with 100% representing the control mean. Single asterisks indicate a significant difference at the 0.05 confidence level; double asterisks indicate a difference at the 0.01 confidence level. Scale bars are identical for **A** and **B**; scale bar = 20 μm. (Modified from Figs. 1 and 2 in Oesterle et al. 1997.)

binds the EGF receptor (EGFR, also known as erbB1), and erbB1 is reported to be expressed in chick inner ear SE via PCR analysis (Lee and Cotanche 1996; Pickles and van Heumen 1997). However, it has been localized to the HCs by immunocytochemical techniques (Lee and Cotanche 1996). As discussed later, EGF family members are involved in regulating proliferation in mammalian inner ear tissues.

In addition to triggering stem/progenitor cell proliferation in other tissues, growth factors also serve to keep the mitotic activity of progenitor cells and stem cells in check. Efforts are currently underway to delineate the growth factors that negatively regulate adult stem cell proliferation in mature tissues and maintain these cells in a quiescent state. Negative regulators (e.g., fibroblast growth factor 18 and members of the transforming growth factor-beta (TGF-β) super-family) are beginning to be identified in other regenerating tissues (e.g., skin, muscle, olfactory epithelium: McCroskery et al. 2003; Wu et al. 2003; Blanpain et al. 2004). It is likely that these growth factors and others also limit division of stem/progenitor cells in adult inner ear SE. The observation that proliferation is increased in cultures of isolated avian inner ear SE relative to that seen in organotypic explants (Stone et al. 1996) supports the hypothesis that stromal cells may be a source of inhibitory factors regulating inner ear stem/progenitor cell proliferation. One negative growth regulator for nonmammalian inner ear SE has been identified; basic FGF-2 inhibits proliferation of cultured auditory and vestibular SE taken from posthatch chickens (Oesterle et al. 2000). This factor is expressed in developing and posthatch chicken inner ear SE and is highly expressed in SCs (Umemoto et al. 1995; Lee and Cotanche 1996; Carnicero et al. 2001). In bullfrog and mammalian vestibular end organs, FGF-2 is reported to be present in HCs (Zheng et al. 1997; Cristobal et al. 2002). Other sources of inhibitory factors remain to be identified.

It is unclear whether levels of FGF-2 change in response to HC damage. One study reported induction of FGF-2 protein expression in the SC layer after noise-induced damage in chicken auditory SE (Umemoto et al. 1995). In contrast, another study noted no differences in FGF-2 expression patterns after noise insult (Lee and Cotanche 1996), and a related study utilizing semiquantitative reverse transcriptase (RT)-PCR techniques demonstrated no quantitative changes of *FGF-2* mRNA expression in the chicken auditory SE after ototoxic insult (Pickles and van Heumen 1997). The latter two studies also noted a redistribution of fibroblast growth factor receptor 1 (FGFR1), a high-affinity receptor for FGF-2, from HCs to SCs after damage. Hence, the binding of FGF-2 to FGFR1 was postulated to lead to SC proliferation in regenerating avian auditory epithelium (Lee and Cotanche 1996). Given the inhibitory effects of FGF-2 on SC division, this growth factor may also be important in promoting differentiation of HC and/or SC precursors (Oesterle et al. 2000). The latter idea is supported by the finding that FGF-2 leads to proliferation-dependent increases in the numbers of cells expressing HC markers in dissociated auditory epithelia taken from developing avian inner ear (Carnicero et al. 2004).

Fibroblast growth factor receptor 3 (FGFR3) also appears to play a role in regulating SC proliferation in nonmammals. FGFR3 is highly expressed in SCs of the mature avian auditory SE. It is rapidly downregulated after damage, and it begins to be expressed again after progenitor cells exit the cell cycle (Bermingham-McDonogh et al. 2001). Thus, a downregulation of signaling via FGFR3 appears to allow a proliferative response to occur that is otherwise suppressed in SCs under normal conditions. These results suggest FGFR3 signaling plays a role in maintaining avian auditory SCs in a state of growth arrest. Interestingly, this same gene was found to be upregulated in mammalian SCs after noise damage to the adult rat organ of Corti (Pirvola et al. 1995), a tissue that has no capacity or limited capacity for regeneration. Thus, a system that does not regenerate regulates this gene in the opposite direction than a system capable of regeneration.

In summary, members of the IGF and FGF superfamilies of growth factors appear to be important regulators of progenitor cell proliferation in nonmammalian inner ear SE. IGFs and FGFs are also important regulators of stem/progenitor cell proliferation in other tissues, including during inner ear development (Richards et al. 1992; Kilpatrick and Bartlett 1993; Gritti et al. 1996, 1999; Aberg et al. 2003, and reviewed in Varela-Nieto et al. 2004 and Barald and Kelley 2004). It should be emphasized that there is no direct evidence (as yet) that IGFs and FGFs are involved in HC regeneration in nonmammalian vertebrates in vivo. Nevertheless, that they appear to be capable of altering SC proliferation in the mature epithelia is harmonious with the idea that regeneration is likely to be regulated by the same factors normally involved in embryonic development. While investigators have begun to identify growth factors that are mitogenic for progenitor cells in nonmammalian vestibular SE, growth factors that have mitogenic effects on the auditory SE remain to be identified.

2.1.2.3.2 Regulation of Proliferation by Direct Cell–Cell Contact. Progression through the cell cycle is regulated by changes in cell–cell contact mediated through cell adhesion molecules (e.g., Levenberg et al. 1999). As discussed in Section 2.1.2.1, the rate of progenitor cell division correlates positively with the degree of HC death. Therefore, it is likely that direct signaling between HCs and SCs is important in maintaining SCs in a mitotically quiescent state (Cotanche 1987a; Corwin et al. 1991; Lewis 1991). Direct SC–HC interactions become altered and then lost as HCs are extruded (e.g., Hirose et al. 2004). To date, few potential candidate molecules for regulating SC division through direct cell–cell contact have been identified. One study by Warchol (2002) showed that disruption of the cell–cell adhesion molecule, N-cadherin, leads to a significant decrease in SC division in cultured chicken utricular epithelia. This finding suggests SC division is regulated by cell–cell contact via N-cadherin, although it is not clear whether this signaling is between SCs or between HCs and SCs.

2.1.2.3.3 Leukocytes as Sources for Signaling. Leukocytes are recruited to sites of tissue injury and play a role in tissue repair. They secrete cytokines and growth factors that stimulate the proliferation of numerous cell types, including some CNS glia (reviewed in Hamilton et al. 1993; Turpin and Lopez-Berestein 1993). It has been hypothesized that leukocytes play a role in triggering HC regeneration, possibly by releasing mitogenic growth factors that initiate progenitor cell proliferation (Corwin et al. 1991; Jones and Corwin 1993, 1996; Warchol 1997). Leukocytes are present in undamaged avian inner ear SE (Warchol 1997; Bhave et al. 1998; Oesterle et al. 2003). Their numbers increase at sites of HC loss in the avian inner ear SE (Warchol 1997; Bhave et al. 1998) and in amphibian lateral-line neuromasts (Jones and Corwin 1993). Leukocyte numbers increase significantly at HC lesion sites *prior* to regenerative proliferation in some nonmammalian SE (Jones and Corwin 1996; Warchol 1997; Bhave et al. 1998). However, the recent report of an absence of leukocytes in lesioned areas of the avian auditory SE suggests they many not be necessary for the initiation of SC division (O'Halloran and Oesterle 2004). Two observations support the idea that leukocytes release factors that alter progenitor cell proliferation in inner ear SE: (1) tumor necrosis factor-alpha (TNF-α), a macrophage secretory product, enhances proliferation of cultured vestibular SCs (Warchol 1999); and (2) dexamethasone, an anti-inflammatory drug that inhibits cytokine production by macrophages, reduces SC proliferation in cultured utricular macula after HC injury (Warchol 1999).

Several sorts of leukocytes have been reported in inner ear SE of birds (Warchol 1997; Bhave et al. 1998). The most common leukocytes are ramified cells of the myeloid lineage. Many of these are major histocompatibility complex (MHC) class II positive, and a small percentage are mature tissue macrophages (O'Halloran and Oesterle 2004). Macrophages have also been reported in lateral line organs of axolotls (Jones and Corwin 1993, 1996), damaged mammalian organ of Corti (Bohne 1971; Fredelius and Rask-Andersen 1990; Roberson and Rubel 1994; Vago et al. 1998), and normal and damaged mammalian vestibular SE (Oesterle et al. 2003). Monocytes have been observed in the damaged organ of Corti (Fredelius and Rask-Andersen 1990), and neutrophils and lymphocytes have been detected in mammalian vestibular SE (Oesterle et al. 2003). Lymphocytes (B and T cells) were not identified in the normal avian SE or in the SE at 1–3 days post aminoglycoside-insult; however, they do appear to reside in blood vessels in the loose connective tissue underneath the SE and in nonsensory tissues adjacent to the SE (O'Halloran and Oesterle 2004). In total, these findings suggest that studies examining functional roles of leukocytes within the SE need to take into account that there may be several leukocyte subtypes present, and the different subtypes may release distinct mitogenic substances with potentially variable effects.

Proliferating leukocytes have been detected in avian and mammalian inner ear SE (Roberson and Rubel 1994; Vago et al. 1998; Warchol 1997; O'Halloran and Oesterle 2004). Hence, studies of mitotic tracer incorporation in the SE following experimental treatment need to take into account that the possibility some mitotically active cells in the SE may be immune cells or other migratory cells derived from outside the SE. It is not known whether leukocytes are capable of dividing within the SE or if they divide outside of the SE and then get recruited to the SE after HC damage.

2.1.2.3.4 Intracellular Signaling Pathways. There is considerable interest in determining the signal transduction pathways that underlie regenerative proliferation. Experimental treatments that raise intracellular cyclic adenosine monophosphate (cAMP) levels lead to increased SC proliferation in the inner ear SE. In explanted auditory end organs, in vitro application of high levels of forskolin, an activator of adenyl cyclase that catalyzes the production of cAMP, results in increased SC proliferation (Navaratnam et al. 1996). Incubations with inhibitors of the cAMP-dependent protein kinase, protein kinase A (PKA), block this effect, suggesting that activation of PKA is required for the forskolin-induced increase in SC division. Forskolin also stimulates proliferation in cultured pieces of vestibular epithelium taken from neonatal mice, but under different conditions than those for the chicken basilar papilla (Montcouquoil and Corwin 2001a). The authors of the study in mice suggest that elevated cAMP induces S-phase entry by increasing the number of growth factor receptors available at the plasma membrane.

Additional signaling cascades are reported to underlie SC proliferation in cultured inner ear SE of mature chickens (Witte et al. 2001) and neonatal mice (Montcouquoil and Corwin 2001b). Cell proliferation is reduced by inhibitors of several important signaling intermediates, including mitogen-activated protein kinase (MAPK), protein kinase C, phosphatidylinositol 3-kinase (PI-3K), and target of rapamycin (TOR). Comparisons between mammal and chick indicate that activated PI-3K and TOR are required for S-phase entry in both avian and mammalian vestibular epithelia, but activation of the MAPK pathway appears to have a more significant role in avians than in mammals. Substantial work remains to be done to characterize the extracellular molecules that directly activate these intracellular pathways and to identify additional signaling pathways of importance.

2.2 Cellular Differentiation

Viewed simplistically, cells derived from progenitor cell division in inner ear SE and lateral line neuromasts acquire two fates: HC or SC. However, as discussed in the preceding text, it is not clear whether distinct subtypes of SCs are present in nonmammalian HC epithelia. Therefore, it is possible that dividing progenitor cells have the capacity to give rise to HCs, terminally mitotic SCs, and SCs with proliferative potential (stem/progenitor cells). Before establishing their mature roles in the SE, postmitotic cells (1) become specified as one cell type or another; (2) undergo molecular and morphologic differentiation, acquiring specific features (structure and function) of a given cell type; and (3) develop mechanisms to maintain their specific phenotype. These processes occur over a prolonged period. For example, in the avian vestibular epithelium, complete differentiation of type I HCs, including elaboration of the calyceal nerve ending surrounding the cell soma, takes as long as 60 days (Weisleder and Rubel 1995). In the next section of the chapter, an overview is provided of what is known about morphological, molecular, and functional changes that occur within regenerated HCs and SCs as they become specified and acquire mature properties.

2.2.1 General Features of Cellular Differentiation During Regeneration

2.2.1.1 Supporting Cells

Mature nonmammalian SCs are distinct from HCs with respect to morphology and function. For example, SCs contact both the lumenal surface of the epithelium and the basal lamina, and they provide structural support to the epithelium via a unique microtubular array (Tanaka and Smith 1978; Ginzberg and Gilula 1979). Unlike HCs, SCs in nonmammals generate the extracellular membranes that sit atop and below the SE. For example, in the basilar papilla, there is steady-state production of tectorial membrane in maturity (Goodyear et al. 1996), and the tectorial membrane is partially regenerated after marked noise-induced HC loss (Cotanche 1987b; Epstein and Cotanche 1995). Unlike HCs, SCs possess connexons, the building blocks of gap junctions, which interconnect SCs and allow rapid flow of ions, such as K^+ and Ca^{2+}. Transport of ions within SCs is thought to be important for cellular signaling and for preventing intoxication of the HCs by K^+ ion buildup in the narrow extracellular spaces surrounding the HCs.

SCs have been shown to express some selective markers that relate to their functional specializations. For example, SCs in the otolithic organs make the extracellular matrix molecule, Gm-2 (Goodyear et al. 1995), and SCs in the basilar papilla make alpha and beta tectorins, components of the tectorial membrane (Goodyear et al. 1996). Connexins and cytokeratins are abundant in SCs, but they are not in HCs (Nickel et al. 2006; Stone et al. 1996). As referred to earlier, very little evidence for specialization among nonmammalian SCs has been collected to date. For instance, it is unclear if subsets of SCs are specialized to continually form the tectorial membrane while others are poised to divide and form new HCs, if necessary. All tolled, there is more evidence in the literature pointing to functional and molecular homogeneity among SCs in nonmammalian inner ears than to specialization.

While there have been analyses of mature and developing SCs (e.g., Tanaka and Smith 1978; Ginzberg and Gilula 1979; Goodyear et al. 1995), studies of regenerated SCs are very limited. This is due in part to the fact that molecular markers distinguishing mature SCs from HCs have only recently begun to emerge, and investigators have primarily been interested in characterizing newly formed HCs. Further, because only a fraction of SCs in a given field are regenerated (Roberson et al. 1996), it is difficult to identify regenerated SCs without double-labeling them with a cell-selective marker and a nucleotide analog such as BrdU.

2.2.1.2 Hair Cells

After HC loss, new HCs are regenerated by progenitor cell division at the lumenal surface of the SE (Raphael 1992; Baird et al. 1993; Stone and Cotanche 1994; Tsue et al. 1994a; Weisleder et al. 1995). Studies in chicken show that newly formed HCs appear round and contact the lumenal surface (Duckert and Rubel 1990; Stone and Rubel 2000a). This contact appears to be maintained over the course of HC differentiation, despite dramatic changes in the cell's shape and relationship to the basal lamina. Studies combining the use of HC markers with nucleotide birth-dating have shown that, by 48 hours after cell division, HC precursors have elaborated a cytoplasmic process that extends away from the lumen and makes contact with the basal lamina (Stone and Rubel 2000a; see Figs. 5.2A, B). Regenerated HCs in adult bullfrogs (*Rana catesbeiana*) show a similar intermediate phenotype (Steyger et al. 1997). This shape is also seen in avian HCs as they develop (Whitehead and Morest 1985), and it is also reminiscent of mature SCs. A few days later, during differentiation in the chicken, the HC's adlumenal process is retracted, and regenerated HCs acquire their mature columnar or flask-like shape and luminal position in the SE.

In the chicken basilar papilla, the largest number of new cells is created on day 2 or day 3 after noise exposure or drug exposure, respectively (Stone and Cotanche 1994; Bhave et al. 1995, 1998; Stone et al. 1999). By 5–7 days after the onset of the damaging stimulus, mitotic activity among progenitor cells has diminished significantly, and by 11 days, dividing cells are rare. By the time this decline in cell division occurs, numerous newly formed HCs and SCs have begun to differentiate in the lesioned area (Cotanche 1987a; Cotanche et al. 1991; Marsh et al. 1990; Lippe et al. 1991; Janas et al. 1995; Stone and Rubel 2000a). Some molecular markers have been used to label postmitotic HC precursors at very early stages of differentiation. These include antibodies against class III ß-tubulin, which label the cytoplasm of HC precursors 3 days after the onset of the damaging stimulus, as early as 14 hours after progenitor cells undergo S phase in the drug-damaged basilar papilla (Stone and Rubel 2000a) and approximately 2 days after S phase in the undamaged utricular macula (Matsui et al. 2000). In addition, antibodies to the homeodomain transcription factor, *Prox1*, label the nuclei of HC precursors around 3 hours after class III ß-tubulin is first detected (Stone et al. 2004). Primitive stereocilia are first detectable on new HCs between 4 and 6 days after the onset of damage (Cotanche 1987a; Stone and Rubel 2000a). They resemble the short, immature stereocilia seen during development of the SE (Cotanche and Sulik 1984). Several molecular markers of stereocilia have been identified, and their expression during HC maturation has been characterized, including the mushroom toxin, phalloidin, and a phosphatase called HC antigen (Bartolami et al. 1991; Goodyear et al. 2003). Other HC markers include unconventional myosins 1c, 6, and 7a (Hasson et al. 1997);

the actin-binding protein fimbrin (Lee and Cotanche 1996); and the calcium sensor, calmodulin (Stone and Rubel 2000a). The physiological maturation of HCs, including the establishment of neural contacts, is discussed in detail in Saunders and Salvi, Chapter 3).

2.2.2 Nonmitotic Hair Cell Regeneration: Transdifferentiation

Studies in a range of nonmammalian animal classes suggest that, after HC damage, some SCs in auditory and vestibular SE convert directly into HCs without an intervening cell division (Fig. 5.2B). This mechanism of forming new HCs, also referred to as direct transdifferentiation, is particularly compelling when one considers that, in most mammalian HC epithelia, progenitor cell division is not robustly increased after damage (Warchol et al. 1993; Roberson and Rubel 1994; Sobkowicz et al. 1997), and the few mitotic events that do occur rarely appear to lead to the production of new HCs (Rubel et al. 1995; Oesterle et al. 2003).

Early evidence for transdifferentiation was purely morphological. In the bullfrog saccule after gentamicin treatment in vivo, Baird et al. (1993) noted that SC nuclei appear to move from the SC layer into the HC layer without a replacement of nuclei in the SC layer, and the number of cells with stereocilia increased concurrently. In the bullfrog saccule (Baird et al. 1996; Steyger et al. 1997) and the chicken basilar papilla (Adler and Raphael 1996), investigators documented the emergence of cells with the morphological appearance of SCs (long somata, direct contact with the basal lamina, and no stereociliary bundles) and also with both molecular and morphological features of HCs (e.g., immunoreactivity for HC-selective calcium binding proteins, electron-dense cytoplasm). These cells were interpreted to be in the process of directly converting from a SC into a HC.

Nucleotide labeling studies in chickens provide additional support for nonmitotic HC regeneration. Roberson et al. (1996, 2004) used an osmotic pump to continually infuse either [³H]thymidine or BrdU into the perilymph of mature chickens, thereby labeling all cells in the SE undergoing DNA synthesis. Chickens were then administered gentamicin to trigger complete HC loss in the proximal portion of the basilar papilla. Via the osmotic pump, nucleotides were available for nearly 2 weeks as HC regeneration proceeded. Roberson et al. (1996) found that 86% of SCs and 50% of regenerated HCs in the damaged region failed to display nuclear nucleotide labeling. These data suggest that only a small percentage of SCs in the damaged basilar papilla divided after complete HC loss and that half of all regenerated HCs were formed by direct SC-to-HC conversion.

Further evidence for direct transdifferentiation is provided by experiments that blocked SC division after HC loss. In cultured bullfrog saccules, HC regeneration is not significantly affected when SC mitoses are inhibited after gentamicin treatment using the DNA polymerase inhibitor aphidicolin (Baird et al. 1996). Numbers of stereociliary bundles (HCs) did not vary significantly between control cultures, cultures in which SC division proceeded as normal, and cultures with highly curtailed cell division (i.e., aphidicolin treatment). These findings suggest that nonmitotic processes lead to the restoration of numerous HCs. A similar uncoupling of HC regeneration and SC proliferation was documented in the chicken basilar papilla after acoustic overstimulation (Adler and Raphael 1996) and in the newt (*Notophthalmus viridescens*) saccule after gentamicin damage in vitro (Taylor and Forge 2005).

While the time course of mitotic activity has been carefully studied in several tissues, data are only beginning to emerge with respect to when and where transdifferentiation occurs during the HC regeneration period. Conversion of SCs to HCs appears to be an early phase of HC regeneration. For example, Roberson et al. (2004) showed, using continual BrdU infusion into the perilymph of live chickens, that nonmitotically generated HCs appear 24–48 hours before mitotically formed HCs. A recent study of the chicken basilar papilla suggests that direct transdifferentiation is initiated in SCs before overt signs of HC damage and is more likely to occur in the abneural half of the SE than in the neural half (Cafaro et al. 2007).

The conversion of one differentiated cell type into another without an intervening mitosis is a highly unusual process for sensorigenesis, and several questions remain regarding this process in inner ear SE. The identity of the cell type that undergoes transdifferentiation is not known. Because few markers for SCs exist in each animal class, it is difficult to investigate whether a particular subpopulation of SCs is specialized for transdifferentiation. The possibility also exists that HC precursors (cells that are undifferentiated but committed to the HC fate) exist in HC epithelia and are triggered to complete their differentiation once nearby mature HCs are lost (Morest and Cotanche 2004). In the undamaged basilar papilla, evidence for undifferentiated HC precursors has not been presented. However, immature HCs are abundant in nonmammalian vestibular epithelia (damaged and undamaged), because HCs are generated on an ongoing basis. For this reason, studies of cellular transdifferentiation in vestibular epithelia must be interpreted carefully. Future studies should strive to identify molecular signals that regulate whether SCs undergo transdifferentiation or mitotic activity. In addition, studies should be aimed at determining whether mechanisms involved in HC specification, differentiation, and maintenance are similar during nonmitotic and mitotic HC regeneration. It has been postulated by Roberson et al. (2004) that, because immature HC precursors in each case demonstrate an intermediate morphology (elongated cell shape, contact with the basal lamina, expression of HC markers), signals regulating differentiation are likely to be the same regardless of whether the HC is derived by SC mitosis or transdifferentiation.

2.2.3 Molecular Regulation of Differentiation

For an undifferentiated cell to become integrated into a tissue and to adopt a specific function, it must become committed to a particular cell fate and then acquire the specific features of that cell type. These processes are regulated by the collective events that take place in the cell's environment (cell-extrinsic) and within the cell itself (cell-intrinsic). Much of what is known about molecular regulation of HC and SC specification, differentiation, and maintenance has been derived from developmental studies in mammals and nonmammals (reviewed in Lewis 1991, 1996; Cantos et al. 2000; Raz and Kelley 1997; Bryant et al. 2002; Fekete and Wu 2002; Barald and Kelley 2004). Because similar cell types are generated in the SE during embryogenesis and regeneration, it is hypothesized that similar molecules are enlisted to play specific roles during both events. In some cases, analysis during regeneration of the expression of genes known to be important for HC development supports this hypothesis (e.g., *Delta1*, *Notch1*; Stone and Rubel 1999; Eddison et al. 2000; Daudet and Lewis 2005). However, there are also examples of genes in which expression patterns are highly divergent during embryogenesis and regeneration (e.g., *BEN*: Goodyear et al. 2001; *Prox1*: Stone et al. 2003, 2004). The challenge for investigators is to piece together genetic cascades that are necessary and/or sufficient for HC/SC production at different ages. The next sections discuss how some molecules function during sensorigenesis, and where possible, address evidence that supports or refutes their role in HC or SC differentiation in mature epithelia, during nonmammalian HC regeneration. (The function of these molecules in mammalian regeneration is discussed in Section 3.2.)

2.2.3.1 Cell-Extrinsic Signals

Signaling through the extracellular receptor, Notch, is without doubt the best studied extrinsic regulatory pathway in inner ear SE development (reviewed in Eddison et al. 2000; Kelley 2003). Signaling through Notch plays critical roles in several tissues in the body, including tissues that are formed throughout life such as the intestine and the hematopoietic system (Yang et al. 2001; Radtke et al. 2004). In the developing nervous system, selection of competent progenitor cells, execution of lineage decisions, and maintenance of stem cell populations depend upon Notch activity (Hartenstein and Posakony 1990; reviewed in Lewis 1996). Notch signaling is critical for tissue development and homeostatis in a wide range of species, including fruit flies (*Drosophila melanogaster*), zebrafish, chickens, mice, and humans. An excellent and well characterized example of Notch's role in regulating cellular patterning is illustrated by the developing fruit fly compound eye (Nagaraj and Banerjee 2004).

The Notch receptor is activated by ligands (Delta or Serrate/Jagged) that are anchored in the membrane of adjacent cells (Fehon et al. 1990; Artavanis-Tsakonas and Simpson 1991), and thus it mediates direct cell–cell interactions (Fig. 5.5). When Notch is bound by ligand, a portion of the receptor (the Notch intracellular domain) enters the nucleus and activates CSL transcription factors, which in turn promote transcription of genes encoding repressor transcription factors in the HES (Hairy and Enhancer of Split) and HRT (Hairy-related Transcription Factor) families (reviewed in Iso et al. 2003; Kageyama et al. 2005). HES inhibits transcription of proneural genes, which direct cells to acquire the neural phenotype by driving expression of genes that promote neural or sensory differentiation. Notch activation also inhibits transcription of

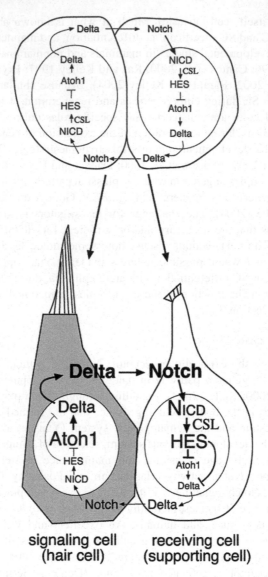

FIGURE 5.5. Regulation of cell fate specification and differentiation via the Notch signaling pathway. A model for the role of Notch signaling pathway in cellular specification within hair-cell epithelia is illustrated here. Two adjacent cells are depicted before specification (**top**) and during specification (**bottom**). The cells on top represent cells shortly after mitosis; they have similar expression levels/activity of Notch ligands (Delta is shown), *HES*, and *Atoh1*. Over time, levels of Notch ligand become altered in one cell such that it becomes the "signaling cell" and adjacent cells become "receiving cells." Activation of the Notch receptor in a receiving cell triggers translocation of the receptor's intracellular domain (NICD) from the cytoplasm to the nucleus. NICD (through CSL) causes HES to become a transcriptional repressor, and this leads to reduced transcription of *Atoh1* and *Delta*. The effect of decreased *Delta* transcription is that the receiving cell

Notch ligands (Skeath and Carroll 1992; Bertrand et al. 2002). Based on this lateral inhibition with feedback, adjacent cells with differing levels of proneural transcription factor and Notch ligand expression will acquire distinct fates (Hoppe and Greenspan 1986; Collier et al. 1996). Elevated levels of Notch ligands and proneural transcription factors in a given cell correlate with acquisition of the sensory/neural cell phenotype. Through Notch signaling, tissues become patterned with alternating or more complex cellular arrangements. Further, activation of Notch in fruit flies has been implicated in regulating mitotic activity (Go et al. 1998; Johnston and Edgar 1998) and in maintaining a proliferative pool of progenitor cells (Henrique et al. 1997). Notch may also directly promote differentiation of non-neural phenotypes (reviewed in Wang and Barres 2000).

Notch function is critical during embryogenesis for specification of sensory areas in the otic epithelium and for proper differentiation of cell types in the inner ear SE. Three ligands are known to be expressed in the developing otocyst (e.g., Lindsell et al. 1996; Adam et al. 1998; Lewis et al. 1998; Morrison et al. 1999; Shailam et al. 1999; Cole et al. 2000; Eddison et al. 2000), with differing distributions. Jagged1 (called Serrate1 in chickens) is the earliest expressed ligand, appearing in the sensory primordium prior to HC differentiation (Adam et al. 1998; Cole et al. 2000; Zine et al. 2000; Zine and de Ribaupierre 2002). Alteration or absence of *Jagged1* function in mice results in partial or total failure of SE formation (Kiernan et al. 2001; Tsai et al. 2001; Kiernan et al. 2006). Further, ectopic activation of Notch signaling in the early otic epithelium of chicken embryos leads to the formation of ectopic sensory patches (Daudet and Lewis 2005). These studies suggest that Notch activation by Jagged1 is required for specification of the early otic epithelium as prosensory in nature.

Later during development, signaling through Notch antagonizes specification and differentiation of cells within prosensory epithelial patches as HCs (Eddison et al. 2000; Kelley 2003; Daudet and Lewis 2005). Both *Jagged2* and *Delta*1/DLL1 are expressed in nascent HCs (Adam et al. 1998; Lanford et al. 1999; Shailam et al. 1999; Zine and de Ribaupierre 2002). Deletion of *Jagged2* leads to overproduction of HCs (Lanford et al. 1999; Kiernan et al. 2005; Brooker et al. 2006), as does inhibition of Notch1 function (Zine et al. 2000). Signaling through Delta ligands is required for lateral inhibition of HC specification in the inner ears of zebrafish and mice. Zebrafish with altered *DeltaA* function show a 5- to 6-fold overproduction of HCs and a deficiency of SCs (Riley et al. 1999). A similar but more dramatic finding is seen in the zebrafish mutant, *mind bomb* (*mib*) (Jiang et al. 1996). Hair cells are overproduced, apparently at the expense of SCs, in inner ear organs (Haddon et al. 1998) and lateral line neuromasts (Itoh and Chitnis 2001). *Mib* encodes an ubiquitin ligase that

◄——

FIGURE 5.5. cannot activate Notch in the sending cell, and as a result, the adjacent cells acquire a strong asymmetry with respect to their expression levels of *Delta* and *Atoh1*. As a result, the signaling cell acquires the hair cell fate, and the receiving cell becomes a supporting cell.

acts in the signaling cell to facilitate endocytosis of the Delta ligand, along with part of the Notch receptor from the nearby receiving cell (Itoh et al. 2003). This process assists in the propagation of the Notch signal in the receiving cell. Thus, precursor cells in *mib* mutants have reduced Notch activation, and as a result, more cells differentiate as HCs. In mice, *DLL1* is required to prevent precocious differentiation of HCs (Kiernan et al. 2005; Brooker et al. 2006), although studies in chickens do not support a critical role for *Delta1* in chicken HC development (Eddison et al. 2000).

These studies support a critical role for cell–cell signaling through the Notch receptor in cell fate acquisition during SE development in the inner ear. However, little is known about Notch function in postembryonic replacement of HCs in normal or injured states. In mature chickens, Notch and its ligands, *Delta1* and *Serrate1*, are expressed in a manner that supports their function in HC regeneration (Stone and Rubel 1999). In mitotically quiescent organs (basilar papillae), *Notch1* is expressed in all SCs and *Serrate1* is expressed in both HCs and SCs, but no *Delta1* expression is seen. In contrast, *Notch1, Delta1,* and *Serrate1* are highly expressed in normal vestibular epithelia, which undergo continual HC regeneration. When HC damage is induced in the basilar papilla, *Delta1* transcription is increased at low levels in dividing progenitors. After mitosis, *Delta1* mRNA is weakly and symmetrically expressed in sibling cells that ultimately acquire distinct fates (HC versus SC), but by 3 days after mitosis, it becomes highly expressed in HC precursors and downregulated in SC precursors. These findings show that *Delta1* expression distinguishes HC precursors from SC precursors shortly after the cells are born, and they suggest that, during regeneration, *Delta1* in nascent HCs laterally inhibits adjacent cells from differentiating as HCs. It is important to restate, however, that while *Delta1* is expressed in developing SE in chickens, experimental augmentation of *Delta1* activity is not sufficient to inhibit HC differentiation during this period (Eddison et al. 2000). A role for *Delta1* in postembryonic HC production remains to be demonstrated.

Additional extrinsic factors believed to regulate HC differentiation during development include steroid hormones (Raz and Kelley 1999), cadherins (Hackett et al. 2002), and Wnts (Dabdoub et al. 2003; Montcouquiol et al. 2003; Stevens et al. 2003). However, a function for these factors in regulating the differentiation of HCs born postembryonically has not been explored.

2.2.3.2 Cell-Intrinsic Signals

Cell fate is also controlled at the intracellular level, by nuclear and cytoplasmic proteins. Several transcription factors, which regulate gene expression within the nucleus, have been shown to be critical for SE differentiation and patterning during inner ear development (reviewed in Fekete and Wu 2002). Some of these factors (e.g., Atoh1, HES1, HES5 and Ngn1) are regulated by Notch signaling (Fritzsch et al. 2000; Hassan and Bellen 2000) and, like Notch, they were originally described in fruit flies. Atonal is a basic helix–loop–helix transcription (bHLH) factor that promotes neural specification and differentiation in fruit flies (Jarman et al. 1993; Ben-Arie et al. 1996). In vertebrates, *atonal*'s homolog, *Atoh1* (Ben-Arie et al. 2000), controls cellular specification in several tissues, including the nervous system (Ben-Arie et al. 1996) and the intestine (Yang et al. 2001). In the developing inner ear of mice and chicks, *Atoh1* is expressed exclusively in developing sensory patches (Bermingham et al. 1999; Shailam et al. 1999; Ben-Arie et al. 2000; Lanford et al. 2000; Chen et al. 2002; Stone et al. 2003). *Atoh1* expression is also seen in developing lateral line neuromasts of zebrafish (Itoh and Chitnis 2001). *Atoh1* expression first emerges in sensory progenitor cells, as in fruit flies, and later becomes elevated in postmitotic HC precursors (Chen et al. 2002; Woods et al. 2004; Matei et al. 2005). Mice with a targeted deletion of *Atoh1* (also called *Math1*) fail to form HCs in any inner ear SE (Bermingham et al. 1999). Thus, *Atoh1* function is necessary for HC specification and/or differentiation during development.

Studies in rodents suggest that this transcription factor triggers nonmitotic HC regeneration in the neonatal and adult rat utricle (Zheng et al. 2000; Shou et al. 2003) and in the adult guinea pig organ of Corti (Kawamoto et al. 2003; Izumikawa et al. 2005; discussed in detail in Section 3.2). A recent study in mature chickens supports a role for *Atoh1* in both mitotic and nonmitotic regeneration in the basilar papilla (Cafaro et al. 2007). While no Atoh1 expression is seen in the normal basilar papilla, it becomes upregulated after damage, initially in transdifferentiating SCs, then in dividing SCs, and later, at a very high level, in regenerated HCs. Remarkably, Atoh1 upregulation in transdifferentiating SCs occurs as early as 15 hours after ototoxin treatment in areas underlying HCs that do not yet show signs of damage but will eventually be lost. This observation demonstrates that SC conversion into HCs is triggered by changes associated with HC damage, not loss, and therefore that critical signals for HC regeneration are activated very shortly after damage is initiated.

In contrast to Atoh1, the bHLH transcription factor, HES1, appears in mice to inhibit cells from differentiating as HCs, pushing them toward a SC fate. *HES1* is a homolog of *Drosophila hairy/enhancer of split* (*h/espl*), which is activated by Notch signaling. In fruit flies, *h/espl* antagonizes neural specification and differentiation by inhibiting transcription of proneural bHLH genes such as *atonal* (Davis and Turner 2001). A similar function for HES family members has been demonstrated in vertebrates (reviewed in Kageyama et al. 2005), and studies in the inner ear suggest this family of transcription factors also inhibits HC specification and differentiation. In mice, *HES1* and *HES5* are strongly expressed in progenitors and then in SCs in developing inner ear SE (Shailam et al. 1999; Lanford et al. 2000; Zheng et al. 2000; Zine et al. 2001). In the mouse organ of Corti, loss of *HES1* leads to overproduction of inner HCs and loss of *HES5* leads to overprduction of outer HCs (Zheng et al. 2000; Zine et al. 2001). HCs are also formed in higher numbers than normal in the utricle and saccule. To date, there has been no investigation of the expression or the role of either HES-type transcription factors in HC regeneration.

Another transcription factor associated with inhibiting neurogenesis in vertebrates, repressor element-1 silencing transcription factor (REST) shows altered expression during HC regeneration. REST functions as a transcriptional silencer of neural genes in non-neuronal cells, and its activity must be downregulated in order for some neurons to develop (Schoenherr and Anderson 1995; Paquette et al. 2000). In the chicken basilar papilla, transcription of *REST* is increased in SCs after ototoxin-induced HC damage (Roberson et al. 2002), suggesting REST helps to establish the SC phenotype in regenerated cells. After ototoxin exposure, *REST* is also increased in native HCs located outside the HC lesion. This finding suggests that mildly damaged HCs may be triggered to regress to a "less developed" state.

There is emerging evidence that the homeodomain transcription factor, Prox1, may also be involved in HC development and regeneration. *Prox1* is a homolog of *Prospero* in fruit flies, which is required for normal development of the embryonic and adult nervous system (reviewed in Doe 1996; Jan and Jan 1998). Prospero regulates production of correct cell types and numbers in the nervous system (Doe et al. 1991; Vaessin et al. 1991) by (1) promoting progenitors to exit the cell cycle (Li and Vaessin 2000) and (2) driving differentiation of neural progenitor cells (Doe et al. 1991). It also regulates cell proliferation, differentiation, and apoptosis in several vertebrate tissues, including the optic lens (Wigle et al. 1999), the retina (Dyer et al. 2003), and the lymphatic vasculature (Wigle et al. 2002). In the developing inner ear of chickens (Stone et al. 2003), *Prox1* is expressed in early sensory primordia, and it is downregulated in HCs and SCs after they have differentiated. In maturity, *Prox1* expresssion is weakly expressed in a subpopulation of SCs in the quiescent basilar papilla. After ototoxin-induced HC damage, *Prox1* expression is upregulated in dividing SCs and in regenerated HCs but not SCs (Stone et al. 2004). Since transcription of *Prospero* in *Drosophila* is regulated by signaling through Notch (Reddy and Rodrigues 1999), it is intriguing to speculate that Notch may play a role in limiting *Prox1* transcription to specified HCs during regeneration. Functions for REST and Prox1 in HC development and regeneration remain to be demonstrated.

Numerous other transcription factors have been implicated in HC differentiation and maintenance during inner ear development (reviewed in Torres and Giraldez 1998; Fekete and Wu 2002; Barald and Kelley 2004). However, the function of most of these factors in HC regeneration has not been explored. The critical role of transcription factors and their regulators in directing HC fate specification and differentiation has garnered increased attention, ensuring a rapid increase in knowledge over the next 5 years. This progress will be hastened by the application of gene microarray technology to identify additional transcription factors that are upregulated or downregulated during HC regeneration (Hawkins et al. 2003).

3. Hair Cell Regeneration in Mammals

Since the discovery of HC regeneration in the warm-blooded avians, several investigations have explored the natural capacity for regeneration in the mammalian SE, as well as the ability to augment the regenerative process through experimental manipulation. The latter studies have been possible largely because of the following advances: (1) development of culture methods for mature mammalian vestibular organs (Warchol et al. 1993; Yamashita and

Oesterle 1995) and neonatal auditory end organs (Sobkowicz et al. 1975; Lefebvre et al. 1993), (2) isolation and maintenance of inner ear stem/progenitor cells from neonatal and mature mammalian SE (Malgrange et al. 2002; Li et al. 2003; Doetzlhofer et al. 2004; White et al. 2006), (3) development of methods to infuse the in vivo inner ear directly with test agents (e.g., Brown et al. 1993; Kuntz and Oesterle 1998a,b; Kopke et al. 2001), (4) gene-manipulation technologies using constitutive and conditional in vivo models, and (5) delivery of exogenous genes into cultured SE or into the SE of intact mature animals using viral and nonviral vectors (Zheng et al. 2000; Kawamoto et al. 2003; Shou et al. 2003; Izumikawa et al. 2005). While some techniques remain an obstacle for investigators in this area (e.g., cultures of the mature organ of Corti), these advances have provided the means to test roles directly for proposed regulators of cell proliferation and differentiation during HC regeneration, and as a result, significant knowledge about HC regeneration in mammals has been gleaned.

3.1 Cell Proliferation

New HCs are not formed normally in postembryonic mammalian organ of Corti, an organ that is mitotically quiescent (Ruben 1967; Roberson and Rubel 1994; Sobkowicz et al. 1997). The picture is less clear in the mammalian vestibular SE. Immature-appearing stereociliary bundles are seen in the normal undamaged vestibular SE of mature guinea pig and bats (Forge et al. 1993, 1998; Rubel et al. 1995; Lambert et al. 1997; Kirkegaard and Jørgensen 2000, 2001), suggesting ongoing regeneration of vestibular HCs may occur in adult mammals. However, the immature bundles may rather reflect the repair of native bundleless HCs (Zheng et al. 1999), fluctuations in bundle structure (Sobkowicz et al. 1996, 1997; Zheng et al. 1999; Baird et al. 2000; Gale et al. 2000), or the presence of an additional HC subtype. Spontaneous cell division is rare or absent in the mature mammalian vestibular SE. A small amount of cell proliferation was reported in the SE in organotypic cultures of adult guinea pig (Warchol et al. 1993) and mice (Lambert 1994) utricles, but virtually no cell division was seen in similar cultures in mice (Yamashita and Oesterle 1995). The limited proliferation reported by Warchol et al. (1993) and Lambert (1994) may be attributable to surgical damage in the SE at the time of explantation, or it may reflect mitotitc activity in leukocytes rather than in SCs (Warchol 1997; Bhave et al. 1998; Oesterle et al. 2003). In vivo studies in mature rodents support the idea that there is normally no ongoing proliferation in the adult vestibular SE and that no new HCs are regenerated on an ongoing basis via a mitotic pathway (Ruben 1967, 1969; Rubel et al. 1995; Li and Forge 1997; Kuntz and Oesterle 1998a).

Regarding the traumatized ear, there is evidence to suggest that, despite a limited proliferative response to HC damage in the mammalian vestibular SE, some HCs may nonetheless spontaneously be replaced after damage/loss. Immature stereociliary bundles were seen in the guinea pig vestibular SE after ototoxic damage (Forge et al. 1993), suggesting the de novo formation of HCs in adult mammals. Further, initial decreases in stereocilia and HC density in drug-damaged chinchilla and guinea pig vestibular organs were followed by significant recovery (Lopez et al. 1997; Forge et al. 1998). The mechanisms of this recovery remain to be determined. They could arise by direct transdifferentiation of SCs (see Section 2.2.2; Li and Forge 1997; Steyger et al. 1997; Forge et al. 1998), by HC dedifferentiation and recovery (Sobkowicz et al. 1996, 1997; Baird et al. 2000), or by regenerative proliferation by progenitor cells. Regarding the latter possibility, damaged mammalian vestibular SE show a small increase in SC division over control levels after HC damage is induced in vitro (Warchol et al. 1993; Lambert 1994; Zheng et al. 1997; Zheng and Gao 1997) and in vivo (Rubel et al. 1995; Li and Forge 1997; Kuntz and Oesterle 1998a; Ogata et al. 1999). More dramatic increases in SC division have been reported in chinchilla ampullary organs (Tanyeri et al. 1995; Lopez et al. 1997, 1998), but this finding remains to be replicated. It remains controversial whether SCs in the mature mammalian vestibular SE possess the ability to produce new HCs by renewed mitotic activity. The absence of labeled HCs in guinea pig or gerbil utricles continuously infused with a cell proliferation marker raises concerns whether any mitotically regenerated cells receive adequate signals to acquire the HC fate (Rubel et al. 1995; Ogata et al. 1999). However, Warchol et al. (1993) demonstrated the presence of postmitotic cells with HC features in cultured adult mammalian utricles. Further, in vivo data in drug-damaged rats, gerbils, and chinchilla showed that postmitotic cells reside in the lumenal compartment of the vestibular SE (Tanyeri et al. 1995; Kuntz and Oesterle 1998a; Ogata et al. 1999), a region typically occupied by HCs. The location of the nucleus, however, is not a reliable phenotypic indicator for cell type in damaged epithelia owing to the disorganization of the tissue, the migration of SC precursor cell nuclei to lumenal portions of the epithelium (Raphael 1992; Tsue et al. 1994a), and the potential presence of proliferating leukocytes in the SE (Warchol 1997; Bhave et al. 1998; Vago et al. 1998). Ultrastrucutral analysis of postmitotic cells in the utricles of adult rats implanted with osmotic pumps and infused with a cell proliferation marker after HC damage suggest some regenerated cells begin to differentiate as HCs, and more differentiate as SCs (Kuntz and Oesterle 1998b; Oesterle et al. 2003). Further, rare cells that double-label for BrdU and a HC-specific protein have been documented in adult gerbil utricles after HC damage in vivo (Ogata et al. 1999).

Unlike the vestibular SE, HC loss in the mammalian auditory SE does not normally lead to HC replacement. The organ of Corti appears unable to repair itself normally via either mitotic or nonmitotic mechanisms (Sobkowicz et al. 1992, 1996, 1997; Roberson and Rubel 1994; Chardin and Romand 1995; Zine and de Ribaupierre 1998). Until recently, there was uncertainty as to whether SCs in the mature organ of Corti retain any capacity to reenter the cell cycle and generate new HCs. Several lines of evidence (discussed in Sections 3.1.1–3.1.2) now demonstrate that postmitotic SCs in the organ of Corti can be experimentally manipulated to reenter the cell cycle and to divide. Hence, organ of Corti

SCs are potential targets for therapeutic manipulation to generate replacement HCs, and this is an active area of research.

In sum, evidence suggests that limited numbers of new vestibular HCs may be regenerated via a mitotic pathway in the adult rodent ear after HC damage, but this remains to be demonstrated definitively. Although cells with features of HC progenitors appear to be present in the adult mammalian vestibular SE, the extremely low rate of cell division under normal conditions and after HC damage underscores the high level of negative growth control exerted upon the stem/progenitor cell population in this tissue. As further discussed in Sections 3.1.1–3.1.2, mechanisms underlying this resistance to mitotic activity are starting to be unraveled. Many critical questions related to mitotic HC replacement in mature mammals must still be addressed, including: (1) Why is mitotic regeneration absent (auditory SE) or limited (vestibular SE)? (2) Is this absence due to the lack of regenerative signals, to the presence of strong inhibitory signals, and/or to a paucity of tissue stem/progenitor cells? (3) How long are regenerated HCs retained in the mammalian vestibular SE, and what are their functional characteristics? (4) Can more significant levels of HC regeneration be stimulated in the inner ear SE of mature mammals? Some partial answers to these questions are provided in the discussion that follows.

3.1.1 Progenitors: Identity, Location, and Behavior

Questions of progenitor cell identity, location, and behavior have been difficult to address experimentally in mammalian inner ear SE because of the low rate, or absence, of inducible proliferation in this tissue. Several lines of evidence do suggest that inner ear stem/progenitor cells are present in the adult mammalian vestibular SE, albeit in a quiescent or restricted state. First, as discussed in the preceding text, limited levels of mitotic activity are seen in mature, mammalian vestibular SC after aminoglycoside-induced HC damage (Warchol et al. 1993; Lambert 1994; Rubel et al. 1995; Oesterle et al. 2003). Second, and as will be discussed further in the text that follows, growth factors that are mitogenic for stem/progenitor cells in other systems (e.g., EGF, TGF-α, IGF-1, FGF-2) stimulate proliferation of mammalian vestibular SCs (Yamashita and Oesterle 1995; Zheng et al. 1997). This proliferation leads to the production of a small numbers of new SCs, and possibly new HCs, in adult mammalian vestibular SE (Lambert 1994; Yamashita and Oesterle 1995; Kuntz and Oesterle 1998a; Oesterle et al. 2003). Third, vestibular SCs of adult mice are positive for nestin (Lopez et al. 2004), an intermediate filament protein that is a marker of stem/progenitor cells in other tissues (Lendahl et al. 1990; Johansson et al. 2002). Lastly, cells isolated from adult mouse utricular SE contain multipotent self-renewing cells that can generate cells with HC characteristics (Li et al. 2003), and this self-renewal is augmented by application of EGF and IGF-1. Taken together, these data suggest the presence of inner ear stem/progenitor cells that depend upon growth factors to survive and proliferate. In recent ultrastructural studies in mature rat utricular SE, proliferating cells were identified with and without SC characteristics (Oesterle et al. 2003). The proliferating cells without

SC characteristics were located adjacent to the basal lamina and were termed "active cells" because they were relatively nondescript but contained massive numbers of polyribosomes in their cytoplasm, signifying active protein synthesis. It is tempting to speculate that these active cells are stem/progenitor cells, but further work is needed to determine their identity.

In contrast to the vestibular SE, a variety of morphologically distinct SC types exist in the organ of Corti. Some of these SCs are highly differentiated (e.g., pillar and Deiters' cells). This high level of differentiation has raised concerns that many SCs in the mature organ of Corti may be too highly differentiated to reenter the cell cycle in response to HC damage. However, several findings suggest stem/progenitor cells may be present in the organ of Corti. Colony-forming cells have been isolated from the auditory SE of newborn rodents, which contains some nestin-positive cells (Malgrange et al. 2002; Lopez et al. 2004). Early postnatal organ of Corti cells, when dissociated and cultured under certain conditions, divide and generate new HCs in vitro (Doetzlhofer et al. 2004). Further, White et al. (2006) showed that cultures of purified SCs from neonatal and postnatal (P14) mouse organ of Corti, when grown in combination with periotic mesenchyme, EGF, and FGF-2, undergo substantial proliferative activity, and some postmitotic cells differentiate into HCs.

Targeted deletions of cell cycle-related genes further demonstrate the capacity of postnatal organ of Corti SCs to reenter the cell cycle and to proliferate. For example, deletion of the cyclin-dependent kinase inhibitor, $p27^{Kip1}$, in mice leads to production of supernumerary cells in the organ of Corti through excessive mitoses that extend significantly past the normal developmental period (Chen and Segil 1999; Löwenheim et al. 1999). Two other cell cycle regulatory genes, $p19^{Ink4d}$ and *retinoblastoma*, are also necessary for normal developmental cell cycle exit in the organ of Corti (Chen et al. 2003; Mantela et al. 2005; Sage et al. 2005). Deletion of either gene leads to highly disregulated DNA synthesis, including DNA synthesis in mature HCs. In sum, these studies demonstrate that proliferative potential is retained among SCs in the postnatal organ of Corti. Future work is needed to delineate the origins and nature of the signals that normally restrict this proliferation and to identify ways to modulate inhibitory signals in order to augment mammalian HC regeneration.

3.1.2 Proliferation Signals

Mitotic production of new HCs is a dominant regenerative strategy in birds and in other species capable of robust HC regeneration. Therefore, there is considerable interest in identifying factors that promote cell cycle reentry in mature mammalian SCs. Further, in the event that direct transdifferentiation of SCs into HCs is developed as a mechanism to induce HC regeneration, the restoration of converted SCs may be critical for maintaining the proper function of the SE. Therefore, effective therapies in humans are likely to require the generation of new SCs, as well as the new HCs.

Several studies have addressed whether addition of classic growth factors can stimulate proliferation of endogenous stem/progenitor cells in mature inner ear SE. Despite extensive testing of several candidates (e.g., Zine and de Ribaupierre 1998; Zheng et al. 1999), growth factors capable of stimulating proliferation in the organ of Corti have not yet been identified. A study by Lefebvre et al. (1993) suggested that retinoic acid may act synergistically with serum to stimulate proliferation in the developing organ of Corti of drug-damaged rats, but this finding has not been replicated (Chardin and Romand 1995). Recent data support a role for EGF (and factors derived from the periotic mesenchyme) in promoting HC production and differentiation in dissociated cultures of embryonic and early postnatal organ of Corti (Doetzlhofer et al. 2004).

In the mammalian vestibular SE, a few growth factors show promise for increasing proliferation and promoting new HC formation (reviewed in Staecker and Van de Water 1998; Oesterle and Hume 1999; see Table 5.1). Some factors (FGF-2, IGF-1, glial growth factor 2 [GGF-2], and heregulin) are mitogenic for developing (neonatal) mammalian vestibular SCs (Zheng et al. 1997, 1999; Montcouquiol and Corwin 2001b; Hume et al. 2003), yet fail to stimulate proliferation in mature SCs (Yamashita and Oesterle 1995; Kuntz and Oesterle 1998b; Hume et al. 2003). Neonatal inner ear tissue may respond differently to mitogens than adult tissue (Hume et al. 2003), or the differences may reflect the range in culture preparations used (e.g., isolated sheets of vestibular SE versus organ-otypic cultures). Increased expression of FGF and IGF-1 receptors has been described in utricular SE sheet cultures relative to that seen in vivo (Zheng et al. 1997). The expression of many other genes is altered by enzymatic treatment and mechanical isolation of the SE, and this may affect growth factor respon-siveness (Chen et al. 2002). In contrast, in organotypic cultures, the SE maintains contact with the extracellular matrix and more closely resembles the in vivo ear architecturally.

In the adult vestibular SE, the most effective mitogenic factors identified to date are EGF (Yamashita and Oesterle 1995; Zheng et al. 1997, 1999) and TGF-α (Lambert 1994; Yamashita and Oesterle 1995; Zheng et al. 1997, 1999; Kuntz and Oesterle 1998a; Oesterle et al. 2000), two structurally related members of the large EGF-ligand family. Both ligands bind the EGF receptor, erbB1. TGF-α stimulates proliferation in cultured utricular SE taken from neonatal rats and adult mice (Lambert 1994; Yamashita and Oesterle 1995; Zheng et al. 1997, 1999). TGF-α's mitogenic effects are potentiated by insulin (Yamashita and Oesterle 1995; Kuntz and Oesterle 1998a), and infusion of TGF-α plus insulin into the adult rat ear stimulates the production of new SCs, and possibly new HCs, in the utricular SE in vivo (Kuntz and Oesterle 1998a; Oesterle et al. 2003) (Fig. 5.6). EGF, when used in combination with insulin, also stimulates SC proliferation in cultured utricles from adult mice (Yamashita and Oesterle 1995) and in isolated sheets of utricular SE from neonatal rats (Zheng et al. 1997, 1999). A cocktail of TGF-α, IGF-1, and retinoic acid was reported to enhance vestibular HC renewal/repair and to improve vestibular function in adult guinea pigs lesioned with an ototoxin (Kopke et al. 2001). Unfortunately, proliferation markers were not used in this study, making it difficult to assess

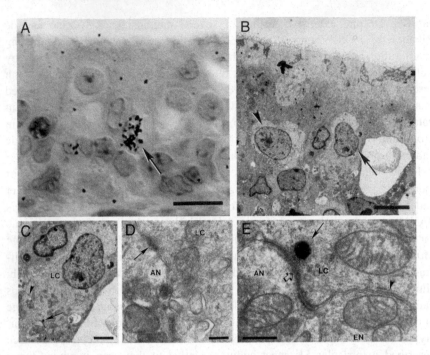

FIGURE 5.6. TGF-α with insulin stimulates the production of new cells in the mature mammalian vestibular SE via a mitotic pathway. TGF-α with insulin was infused, along with the cell-proliferation marker [³H]thymidine, into the inner ears of adult rats via mini-osmotic pumps, and the effects were assessed on normal and gentamicin-damaged utricles. Labeled cells were identified at the light microscope level. Cells were photographed prior to reembedding the thick sections in plastic and processing them for transmission electron microscopy (TEM). The light micrographs were used to precisely identify the [³H]thymidine-labeled cells at the TEM level. Ultrastructural characteristics of newly generated SE cells were examined. Five classes of [³H]thymidine-labeled cells were identified on the basis of their ultrastructural characteristics: (1) cells with synaptic specializations that appeared to be newly generated HCs (A–C), (2) SCs, (3) leukocytes, (4) cells classified as "active cells" that are relatively nondescript but contain massive numbers of polyribosomes, and (5) degenerating HCs. This figure shows a labeled cell with synaptic specializations with afferent and efferent terminals. (A) Light photomicrograph taken from a gentamicin-damaged rat ear that was infused with [³H]thymidine and TGF-α plus insulin for 3 days and fixed 4 days after pump implantation. A clearly labeled cell (arrow) is located in the lumenal half of the sensory epithelium. The focus is on the silver grains. (B) Same section shown in (A), processed for TEM. The arrow points to the labeled cell in (A), a putative HC. The cytoplasm of the labeled cell is electronlucent like that of nearby unlabeled HCs (arrowhead) and lighter than that of adjacent unlabeled SCs. (C–E) Higher magnification of synaptic specializations on the labeled cell shown in (A) and (B). (C) Numerous neural elements abut the labeled cell (LC). The arrowhead and arrow point to regions of the labeled cell with synaptic specializations. (D) Higher magnification of the region indicated by the arrowhead in (C). A membrane density (arrow) can be seen on a presumed afferent neural element (AN) at a contact with the labeled cell (LC). (E) Higher magnification of the region indicated by the arrow in (C).

the nature of the repair process and whether a proliferation-mediated mechanism was involved.

ErbB1 is expressed by the majority of SCs in the mature mammalian vestibular SE (Matsunaga et al. 2001; Zhang et al. 2002; Hume et al. 2003). However, mitogenic effects of TGF-α or EGF are limited to small numbers of SCs (Yamashita and Oesterle 1995; Kuntz and Oesterle 1998a). These findings suggest that the lack of responsiveness seen in most SCs is not due to the lack of receptors for EGF/TGF-α but rather to the activities of additional regulators of cell proliferation.

In mammalian inner ear SE, the gradual slowing and eventual cessation of ongoing progenitor cell proliferation toward the end of embryonic development demonstrates that HC genesis is under strong negative control (Ruben 1967). It makes teleological sense that numbers of cells in the auditory SE in particular must be tightly regulated, because of the importance of vibratory characteristics in the SE for the encoding of sound information. One working hypothesis is that differentiating and/or mature HCs produce signals that inhibit stem/progenitor cell division. One family of growth factors known to inhibit stem/progenitor cell division is transforming growth factor-beta (TGF-β) (e.g., Kawauchi et al. 2004). Addition of TGF-ß1, 2, 3, or 5 has been shown to reduce proliferation in isolated vestibular SCs from neonatal rats (Zheng et al. 1997). However, growth factors that have modulatory effects in intact SE in vitro and in the SE in vivo remain to be identified. The identification of negative growth signals will be very important, not only for understanding HC development, but also for devising strategies to deal with HC injury and loss, in which persistent growth-inhibitory signals could thwart attempts to promote regeneration. Robust proliferation of stem/progenitor cells in mature mammalian inner ear SE may require mitogenic stimulation to be coupled with release from tonic inhibition.

In summary, while studies on mitogenic factors such as TGF-α provide encouraging evidence that stem/progenitor cell division in the mature mammalian SE can be augmented, the number of cells that are triggered to reenter the cell cycle is extremely small relative to that seen in the avian SE. It is possible that additional growth factors can have more dramatic mitogenic effects. Many growth factors remain to be tested individually, in combination, and in specific sequences, and the mechanisms of growth factors with strong mitogenic effects need to be further explored as a treatment to augment mammalian HC regeneration. However, before any growth factor can be seriously considered for treatment of clinical problems related to HC loss, its effects in nearby tissues

---

FIGURE 5.6. Note the round synaptic body (arrow) in the labeled cell (LC) opposite an afferent nerve (AN) terminal and the membrane thickenings (open arrow) of the synaptic membranes. A presumed efferent terminal (EN) also apposes the labeled cell, adjacent to subsurface cisternae (arrowhead) in the labeled cell. Scale bars = 10 μm in **A**, 5 μm in **B**, 2 μm in **C**; 250 nm in **D**; 200 nm in **E**. Modified from Figs. 4 and 5 in Oesterle et al. 2003.)

must also be fully explored. Kuntz and Oesterle (1998a) showed that TGF-α plus insulin stimulates cell proliferation in many regions of the extrasensory epithelium (e.g., transitional epithelium, stroma, and squamous cells lining the portion of the stroma facing the perilymph), in addition to the vestibular SE. Such effects could be deleterious to inner ear function.

3.2 Differentiation

Lost HCs are not replaced in the adult organ of Corti, and the SCs involved in the repair process develop into a permanent epithelial scar that prevents further damage to the epithelium (Hawkins 1976; Johnson et al. 1981; Forge 1985). However, in the developing organ of Corti, undifferentiated cells can serve as progenitors for new HCs. Hair cell ablation experiments in the embryonic organ of Corti demonstrated that lost HCs can be replaced in vitro, possibly through direct transformation of uncommitted nonsensory cells that change their normal developmental fate (Kelley et al. 1995). Hair cell replacement might be possible in the mature organ of Corti, as long as cells with the capacity to transdifferentiate are retained in the epithelium. As development proceeds, nonsensory cells within the organ of Corti specialize into a variety of morphologically distinct SC types (Hensen's cells, Deiters' cells, pillar cells, inner phalangeal cells, and border cells), possibly limiting any regenerative process. Nevertheless, some observations suggest that SCs in the mature organ of Corti may have conserved some potential to differentiate into sensory cells. In the aminoglycoside-damaged rat organ of Corti, some Deiters' cells undergo atypical differentiation at their apical poles, forming a distinct tuft of actin-rich microvilli reminiscent of immature stereociliary bundles on nascent HCs (Lenoir and Vago 1996, 1997; Romand et al. 1996; Daudet et al. 1998; Parietti et al. 1998). These cells have neither a cuticular plate nor stereocilia, they do not express typical HC markers, and their basal pole is not contacted by nerve fibers. However, the unique features of these cells suggest that they could be SCs engaged in the process of direct transdifferentiation into HCs (Lenoir and Vago 1996, 1997; Romand et al. 1996; Daudet et al. 1998; Parietti et al. 1998). The atypical cells disappear from the scarring epithelium after a few weeks in vivo (Daudet et al. 1998). Exogenous application of EGF in vitro had no effects on their differentiation (Daudet et al. 2002), though Lefebvre et al. (2000) report that EGF stimulates production of supernumerary HCs in neonatal rat organ of Corti explants. Application of TGF-α appears to promote reorganization of the actin cytoskeleton in the atypical cells, but it does not coax them to progress further toward a definitive HC phenotype (Daudet et al. 2002). It is conceivable that atypical cells could be stimulated toward adapting a HC phenotype by other unidentified factors, but these factors remain to be identified. As discussed earlier, histological studies of the vestibular SE of adult guinea pigs provide evidence that SCs may undergo direct transdifferentiation into HCs via a nonproliferative pathway (Li and Forge 1997; Forge et al. 1998). Support for this interpretation is provided by the observation that the number of supporting cells shows a small but significant reduction during the course of recovery (Forge et al. 1998).

Based on the potential capacity of mature SCs for phenotypic conversion, several recent studies have begun to identify molecules that are able to activate or suppress HC fate determination. As discussed in depth earlier (Sections 2.2.3.12), recent experiments have implicated the basic helix–loop–helix (HLH) family of transcription factors, including *Atoh1* (also called *Math1*), *HES1*, and *HES5*, as central controllers of cellular specification and differentiation in the SE (reviewed in Gao 2003). *HES1* and *HES5* act as negative regulators of HC differentiation. The deletion of *HES1* or *HES5* in mutant mice leads to an overproduction of HCs in auditory and vestibular SE (Zine et al. 2001). *Math1*, the mouse homolog of the *Drosophila* gene *atonal*, acts as a positive regulator of HC differentiation, and is necessary and sufficient for HC generation (Bermingham et al. 1999; Zheng et al. 2000; Zine et al. 2001; Woods et al. 2004). After maturation of HCs, the expression of *Math1* is downregulated (Zheng et al. 2000). Virally delivered transgenes for *Math1* or *Hath1* (the human *atonal* homolog) have been introduced into cultures of immature rat organ of Corti and adult rat utricular SE. This misexpression leads to the production of ectopic HCs, as well as HCs in the SE, presumably by triggering nonsensory cells to convert into HCs without intervening mitoses (Zheng et al. 2000; Shou et al. 2003; Woods et al. 2004). Even more striking is the finding that misexpression of *Math1* in mature guinea pig organ of Corti in vivo causes the appearance of new HCs in the organ of Corti and in adjacent nonsensory epithelium (Kawamoto et al. 2003; Izumikawa et al. 2005). In *Math1*-transfected animals, auditory nerve axons appear to regrow to the new HCs (Kawamoto et al. 2003), and hearing thresholds in ototoxin-damage ears appear to improve (Izumikawa et al. 2005). These exciting findings suggest that the reactivation of developmental regulatory genes in mature tissues is a potential strategy for HC replacement therapy in the adult mammalian inner ear.

Another promising potential therapeutic strategy for HC replacement involves the use of a small molecule, the gamma secretase inhibitor *N*-[*N*-(3,5-difluorophenacetyl-L-alanyl)]-*S*-phenylglycine *t*-butyl ester (DAPT), to block endogenous Notch signaling. Supporting cells in the mammalian inner ear SE express the Notch receptor and Notch effector genes, which antagonize the ability of the *Atoh1* (*Math1*) transcription factor to induce HC differentiation (Landford et al. 1999, 2000; Morrison et al. 1999; Zheng et al. 2000; Zine et al. 2001). The reduction of Notch signaling in inner ear SE in a variety of mice mutants results in the production of supernumerary HCs (Landford et al. 1999; Zhang et al. 2000; Zheng et al. 2000; Kiernan et al. 2001; Zine 2003). Notch signaling can be blocked by DAPT, which prevents gamma secretase-mediated cleavage and activation of the Notch receptor. Several groups of investigators have shown that DAPT treatment causes robust supernumerary HC production in cultured organ of Corti from embryonic or neonatal mice (Woods et al. 2004; Tang et al. 2006; Yamamoto et al. 2006). The new HCs are thought to arise via direct transdifferentiation from SCs or cells outside the organ of Corti. The remarkable effectiveness of DAPT in stimulating transdifferentiation raises the possibility that inhibition of Notch signaling in the mature damaged organ of Corti might be an effective therapeutic treatment for hearing loss. For now, it remains to be determined whether DAPT can trigger SC-to-HC conversion in adult SE, whether HCs regenerated through this mechanism can fully differentiate and survive to restore auditory function after damage, and if DAPT can be effectively delivered to the inner ear in vivo to trigger HC regeneration.

4. Summary and Future Directions

In addition to the fundamental importance of unraveling cellular and molecular processes leading to regeneration of HCs from stem/progenitor cells in the inner ear SE, the ability to induce and control the proliferation and differentiation of cells in the mature SE may have important practical applications for the treatment of human hearing and balance disorders. It is impossible to predict when research efforts to delineate these processes will lead to a therapy for these impairments. However, great progress has been made in understanding HC regeneration since its discovery in mature birds in the late 1980s. It is anticipated that future progress will be hastened by technical advances, including improved methods for in vitro and in vivo gene/drug delivery to the inner ear, development of SE cells lines for easier in vitro studies (e.g., Rivolta et al. 1998; Kalinec et al. 1999), and genomics analyses that reveal molecular profiles for stem/progenitors as well as genes that regulate progenitor cell behavior and specification/differentiation of cell progeny (Hawkins et al. 2003).

Acknowledgments. We thank the National Institute on Deafness and Other Communication Disorders (DC03944, DC03696, DC04661), the National Organization for Hearing Research, and the Deafness Research Foundation for supporting our research and other research programs aimed at understanding cellular and molecular mechanisms of hair cell regeneration.

References

Aberg MAI, Aberg ND, Palmer TD, Alborn A-M, Carlsson-Skwirut C, Bang P, Rosengren LE, Olsson T, Gage FH, Ericksson PS (2003) IGF-1 has a direct proliferative effect in adult hippocampal progenitor cells. Mol Cell Neurosci. 24:23–40.
Adam J, Myat A, Le Roux I, Eddison M, Henrique D, Ish-Horowicz D, Lewis J (1998) Cell fate choices and the expression of Notch, Delta and Serrate homologues in the chick inner ear: parallels with *Drosophila* sense-organ development. Development 125:4645–4654.
Adler HJ, Raphael Y (1996) New hair cells arise from supporting cell conversion in the acoustically damaged chick inner ear. Neurosci Lett 205:17–20.
Adler HJ, Saunders JC (1995) Hair cell replacement in the avian inner ear following two exposures to intense sound. J Neurocytol 24:111–116.

Artavanis-Tsakonas S, Simpson P (1991) Choosing a cell fate: a view from the Notch locus. Trends Genet 7:403–408.

Avallone B, Balsamo G, Trapani S, Marmo F (2002) Apoptosis during chick inner ear development: some observations by TEM and TUNEL techniques. Eur J Histochem 46:53–59.

Avallone B, Porritiello M, Esposito D, Mutone R, Balsamo G, Marmo F (2003) Evidence for hair cell regeneration in the crista ampullaris of the lizard *Podarcis sicula*. Hear Res 178:79–88.

Baird RA, Torres MA, Schuff NR (1993) Hair cell regeneration in the bullfrog vestibular otolith organs following aminoglycoside toxicity. Hear Res 65:164–174.

Baird RA, Steyger PS, Schuff N (1996) Mitotic and nonmitotic hair cell regeneration in the bullfrog vestibular otolith organs. Ann NY Acad Sci 781:59–70.

Baird RA, Burton MD, Fashena DS, Naeger RA (2000) Hair cell recovery in mitotically blocked cultures of the bullfrog saccule. Proc Natl Acad Sci USA 97:11722–11729.

Balak KJ, Corwin JT, Jones JE (1990) Regenerated hair cells can originate from supporting cell progeny: evidence from phototoxicity and laser ablation experiments in the lateral line system. J Neurosci 10:2502–2512.

Barald KF, Kelley MW (2004) From placode to polarization: new tunes in inner ear development. Development 131:4119–4130.

Bartolami S, Goodyear R, Richardson G (1991) Appearance and distribution of the 275 kD hair-cell antigen during development of the avian inner ear. J Comp Neurol 314:777–788.

Beites CL, Kawauchi S, Crocker CE, Calof AL (2005) Identification and molecular regulation of neural stem cells in the olfactory epithelium. Exp Cell Res 306:309–316.

Ben-Arie N, McCall AE, Berkman S, Eichele G, Bellen HJ, Zoghbi HY (1996) Evolutionary conservation of sequence and expression of the bHLH protein Atonal suggests a conserved role in neurogenesis. Hum Mol Genet 5:1207–1216.

Ben-Arie N, Hassan BA, Bermingham NA, Malicki DM, Armstrong D, Matzuk M, Bellen HJ, Zoghbi HY (2000) Functional conservation of atonal and Math1 in the CNS and PNS. Development 127:1039–1048.

Bermingham NA, Hassan BA, Price SD, Vollrath MA, Ben-Arie N, Eatock RA, Bellen HJ, Lysakowski A, Zoghbi HY (1999) Math1: an essential gene for the generation of inner ear hair cells. Science 284:1837–1841.

Bermingham-McDonogh O, Stone JS, Reh TA, Rubel EW (2001) FGFR3 expression during development and regeneration of the chick inner ear sensory epithelia. Dev Biol 238:247–259.

Bertrand N, Castro DS, Guillemot F (2002) Proneural genes and the specification of neural cell types. Nat Rev Neurosci 3:517–530.

Bever MM, Fekete DM (1999) Ventromedial focus of cell death is absent during development of *Xenopus* and zebrafish inner ears. J Neurocytol 28:781–793.

Bhave SA, Stone JS, Rubel EW, Coltrera MD (1995) Cell cycle progression in gentamicin-damaged avian cochleas. J Neurosci 15:4618–4628.

Bhave SA, Oesterle EC, Coltrera MD (1998) Macrophage and microglia-like cells in the avian inner ear. J Comp Neurol 398:241–256.

Blanpain C, Lowry WE, Geoghegan A, Polak L, Fuchs E (2004) Self-renewal, multipotency, and the existence of two cell populations within an epithelial stem cell niche. Cell 118:635–648.

Bohne B (1971) Scar Formation in the Inner Ear Following Acoustic Injury: Sequence of Changes from Early Signs of Damage to Healed Lesion. Ph. D. Thesis. St. Louis, MO: Washington University.

Bone RC, Ryan AF (1978) Audiometric and histologic correlates of the interaction between kanamycin and subtraumatic levels of noise in the chinchilla. Otolaryngology 86: ORL400–4.

Brooker R, Hozumi K, Lewis J (2006) Notch ligands with contrasting functions: Jagged1 and Delta1 in the mouse inner ear. Development 133:1277–1286.

Brown JN, Miller JM, Altschuler RA, Nuttall AL (1993) Osmotic pump implant for chronic infusion of drugs into the inner ear. Hear Res 70:167–172.

Bryant J, Goodyear RJ, Richardson GP (2002) Sensory organ development in the inner ear: molecular and cellular mechanisms. Br Med Bull 63:39–57.

Cafaro J, Lee GS, Stone JS (2007) Atoh1 expression defines activated progenitors and differentiating hair cells during avian hair cell regeneration. Dev Dyn. 236 (1): 156–170.

Calof AL, Chikaraishi DM (1989) Analysis of neurogenesis in a mammalian neuroepithelium: proliferation and differentiation of an olfactory neuron precursor in vitro. Neuron 3:115–127.

Camarero G, Avendano C, Fernandez-Moreno C, Villar A, Contreras J, de Pable F, Pichel JG, Varela-Nieto I (2001) Delayed inner ear maturation and neuronal loss in postnatal Igf-1–deficient mice. J Neurosci 21:7630–7641.

Cantos R, Cole LK, Acampora D, Simeone A, Wu DK (2000) Patterning of the mammalian cochlea. Proc Natl Acad Sci USA 97:11707–11713.

Carnicero E, Garrido JJ, Alonso MT, Schimmang T (2001) Roles of fibroblast growth factor 2 during innervation of the avian inner ear. J Neurochem 77:786–795.

Carnicero E, Zelarayan LC, Rüttiger L, Knipper M, Alvarez Y, Alonso MT, Schimmang T (2004) Differential roles of fibroblast growth factor-2 during development and maintenance of auditory sensory epithelial. J Neurosci Res 77:787–797.

Chardin S, Romand R (1995) Regeneration and mammalian auditory hair cells. Science 267:707–711.

Chen P, Segil N (1999) p27^{Kip1} links cell proliferation to morphogenesis in the developing organ of Corti. Development 126: 1581–1590.

Chen ZY, Corey DP (2002) Understanding inner ear development with gene expression profiling. J Neurobiol. 5:276–285.

Chen P, Johnson JE, Zoghbi HY, Segil N (2002) The role of Math1 in inner ear development: uncoupling the establishment of the sensory primordium from hair cell fate determination. Development 129:2495–2505.

Chen P, Zindy F, Abdala C, Liu F, Li X, Roussel MF, Segil N (2003) Progressive hearing loss in mice lacking the cyclin-dependent kinase inhibitor Ink4d. Nature 5:422–426.

Cho Y, Gong T-WL, Stöver T, Lomax MI, Altschuler RA (2001) Gene expression profiles of the rat cochlea, cochlear nucleus, and inferior colliculus. J Assoc Res Otolaryngol 3:54–67.

Cole LK, Le Roux I, Nunes F, Laufer E, Lewis J, Wu DK (2000) Sensory organ generation in the chicken inner ear: contributions of bone morphogenetic protein 4, serrate1, and lunatic fringe. J Comp Neurol 424:509–520.

Collier JR, Monk NA, Maini PK, Lewis JH (1996) Pattern formation by lateral inhibition with feedback: a mathematical model of delta-notch intercellular signaling. J Theor Biol 183:429–446.

Corwin JT (1981) Postembryonic production and aging in inner ear hair cells in sharks. J Comp Neurol 201:541–553.

Corwin JT (1983) Postembryonic growth of the macula neglecta auditory detector in the ray, *Raja clavata*: continual increases in hair cell number, neural convergence, and physiological sensitivity. J Comp Neurol 217:345–345.

Corwin JT (1985) Perpetual production of hair cells and maturational changes in hair cell ultrastructure accompany postembryonic growth in an amphibian ear. Proc Natl Acad Sci USA 82:3911–3915.

Corwin JT, Cotanche DA (1988) Regeneration of sensory hair cells after acoustic trauma. Science 240:1772–1774.

Corwin JT, Jones JE, Katayama A, Kelley MW, Warchol ME (1991) Hair cell regeneration: the identities of progenitor cells, potential triggers and instructive cues. Ciba Found Symp 160:103–120.

Cotanche DA (1987a) Regeneration of hair cell stereociliary bundles in the chick cochlea following severe acoustic trauma. Hear Res 30:181–195.

Cotanche DA (1987b) Regeneration of the tectorial membrane in the chick cochlea following severe acoustic trauma. Hear Res 30:197–206.

Cotanche DA (1999) Structural recovery from sound and aminoglycoside damage in the avian cochlea. Audiol Neurootol 4:271–285.

Cotanche DA, Sulik KK (1984) The development of stereociliary bundles in the cochlear duct of chick embryos. Brain Res 318(2):181–193.

Cotanche DA, Petrell A, Picard DA (1991) Structural reorganization of hair cells and supporting cells during noise damage, recovery and regeneration in the chick cochlea. Ciba Found Symp 160:131–142.

Cotanche DA, Messana EP, Ofsie MS (1995) Migration of hyaline cells into the chick basilar papilla during severe noise damage. Hear Res 91:148–159.

Cristobal R, Popper P, Lopez I, Micevych P, De Vellis J, Honrubia V (2002) *In vivo* and *in vitro* localization of brain-derived neurotrophic factor, fibroblast growth factor-2 and their receptors in the bullfrog vestibular end organs. Mol Brain Res 102:83–99.

Cruz RM, Lambert PR, Rubel EW (1987) Light microscopic evidence of hair cell regeneration after gentamicin toxicity in chick cochlea. Arch Otolaryngol Head Neck Surg 113:1058–1062.

Cryns K, Van Camp G (2004) Deafness genes and their diagnostic applications. Audiol Neuro-Otol 9:2–22.

Dabdoub A, Donohue MJ, Brennan A, Wolf V, Montcouquiol M, Sassoon DA, Hseih JC, Rubin JS, Salinas PC, Kelley MW (2003) Wnt signaling mediates reorientation of outer hair cell stereociliary bundles in the mammalian cochlea. Development 130:2375–2384.

Daudet N, Lewis J (2005) Two contrasting roles for Notch activity in chick inner ear development: specification of prosensory patches and lateral inhibition of hair-cell differentiation. Development 132:541–551.

Daudet N, Vago P, Ripoll C, Humbert G, Pujol R, Lenoir M (1998) Characterization of atypical cells in the juvenile rat organ of Corti after aminoglycoside ototoxicity. J Comp Neurol 401:145–162.

Daudet N, Ripoll C, Lenoir M (2002) Transforming growth factor-α-induced cellular changes in organotypic cultures of juvenile, amikacin-treated rat organ of Corti. J Comp Neurol 442:6–22.

Davis RL, Turner DL (2001) Vertebrate hairy and Enhancer of split related proteins: transcriptional repressors regulating cellular differentiation and embryonic patterning. Oncogene 20:8342–8357.

Ding-Pfennigdorff D, Smolders JW, Muller M, Klinke R (1998) Hair cell loss and regeneration after severe acoustic overstimulation in the adult pigeon. Hear Res. 120:109–120.

Doe CQ (1996) Asymmetric cell division and neurogenesis. Opin Genet Dev 6:562–566.

Doe CQ, Chu-LaGraff Q, Wright DM, Scott MP (1991) The prospero gene specifies cell fates in the *Drosophila* central nervous system. Cell 65:451–464.

Doetzlhofer A, White PM, Johnson JE, Segil N, Groves AK (2004) *In vitro* growth and differentiation of mammalian sensory hair cell progenitors for EGF and periotic mesenchyme. Dev Biol 272:432–447.

Duckert LG, Rubel EW (1990) Ultrastructural observations on regenerating hair cells in the chick basilar papilla. Hear Res 48:161–182.

Dye BJ, Frank TC, Newlands SD, Dickman JD (1999) Distribution and time course of hair cell regeneration in the pigeon utricle. Hear Res 13:17–26.

Dyer MA, Livesey FJ, Cepko CL, Oliver G (2003) Prox1 function controls progenitor cell proliferation and horizontal cell genesis in the mammalian retina. Nat Genet 34:53–58.

Eddison M, Le Roux I, Lewis J (2000) Notch signaling in the development of the inner ear: lessons from *Drosophila*. Proc Natl Acad Sci USA 97:11692–11699.

Epstein JE, Cotanche DA (1995) Secretion of a new basal layer of tectorial membrane following gentamicin-induced hair cell loss. Hear Res 90:31–43.

Fehon RG, Johansen K, Rebay I, Artavanis-Tsakonas S (1990) Complex cellular and subcellular regulation of notch expression during embryonic and imaginal development of *Drosophila*: implications for notch function. J Cell Biol 113:657–669.

Fekete DM, Wu DK (2002) Revisiting cell fate specification in the inner ear. Curr Opin Neurobiol 12:35–42.

Fekete DM, Muthukumar S, Karagogeos D (1998) Hair cells and supporting cells share a common progenitor in the avian inner ear. J Neurosci 18:7811–7821.

Fischer AJ, Reh TA (2000) Identification of a proliferating marginal zone of retinal progenitors in postnatal chickens. Dev Biol 2000 220:197–210.

Fischer AJ, Reh TA (2001) Muller glia are a potential source of neural regeneration in the postnatal chicken retina. Nat Neurosci 4:247–252.

Forge A (1985) Outer hair cell loss and supporting cell expansion following chronic gentamicin treatment. Hear Res 19:171–182.

Forge A, Schacht J (2000) Aminoglycoside antibiotics. Audiol Neurootol 5:3–22.

Forge A, Li L, Corwin JT, Nevill G (1993) Ultrastructural evidence for hair cell regeneration in the mammalian inner ear [see comments] Science 259:1616–1619.

Forge A, Li L, Nevill G (1998) Hair cell recovery in the vestibular sensory epithelia of mature guinea pigs. J Comp Neurol 397:69–88.

Fredelius L, Rask-Andersen H (1990) The role of macrophages in the disposal of degeneration products within the organ of corti after acoustic overstimulation. Acta Otolaryngol Stockh 109:76–82.

Fritzsch B, Beisel KW, Bermingham NA (2000) Developmental evolutionary biology of the vertebrate ear: conserving mechanoelectric transductionand developmental pathways in diverging morphologies. NeuroReport 11:R35–44.

Galderisi U, Jori FP, Giordano A (2003) Cell cycle regulation and neural differentiation. Oncogene 22:5208–5219.

Gale JE, Meyers JR, Periasamy A, Corwin JT (2000) Survival of bundleless hair cells and subsequent bundle replacement in the bullfrog's saccule. J Neurobiol 50:81–92.

Gao W (2003) Hair cell development in higher vertebrates. Curr Top Dev Biol 57:293–319.

Ginzberg RD, Gilula NB (1979) Modulation of cell junctions during differentiation of the chicken otocyst sensory epithelium. Dev Biol 68:110–129.

Girod DA, Duckert LG, Rubel EW (1989) Possible precursors of regenerated hair cells in the avian cochlea following acoustic trauma. Hear Res 42:175–194.

Girod DA, Tucci DL, Rubel EW (1991) Anatomical correlates of functional recovery in the avian inner ear following aminoglycoside ototoxicity. Laryngoscope 101: 1139–1149.

Gleich O, Dooling RJ, Manley GA (1994) Inner-ear abnormalities and their functional consequences in Belgian Waterslager canaries (*Serinus canarius*). Hear Res 79:123.

Gleich O, Dooling RJ, Presson JC (1997) Evidence for supporting cell proliferation and hair cell differentiation in the basilar papilla of adult Belgian Waterslager canaries (*Serinus canarius*). J Comp Neurol 377:5–14.

Go MJ, Eastman DS, Artavanis-Tsakonas S (1998) Cell proliferation control by Notch signaling in *Drosophila* development. Development 125:2031–2040.

Goldstein BJ, Fang H, Youngentob SL, Schwob JE (1998) Transplantation of multipotent progenitors from the adult olfactory epithelium. NeuroReport 9:1611–1617.

Goodyear RJ, Richardson GP (2002) Extracellular matrices associated with the apical surfaces of sensory epithelia in the inner ear: molecular and structural diversity. J Neurobiol 53:212–227.

Goodyear R, Holley M, Richardson G (1995) Hair and supporting-cell differentiation during the development of the avian inner ear. J Comp Neurol 351:81–93.

Goodyear R, Killick R, Legan PK, Richardson GP (1996) Distribution of beta-tectorin mRNA in the early posthatch and developing avian inner ear. Hear Res 96:167–178.

Goodyear RJ, Gates R, Lukashkin AN, Richardson GP (1999) Hair-cell numbers continue to increase in the utricular macula of the early posthatch chick. J Neurocytol 28:851–861.

Goodyear RJ, Kwan T, Oh SH, Raphael Y, Richardson GP (2001) The cell adhesion molecule BEN defines a prosensory patch in the developing avian otocyst. J Comp Neurol 434:275–288.

Goodyear RJ, Legan PK, Wright MB, Marcotti W, Oganesian A, Coats SA, Booth CJ, Kros CJ, Seifert RA, Bowen-Pope DF, Richardson GP (2003) A receptor-like inositol lipid phosphatase is required for the maturation of developing cochlear hair bundles. J Neurosci 23:9208–9219.

Gritti A, Parati EA, Cova L, Frolichsthal P, Galli R, Wanke E, Faravelli L, Morassutti DJ, Roisen F, Nickel DD et al. (1996) Multipotential stem cells from the adult mouse brain proliferate and self-renew in response to basic fibroblast growth factor. J. Neurosci 16:1091–1100.

Gritti A, Frolichsthal-Schoeller P, Galli R, Parati EA, Cova L, Pagano SF, Bjornson CR and Vescovi AL (1999) Epidermal and fibroblast growth factors behave as mitogenic regulators for a single multipotent stem cell-like population from the subventricular region of the adult mouse forebrain. J Neurosci 19:3287–3297.

Hackett L, Davies D, Helyer R, Kennedy H, Kros C, Lawlor P, Rivolta MN, Holley M (2002) E-cadherin and the differentiation of mammalian vestibular hair cells. Exp Cell Res 278:19–30.

Haddon C, Jiang YJ, Smithers L, Lewis J (1998) Delta-Notch signalling and the patterning of sensory cell differentiation in the zebrafish ear: evidence from the mind bomb mutant. Development125:4637–4644.

Hall PA, Watt FM (1989) Stem cells: the generation and maintenance of cellular diversity. Development 106:619–633.

Hamilton TA, Ohmori Y, Narumi S, Tannenbaum CS, eds (1993) Regulation of Diversity of Macrophage Activation. Ann Arbor: CRC Press.

Harris JA, Cheng AG, Cunningham LL, MacDonald G, Raible DW, Rubel EW (2003) Neomycin-induced hair cell death and rapid regeneration in the lateral line of zebrafish (*Danio rerio*). J Assoc Res Otolaryngol 4:219–234.

Hartenstein V, Posakony JW (1990) A dual function of the Notch gene in *Drosophila* sensillum development. Dev Biol 142:13–30.

Hashino E, Salvi R (1993) Changing patterns of DNA replication in the noise-damaged chick cochlea. J Cell Sci 105:23–31.

Hashino E, TinHan EK, Salvi RJ (1995) Base-to-apex gradient of cell proliferation in the chick cochlea following kanamycin-induced hair cell loss. Hear Res 88:156–168.

Hashino E, Tanaka Y, Sokabe M (1991) Hair cell damage and recovery following chronic application of kanamycin in the chick cochlea. Hear Res 52:356–368.

Hassan BA, Bellen HJ (2000) Doing the MATH: is the mouse a good model for fly development? Genes Dev 14:1852–1865.

Hasson T, Gillespie PG, Garcia JA, MacDonald RB, Zhao Y, Yee AG, Mooseker MS, Corey DP (1997) Unconventional myosins in inner-ear sensory epithelia. J Cell Biol 137:1287–1307.

Hawkins E (1976) Drug ototoxicity. In Keidel WD, Neff WD (eds) Auditory System. New York: Springer-Verlag, pp. 707–748.

Hawkins RD, Bashiardes S, Helms CA, Hu L, Saccone NL, Warchol ME, Lovett M (2003) Gene expression differences in quiescent versus regenerating hair cells of avian sensory epithelia: implications for human hearing and balance disorders. Hum Mol Genet 12:1261–1272.

Henrique D, Hirsinger E, Adam J, Le Roux I, Pourquie O, Ish-Horowicz D, Lewis J (1997) Maintenance of neuroepithelial progenitor cells by Delta-Notch signalling in the embryonic chick retina. Curr Biol 7:661–670.

Higgs DM, Souza MJ, Wilkins HR, Presson JC, Popper AN (2002) Age- and size-related changes in the inner ear and hearing ability of the adult zebrafish (*Danio rerio*). J Assoc Res Otolaryngol 3:222.

Hirose K, Westrum LE, Cunningham DE, Rubel EW (2004) Electron microscopy of degenerative changes in the chick basilar papilla after gentamicin exposure. J Comp Neurol 470:164–180.

Hong R, Chakravarti D (2003) The human proliferating cell nuclear antigen regulates transcriptional coactivator p300 activity and promotes transcriptional repression. J Biol Chem 278:44505–44513.

Hoppe PE, Greenspan RJ (1986) Local function of the Notch gene for embryonic ectodermal pathway choice in *Drosophila*. Cell 46:773–783.

Huard JM, Youngentob SL, Goldstein BJ, Luskin MB, Schwob JE (1998) Adult olfactory epithelium contains multipotent progenitors that give rise to neurons and non-neural cells. J Comp Neurol 400:469–486.

Hume CR, Kirkegaard M, Oesterle EC (2003) ErbB expression: the mouse inner ear and maturation of the mitogenic response to heregulin. J Assoc Res Otolaryngol 4:422–443.

Inglis-Broadgate SL, Thomson RE, Pellicano F, Tartaglia MA, Pontikis CC, Cooper JD, Iwata T (2005) FGFR3 regulates brain size by controlling progenitor cell proliferation and apoptosis during embryonic development. Dev Biol 279:73–85.

Iso T, Kedes L, Hamamori Y (2003) HES and HERP families: multiple effectors of the Notch signaling pathway. J Cell Physiol 194:237–255.

Itoh M, Chitnis AB (2001) Expression of proneural and neurogenic genes in the zebrafish lateral line primordium correlates with selection of hair cell fate in neuromasts. Mech Dev 102:263–266.

Itoh M, Kim CH, Palardy G, Oda T, Jiang YJ, Maust D, Yeo SY, Lorick K, Wright GJ, Ariza-McNaughton L, Weissman AM, Lewis J, Chandrasekharappa SC, Chitnis AB (2003) Mind bomb is a ubiquitin ligase that is essential for efficient activation of Notch signaling by Delta. Dev Cell 4:67–82.

Izumikawa M, Minoda R, Kawamoto K, Abrashkin KA, Swiderski DL, Dolan DF, Brough DE, Raphael Y (2005) Auditory hair cell replacement and hearing improvement by *Atoh1* gene therapy in deaf mammals. Nat Med 11:271–276.

Jan YN, Jan LY (1998) Asymmetric cell division. Nature 392:775–778.

Janas JD, Cotanche DA, Rubel EW (1995) Avian cochlear hair cell regeneration: stereological analyses of damage and recovery from a single high dose of gentamicin. Hear Res 92:17–29.

Jarman AP, Grau Y, Jan LY, Jan YN (1993) atonal is a proneural gene that directs chordotonal organ formation in the *Drosophila* peripheral nervous system. Cell 73:1307–1321.

Jiang YJ, Brand M, Heisenberg CP, Beuchle D, Furutani-Seiki M, Kelsh RN, Warga RM, Granato M, Haffter P, Hammerschmidt M, Kane DA, Mullins MC, Odenthal J, van Eeden FJ, Nusslein-Volhard C (1996) Mutations affecting neurogenesis and brain morphology in the zebrafish, *Danio rerio*. Development123:205–216.

Johansson CB, Lothian C, Molin M, Okano H, Lendhal U (2002) Nestin enhance requirements for expression in normal and injured adult CNS. J Neurosci Res 69:784–794.

Johnson LG, Hawkins JE Jr, Kingsley TC, Black FO, Matz GJ (1981) Aminoglycoside-induced cochlear pathology in man. Acta Otolaryngol 383(Suppl):1–19.

Johnston LA, Edgar BA (1998) Wingless and Notch regulate cell-cycle arrest in the developing *Drosophila* wing. Nature 394:82–84.

Jones JE, Corwin JT (1993) Replacement of lateral line sensory organs during tail regeneration in salamanders: identification of progenitor cells and analysis of leukocyte activity. J Neurosci 13:1022–1034.

Jones JE, Corwin JT (1996) Regeneration of sensory cells after laser ablation in the lateral line system: hair cell lineage and macrophage behavior revealed by time-lapse video microscopy. J Neurosci 16:649–662.

Jørgensen JM (1991) Regeneration of lateral line and inner ear vestibular cells. Ciba Found Symp 160:151–170.

Jørgensen JM, Flock A (1976) Non-innervated sense organs of the lateral line: development in the regenerating tail of the salamander *Ambystoma mexicanum*. J Neurocytol 5:33–41.

Jørgensen JM, Mathiesen C (1988) The avian inner ear. Continuous production of hair cells in vestibular sensory organs, but not in the auditory papilla. Naturwissenschaften 75:319–320.

Jung KH, Chu K, Kim M, Jeong SW, Song YM, Lee ST, Kim JY, Lee SK, Roh JK (2004) Continuous cytosine-b-D-arabinofuranoside infusion reduces ectopic granule cells in adult rat hippocampus with attenuation of spontaneous recurrent seizures following pilocarpine-induced status epilepticus. Eur J Neurosci 19:3219–3226.

Kageyama R, Ohtsuka T, Hatakeyama J, Ohsawa R (2005) Roles of bHLH genes in neural stem cell differentiation. Exp Cell Res. 306:343–348.

Kalinec F, Kalinec G, Boukhvalova M, Kachar B (1999) Establishment and characterization of conditionally immortalized organ of corti cell lines. Cell Biol Int 23:175–184.

Katayama A, Corwin JT (1989) Cell production in the chicken cochlea. J Comp Neurol 281:129–135.

Katayama A, Corwin JT (1993) Cochlear cytogenesis visualized through pulse labeling of chick embryos in culture. J Comp Neurol 333:28–40.

Kawamoto K, Ishimoto S-I, Minoda R, Brough DE, Raphael Y (2003) Math1 gene transfer generates new cochlear hair cells in mature guinea pigs in vivo. J Neurosci 23:4395–4400.

Kawauchi S, Beites CL, Crocker CE, Wu HH, Bonnin A, Murray R, Calof AL (2004) Molecular signals regulating proliferation of stem and progenitor cells in mouse olfactory epithelium. Dev Neurosci 26:166–180.

Kelley MW (2003) Cell adhesion molecules during inner ear and hair cell development, including notch and its ligands. Curr Top Dev Biol 57:321–356.

Kelley MW, Talreja DR, Corwin JT (1995) Replacement of hair cells after laser microbeam irradiation in cultured organs of Corti from embryonic and neonatal mice. J Neurosci 15:3013–3026.

Kevetter GA, Blumberg KR, Correia MJ (2000) Hair cell and supporting cell density and distribution in the normal and regenerating posterior crista ampullaris of the pigeon. Int J Dev Neurosci 18:855–867.

Kiernan AE, Ahituv N, Fuchs H, Balling R, Avraham KB, Steel KP, Hrabe de Angelis (2001) The Notch ligand Jagged1 is required for inner ear sensory development. Proc Natl Acad Sci USA 98:3873–3878.

Kiernan AE, Cordes R, Kopan R, Gossler A, Gridley T (2005) The Notch ligands DLL1 and JAG2 act synergistically to regulate hair cell development in the mammalian inner ear. Development 132:4353–4362. Epub 2005 Sept 1

Kiernan AE, Xu J, Gridley T (2006) The Notch ligand JAG1 is required for sensory progenitor development in the mammalian inner ear. PLoS Genet 2(1):e4. Epub 2006 Jan 13.

Kil J, Warchol ME, Corwin JT (1997) Cell death, cell proliferation, and estimates of hair cell life spans in the vestibular organs of chicks. Hear Res 114:117–126.

Kilpatrick TJ, Bartlett PF (1993) Cloning and growth of multipotential neural precursors: requirements for proliferation and differentiation. Neuron 10:255–265.

Kirkegaard M, Jørgensen JM (2000) Continuous hair cell turnover in the inner ear vestibular organs of a mammal, the Daubenton's bat (Myotis daubentonii). Naturwissenschaften 87:83–86.

Kirkegaard M, Jørgensen JM (2001) The inner ear macular sensory epithelia of the Daubenton's bat. J Comp Neurol 438:433–444.

Klockars T, Perheentupa T, Dahl H-HM (2002) In silico analyses of mouse inner-ear transcripts. JARO 4:24–40.

Kopke RD, Jackson RL, Li G, Rasmussen MD, Hoffer ME, Frenz DA, Costello M, Schultheiss P, Van de Water TR (2001) Growth factor treatment enhances vestibular hair cell renewal and results in improved vestibular function. Proc Natl Acad Sci USA 98:5886–5891.

Kruger RP, Goodyear RJ, Legan PK, Warchol ME, Raphael Y, Cotanche DA, Richardson GP (1999) The supporting-cell antigen: a receptor-like protein tyrosine phosphatase expressed in the sensory epithelia of the avian inner ear. J Neurosci 19:4815–4827.

Kuntz AL, Oesterle EC (1998a) Transforming growth factor alpha with insulin stimulates cell proliferation in vivo in adult rat vestibular sensory epithelium. J Comp Neurol 399:413–423.

Kuntz AL, Oesterle EC (1998b) Transforming growth factor-α with insulin induces proliferation in rat utricular extrasensory epithelia. Otolaryngol Head Neck Surg 118:816–824.

Lambert PR (1994) Inner ear hair cell regeneration in a mammal: identification of a triggering factor. Laryngoscope 104:701–718.

Lambert PR, Gu R, Corwin JT (1997) Analysis of small hair bundles in the utricles of mature guinea pigs. Am J Otol 18:637–643.

Lanford PJ, Lan Y, Jiang R, Lindsell C, Weinmaster G, Gridley T, Kelley MW (1999) Notch signalling pathway mediates hair cell development in mammalian cochlea. Nat Genet 21:289–292.

Lanford PJ, Shailam R, Norton CR, Gridley T, Kelley MW (2000) Expression of Math1 and HES5 in the cochleae of wildtype and Jag2 mutant mice. J Assoc Res Otolaryngol 1:161–171.

Lang H, Bever MM, Fekete DM (2000) Cell proliferation and cell death in the developing chick inner ear: spatial and temporal patterns. J Comp Neurol 417:205–220.

Lee KH, Cotanche DA (1996) Potential role of bFGF and retinoic acid in the regeneration of chicken cochlear hair cells. Hear Res 94:1–13.

Lefebvre PP, Malgrange B, Staecker H, Moonen G, Van de Water TR (1993) Retinoic acid stimulates regeneration of mammalian auditory hair cells [see comments]. Science 260:692–695.

Lefebvre PP, Malgrange B, Thiry M, Van de Water TR, Moonen G (2000) Epidermal growth factor upregulates production of supernumerary hair cells in neonatal rat organ of Corti explants. Acta Otolaryngol 120:142–145.

Lendahl U, Zimmerman LB, McKay RDG (1990) CNS stem cells express a new class of intermediate filament protein. Cell 60:585–595.

Lenoir M, Vago P (1996) Morphological indications of hair cell neodifferentiation in the organ of Corti of amikacin treated rat pups. CR Acad Sci 319:269–276.

Lenoir M, Vago P (1997) Does the organ of Corti attempt to differentiate new hair cells after antibiotic intoxication in rat pups? Int J Dev Neurosci 15:487–495.

León Y, Vazquez E, Sanz C, Vega JA, Mato JM, Giraldez F, Represa J, Varela-Nieto I (1995) Insulin-like growth factor-I regulates cell proliferation in the developing inner ear, activating glycosylphosphatidylinositol hydrolysis and Fos expression. Endocrinology 136:3494–3503.

León Y, Sanz C, Giráldez F, Varela-Nieto I (1998) Induction of cell growth by insulin and insulin-like growth factor-I is associated with Jun expression in the otic vesicle. J Comp Neurol 398:323–332.

León Y, Sanz C, Frago LM, Camarero G, Cañón S, Vazrela-Nieto I, Giráldez F (1999) Involvement of insulin-like growth factor-1 in inner ear organogenesis and regeneration. Horm Metab Res 31:126–132.

Levenberg S, Yarden A, Kam Z, Geiger B (1999) p27 is involved in N-cadherin-mediated contact inhibition of cell growth and S-phase entry. Oncogene 18:869–876.

Lewis J (1991) Rules for the production of sensory cells. Ciba Found Symp 160:25–39.

Lewis J (1996) Neurogenic genes and vertebrate neurogenesis. Curr Opin Neurobiol 6:3–10.

Lewis AK, Frantz GD, Carpenter DA, de Sauvage FJ, Gao WQ (1998) Distinct expression patterns of notch family receptors and ligands during development of the mammalian inner ear. Mech Dev 78:159–163.

Li L, Forge A (1997) Morphological evidence for supporting cell to hair cell conversion in the mammalian utricular macula. Int J Dev Neurosci 15:433–446.

Li L, Vaessin H (2000) Pan-neural Prospero terminates cell proliferation during *Drosophila* neurogenesis. Genes Dev 14:147–151.

Li H, Liu H, Heller S (2003) Pluripotent stem cells from the adult mouse inner ear. Nat med 9:1293–1299.

Lindsell CE, Boulter J, diSibio G, Gossler A, Weinmaster G (1996) Expression patterns of Jagged, Delta1, Notch1, Notch2, and Notch3 genes identify ligand-receptor pairs that may function in neural development. Mol Cell Neurosci 8:14–27.

Lippe WR, Westbrook EW, Ryals BM (1991) Hair cell regeneration in the chicken cochlea following aminoglycoside toxicity. Hear Res 56:203–210.

Lopez I, Honrubia V, Lee SC, Schoeman G, Beykirch K (1997) Quantification of the process of hair cell loss and recovery in the chinchilla crista ampullaris after gentamicin treatment. Int J Dev Neurosci 15:447–461.

Lopez I, Honrubia V, Lee SC, Li G, Beykirch K (1998) Hair cell recovery in the chinchilla crista ampullaris after gentamicin treatment: a quantitative approach. Otolaryngol Head Neck Surg 119:255–262.

Lopez IA, Zhao PM, Yamaguchi M, de Vellis J, Espinosa-Jeffrey A (2004) Stem/progenitor cells in the postnatal inner ear of the GFP-nestin transgenic mouse. Int J Dev Neurosci 22:205–213.

Löwenheim H, Furness DN, Kil J, Zinn C, Gültig K, Fero ML, Frost D, Gummer AW, Roberts JM, Rubel EW, Hackney CM, Zenner HP (1999) Proc Natl Acad Sci USA 96: 4084–4088.

Malgrange B, Belachew S, Thiry M, Nguyen L, Rogister B, Alvarez M, Rigo J-M, Van De Water TR, Moonen G, Lefebvre PP (2002) Proliferative generation of mammalian auditory hair cells in culture. Mech Dev 112:79–88.

Mantela J, Jiang Z, Ylikoski J, Fritzsch B, Zacksenhaus E, Pirvola U (2005) The retinoblastoma gene pathway regulates the postmitotic state of hair cells of the mouse inner ear. Development 132:2377–2388.

Marean GC, Cunningham D, Burt JM, Beecher MD, Rubel EW (1995) Regenerated hair cells in the European starling: are they more resistant to kanamycin ototoxicity than original hair cells? Hear Res 82:267–276.

Marsh RR, Xu LR, Moy JP, Saunders JC (1990) Recovery of the basilar papilla following intense sound exposure in the chick. Hear Res 46:229–237.

Matei V, Pauley S, Kaing S, Rowitch D, Beisel KW, Morris K, Feng F, Jones K, Lee J, Fritzsch B (2005) Smaller inner ear sensory epithelia in Neurog 1 null mice are related to earlier hair cell cycle exit. Dev Dyn 234:633–650.

Matsui JI, Oesterle EC, Stone JS, Rubel EW (2000) Characterization of damage and regeneration in cultured avian utricles. J Assoc Res Otolaryngol 1:46–63.

Matsui JI, Ogilvie JM, Warchol ME (2002) Inhibition of caspases prevents ototoxic and ongoing hair cell death. J Neurosci 22:1218–1227.

Matsunaga T, Davis JG, Greene MI (2001) Adult rat otic placode-derived neurons and sensory epithelium express all four erbB receptors: a role in regulating vestibular ganglion neuron viability. DNA Cell Biol 20:307–319.

McCroskery S, Thomas M, Maxwell L, Sharma M, Kambadur R (2003) Myostatin negatively regulates satellite cell activation and self-renewal. J Cell Biol 162: 1135–1147.

Montcouquiol M, Corwin JT (2001a) Brief treatments with forskolin enhance S-phase entry in balance epithelia from the ears of rats. J Neurosci 21:974–982.

Montcouquiol M, Corwin JT (2001b) Intracellular signals that control cell proliferation in mammalian balance epithelia: key roles for phosphatidylinositol-3 kinase, mammalian target of rapamycin, and S6 kinases in preference to calcium, protein kinase C, and mitogen-activated protein kinase. J Neurosci 21:570–580.

Montcouquiol M, Rachel RA, Lanford PJ, Copeland NG, Jenkins NA, Kelley MW (2003) Identification of Vangl2 and Scrb1 as planar polarity genes in mammals. Nature 423:173–177.

Morest DK, Cotanche DA (2004) Regeneration of the inner ear as a model of neural plasticity. J Neurosci Res 78:455–460.

Morrison A, Hodgetts C, Gossler A, Hrabe de Angelis M, Lewis J (1999) Expression of Delta1 and Serrate1 (Jagged1) in the mouse inner ear. Mech Dev 84:169–172.

Mumm JS, Shou J, Calof AL (1996) Colony-forming progenitors from mouse olfactory epithelium: evidence for feedback regulation of neuron production. Proc Natl Acad Sci USA 93:11167–11172.

Murray RC, Calof AL (1999) Neuronal regeneration: lessons from the olfactory system. Semin Cell Dev Biol 10:421–431.

Nagaraj R, Banerjee U (2004) The little R cell that could. Int J Dev Biol 48:755–760.

Navaratnam DS, Su HS, Scott S, Oberholtzer JC (1996) Proliferation in the auditory receptor epithelium mediated by a cyclic AMP-dependent signaling pathway. Nat Med 2:1136–1139.

Nickel R, Becker D, Forge A (2006) Molecular and functional characterization of gap junctions in the avian inner ear. J Neurosci 26:6190–6199.

Niemiec AJ, Raphael Y, Moody DB (1994) Return of auditory function following structural regeneration after acoustic trauma: behavioral measures from quail. Hear Res 79:1–16.

Oesterle EC, Hume CR (1999) Growth factor regulation of the cell cycle in developing and mature inner ear sensory epithelia. J Neurocytol 28:877–887.

Oesterle EC, Rubel EW (1993) Postnatal production of supporting cells in the chick cochlea. Hear Res 66:213–224.

Oesterle EC, Cunningham DE, Rubel EW (1992) Ultrastructure of hyaline, border, and vacuole cells in chick inner ear. J Comp Neurol 318:64–82.

Oesterle EC, Tsue TT, Reh TA, Rubel EW (1993) Hair-cell regeneration in organ cultures of the postnatal chicken inner ear. Hear Res 70:85–108.

Oesterle EC, Tsue TT, Rubel EW (1997) Induction of cell proliferation in avian inner ear sensory epithelia by insulin-like growth factor-I and insulin. J Comp Neurol 380:262–274.

Oesterle EC, Bhave SA, Coltrera MD (2000) Basic fibroblast growth factor inhibits cell proliferation in cultured avian inner ear sensory epithelia. J Comp Neurol 424:307–326.

Oesterle EC, Cunningham DE, Westrum LE, Rubel EW (2003) Ultrastructural analysis of (^{3}H)thymidine-labeled cells in the rat utricular macula. J Comp Neurol 463:177–195.

Ogata Y, Slepecky NB, Takahashi M (1999) Study of the gerbil utricular macula following treatment with gentamicin, by use of bromodeoxyuridine and calmodulin immunohistochemical labelling. Hear Res 133:53–60.

O'Halloran EK, Oesterle EC (2004) Characterization of leukocyte subtypes in chicken inner ear sensory epithelia. J Comp Neurol 475:340–360.

Paquette AJ, Perez SE, Anderson DJ (2000) Constitutive expression of the neuron-restrictive silencer factor (NRSF)/REST in differentiating neurons disrupts neuronal gene expression and causes axon pathfinding errors in vivo. Proc Natl Acad Sci USA 97:12318–12323.

Parietti C, Vago P, Humbert G, Lenoir M (1998) Attempt at hair cell neodifferentiation in developing and adult amikacin intoxicated rat cochleae. Brain Res 813:57–66.

Parker MA, Cotanche DA (2004) The potential use of stem cells for cochlear repair. Audiol Neurootol 9:72–80.

Pickles JO, van Heumen WRA (1997) The expression of messenger RNAs coding for growth factors, their receptors, and eph-class receptor tyrosine kinases in normal and ototoxically damaged chick cochleae. Developmental Neuroscience 19: 476–487.

Pirvola U, Cao Y, Oellig C, Suoqiang Z, Pettersson RF, Ylikoski J (1995) The site of action of neuronal acidic fibroblast growth factor is the organ of Corti of the rat cochlea. Proc Natl Acad Sci USA 92:9269–9273.

Platt C (1977) Hair cell distribution and orientation in goldfish otolith organs. J Comp Neurol 172:283–287.

Popper AN, Hoxter B (1984) Growth of a fish ear: 1. Quantitative analysis of hair cell and ganglion cell proliferation. Hear Res 15:133–142.

Presson JC, Popper AN (1990) Possible precursors to new hair cells, support cells, and Schwann cells in the ear of a post-embryonic fish. Hear Res 46:9–22.

Presson JC, Smith T, Mentz L (1995) Proliferating hair cell precursors in the ear of a postembryonic fish are replaced after elimination by cytosine arabinoside. J Neurobiol 26:579–584.

Presson JC, Lanford PJ, Popper AN (1996) Hair cell precursors are ultrastructurally indistinguishable from mature support cells in the ear of a postembryonic fish. Hear Res 100:10–20.

Radtke F, Wilson A, MacDonald HR (2004) Notch signaling in T- and B-cell development. Curr Opin Immunol 6:174–179.

Raphael Y (1992) Evidence for supporting cell mitosis in response to acoustic trauma in the avian inner ear. J Neurocytol 21:663–671.

Raz Y, Kelley MW (1997) Effects of retinoid and thyroid receptors during development of the inner ear. Semin Cell Dev Biol 8:257–264.

Raz Y, Kelley MW (1999) Retinoic acid signaling is necessary for the development of the organ of Corti. Dev Biol 213:180–193.

Reddy GV, Rodrigues V (1999) Sibling cell fate in the *Drosophila* adult external sense organ lineage is specified by prospero function, which is regulated by Numb and Notch. Development126:2083–2092.

Richards LJ, Kilpatrick TJ, Bartlett PF (1992) De novo generation of neuronal cells from the adult mouse brain. Proc Natl Acad Sci USA 89:8591–8595.

Riley BB, Chiang M, Farmer L, Heck R (1999) The deltaA gene of zebrafish mediates lateral inhibition of hair cells in the inner ear and is regulated by pax2.1. Development 126:5669–5678.

Rivolta MN, Grix N, Lawlor P, Ashmore JF, Jagger DJ, Holley MC (1998) Auditory hair cell precursors immortalized from the mammalian inner ear. Proc Biol Sci 265(1406):1595–1603.

Rivolta MN, Grix N, Lawlor P, Ashmore JF, Jagger DJ, Holley MC (2003) Auditory hair cell precursors immortalized from the mammalian inner ear. Proc Biol Sci 265:1595–1603.

Roberson DW, Rubel EW (1994) Cell division in the gerbil cochlea after acoustic trauma. Am J Otol 15:28–34.

Roberson DW, Weisleder P, Bohrer PS, Rubel EW (1992) Ongoing production of sensory cells in the vestibular epithelium of the chick. Hear Res 57:166–174.

Roberson DW, Kreig CS, Rubel EW (1996) Light microscopic evidence that direct transdifferentiation gives rise to new hair cells in regenerating avian auditory epithelium. Aud Neurosci 2:195–205.

Roberson DW, Alosi JA, Mercola M, Cotanche DA (2002) REST mRNA expression in normal and regenerating avian auditory epithelium. Res 172:62–172.

Roberson DW, Alosi JA, Cotanche DA (2004) Direct transdifferentiation gives rise to the earliest new hair cells in regenerating avian auditory epithelium. J Neurosci Res 78:461–471.

Romand R, Chardin S, Le Calvez S (1996) The spontaneous appearance of hair cell-like cells in the mammalian cochlea following aminoglycoside ototoxicity. NeuroReport 8:133–137.

Rosenfeld RG, Roberts CT Jr, eds (1999)The IGF System. Totowa, NJ: Humana Press.

Rubel EW, Dew LA, Roberson DW (1995) Mammalian vestibular hair cell regeneration [letter; comment]. Science 267:701–707.

Ruben RJ (1967) Development of the inner ear of the mouse: a radioautographic study of terminal mitoses. Acta Otolaryngol Suppl 220:1–43.

Ruben RJ (1969) The synthesis of DNA and RNA in the developing inner ear. Laryngoscope 79:1546–1556.

Ryals BM, Rubel EW (1988) Hair cell regeneration after acoustic trauma in adult Coturnix quail. Science 240:1774–1776.

Saffer LD, Gu R, Corwin JT (1996) An RT-PCR analysis of mRNA for growth factor receptors in damaged and control sensory epithelia of rat utricles. Hear Res 94:14–23.

Sage C, Huang M, Karimi K, Gutierrez G, Vollrath MA, Zhang D-S, García-Añoveros J, Hinds PW, Corwin JT, Corey DP, Chen Z-Y (2005) Proliferation of functional hair cells in vivo in the absence of the retinoblastoma protein. Science 307:1114–1118.

Schoenherr CJ, Anderson DJ (1995) Silencing is golden: negative regulation in the control of neuronal gene transcription. Curr Opin Neurobiol 5(5): 566–571.

Schwartz Levey M, Chikaraishi DM, Kauer JS (1991) Characterization of potential precursor populations in the mouse olfactory epithelium using immunocytochemistry and autoradiography. J Neurosci 11:3556–3564.

Schwob JE (2002) Neural regeneration and the peripheral olfactory system. Anat Rec 269:33–49.

Severinsen SA, Jørgensen JM, Nyengaard JR (2003) Structure and growth of the utricular macula in the inner ear of the slider turtle Trachemys scripta. J Assoc Res Otolaryngol 4:505–520.

Shailam R, Lanford PJ, Dolinsky CM, Norton CR, Gridley T, Kelley MW (1999) Expression of proneural and neurogenic genes in the embryonic mammalian vestibular system. J Neurocytol 28:809–819.

Shou J, Zheng JL, Gao WQ (2003) Robust generation of new hair cells in the mature mammalian inner ear by adenoviral expression of Hath1. Mol Cell Neurosci 23: 169–179.

Skeath JB, Carroll SB (1992) Regulation of proneural gene expression and cell fate during neuroblast segregation in the Drosophila embryo. Development 114:939–946.

Sliwinska-Kowalska M, Rzadzinska A, Jedlinska U, Rajkowska E (2000) Hair cell regeneration in the chick basilar papilla after exposure to wide-band noise: evidence for ganglion cell involvement. Hear Res 148:197–212.

Sobkowicz HM, Bereman B, Rose JE (1975) Organotypic development of the organ of Corti in culture. J Neurocytol 4:543–572.

Sobkowicz HM, August BK, Slapnick SM (1992) Epithelial repair following mechanical injury of the developing organ of Corti in culture: an electron microscopic and autoradiographic study. Exp Neurol 115:44–49.

Sobkowicz HM, August BK, Slapnick SM (1996) Post-traumatic survival and recovery of the auditory sensory cells in culture. Acta Oto-Laryngologica 116:257–262.

Sobkowicz HM, August BK, Slapnick SM (1997) Cellular interactions as a response to injury in the organ of Corti in culture. Int J Dev Neurosci 15:463–485.

Staecker H, Van de Water TR (1998) Factors controlling hair-cell regeneration/repair in the inner ear. Curr Opin Neurobiol 8:480–487.

Stevens CB, Davies AL, Battista S, Lewis JH, Fekete DM (2003) Forced activation of Wnt signaling alters morphogenesis and sensory organ identity in the chicken inner ear. Dev Biol 261:149–164.

Steyger PS, Burton M, Hawkins JR, Schuff NR, Baird RA (1997) Calbindin and parvalbumin are early markers of non-mitotically regenerating hair cells in the bullfrog vestibular otolith organs. Int J Dev Neurosci 15:417–432.

Stone JS, Cotanche DA (1994) Identification of the timing of S phase and the patterns of cell proliferation during hair cell regeneration in the chick cochlea. J Comp Neurol 341:50–67.

Stone JS, Rubel EW (1999) Delta1 expression during avian hair cell regeneration. Development 126:961–973.

Stone JS, Rubel EW (2000a) Temporal, spatial, and morphologic features of hair cell regeneration in the avian basilar papilla. J Comp Neurol 417:1–16.

Stone JS, Rubel EW (2000b) Cellular studies of auditory hair cell regeneration in birds. Proc Natl Acad Sci USA 97:11714–11721.

Stone JS, Leano SG, Baker LP, Rubel EW (1996) Hair cell differentiation in chick cochlear epithelium after aminoglycoside toxicity: *in vivo* and *in vitro* observations. J Neurosci 16:6157–6174.

Stone JS, Choi YS, Woolley SM, Yamashita H, Rubel EW (1999) Progenitor cell cycling during hair cell regeneration in the vestibular and auditory epithelia of the chick. J Neurocytol 28:863–876.

Stone JS, Shang JL, Tomarev S (2003) Expression of Prox1 defines regions of the avian otocyst that give rise to sensory or neural cells. J Comp Neurol 460:487–502.

Stone JS, Shang JL, Tomarev S (2004) cProx1 immunoreactivity distinguishes progenitor cells and predicts hair cell fate during avian hair cell regeneration. Dev Dyn 230:597–614.

Tanaka K, Smith CA (1978) Structure of the chicken's inner ear: SEM and TEM study. Am J Anat 153:251–271.

Tang LS, Alger HM, Pereira FA (2006) COUP-TFI controls Notch regulation of hair cell and support cell differentiation. Development 133:3683–3693.

Tanyeri H, Lopez I, Honrubia V (1995) Histological evidence for hair cell regeneration after ototoxic cell destruction with local application of gentamicin in the chinchilla crista ampullaris. Hear Res 89:194–202.

Taylor RR, Forge A (2005) Hair cell regeneration in sensory epithelia from the inner ear of a urodele amphibian. J Comp Neurol 484:105–120.

Torres M, Giraldez F (1998) The development of the vertebrate inner ear. Mech Dev 71:5–21.

Tsai H, Hardisty RE, Rhodes C, Kiernan AE, Roby P, Tymowska-Lalanne Z, Mburu P, Rastan S, Hunter AJ, Brown SD, Steel KP (2001) The mouse slalom mutant demonstrates a role for Jagged1 in neuroepithelial patterning in the organ of Corti. Hum Mol Genet10:507–512.

Tsue TT, Watling DL, Weisleder P, Coltrera MD, Rubel EW (1994a) Identification of hair cell progenitors and intermitotic migration of their nuclei in the normal and regenerating avian inner ear. J Neurosci 14:140–152.

Tsue TT, Oesterle EC, Rubel EW (1994b) Diffusible factors regulate hair cell regeneration in the avian inner ear. Proc Natl Acad Sci USA 91:1584–1588.

Turpin JA, Lopez-Berestein G, eds (1993) Differentiation, maturation, and activation of monocytes and macrophages: functional activity is controlled by a continum of maturation. In Lopez-Berestein G, Klostergaard J (eds) Mononuclear Phagocytes in Cell Biology. Ann Arbor: CRC Press, pp. 72–99.

Umemoto M, Sakagam M, Fukazawa K, Ashida K, Kubo T, Senda T, Yoneda Y (1995) Hair cell regeneration in the chick inner ear following acoustic trauma: ultrastructural and immunohistochemical studies. Cell Tissue Res 281:435–443.

Vaessin H, Grell E, Wolff E, Bier E, Jan LY, Jan YN (1991) prospero is expressed in neuronal precursors and encodes a nuclear protein that is involved in the control of axonal outgrowth in *Drosophila*. Cell 67:941–953.

Vago P, Humbert G, Lenoir M (1998) Amikacin intoxication induces apoptosis and cell proliferation in rat organ of Corti. NeuroReport 9:431–436.

Varela-Nieto I, Morales-Garcia JA, Vigil P, Diaz-Casares A, Gorospe I, Sánchez-Galiano, Cañon S, Camarero G, Contreras J, Cediel R, Leon Y (2004) Trophic effects of insulin-like growth factor-I (IGF-1) in the inner ear. Hear Res 196:19–25.

Wagers AJ, Weissman IL (2004) Plasticity of adult stem cells. Cell 116:639–648.

Wang S, Barres BA (2000) Up a notch: instructing gliogenesis. Neuron 27:197–200.

Warchol ME (1995) Supporting cells in isolated sensory epithelia of avian utricles proliferate in serum-free culture. NeuroReport 6:981–984.

Warchol ME (1997) Macrophage activity in organ cultures of the avian cochlea: Demonstration of a resident population and recruitment to sites of hair cell lesions. J Neurobiol 33:724–734.

Warchol ME (1999) Immune cytokines and dexamethasone influence sensory regeneration in the avian vestibular periphery. J Neurocytol 28:889–900.

Warchol ME (2002) Cell density and N-cadherin interactions regulate cell proliferation in the sensory epithelia of the inner ear. J Neurosci 22:2607–2616.

Warchol ME, Corwin JT (1996) Regenerative proliferation in organ cultures of the avian cochlea: identification of the initial progenitors and determination of the latency of the proliferative response. J Neurosci 16:5466–5477.

Warchol ME, Lambert PR, Goldstein BJ, Forge A, Corwin JT (1993) Regenerative proliferation in inner ear sensory epithelia from adult guinea pigs and humans. Science 259:1619–1622.

Watt FM, Hogan BL (2000) Out of Eden: stem cells and their niches. Science 287:1427–1430.

Weisleder P, Rubel EW (1992) Hair cell regeneration in the avian vestibular epithelium. Exp Neurol 115:2–6.

Weisleder P, Rubel EW (1993) Hair cell regeneration after streptomycin toxicity in the avian vestibular epithelium. J Comp Neurol 331:97–110.

Weisleder P, Tsue TT, Rubel EW (1995) Hair cell replacement in avian vestibular epithelium: supporting cell to type I hair cell. Hear Res 82:125–133.

Weissman IL, Anderson DJ, Gage F (2001) Stem and progenitor cells: origins, phenotypes, lineage commitments, and transdifferentiations. Annu Rev Cell Dev Biol 17:387–403.

White PM, Doetzlhofer A, Lee YS, Groves AK, Segil N (2006) Mammalian cochlear supporting cells can divide and trans-differentiate into hair cells. Nature 441:984–987.

Whitehead MC, Morest DK (1985) The development of innervation patterns in the avian cochlea. Neuroscience 14:255–276.

Wigle JT, Chowdhury K, Gruss P, Oliver G (1999) Prox1 function is crucial for mouse lens-fibre elongation. Nat Genet 21:318–322.

Wigle JT, Harvey N, Detmar M, Lagutina I, Grosveld G, Gunn MD, Jackson DG, Oliver G (2002) An essential role for Prox1 in the induction of the lymphatic endothelial cell phenotype. EMBO J 21:1505–1513.

Wilkins HR, Presson JC, Popper AN (1999) Proliferation of vertebrate inner ear supporting cells. J Neurobiol 39:527–535.

Williams JA, Holder N (2000) Cell turnover in neuromasts of zebrafish larvae. Hear Res 143:171–181.

Witte MC, Montcouquiol M, Corwin JT (2001) Regeneration in avian hair cell epithelia: identification of intracellular signals required for S-phase entry. Eur J Neurosci 14:829–838.

Woods C, Montcouquiol M, Kelley MW (2004) Math1 regulates development of the sensory epithelium in the mammalian cochlea. Nat Neurosci 1–9.

Woolley SM, Rubel EW (1999) High-frequency auditory feedback is not required for adult song maintenance in Bengalese finches. J Neurosci 19:358–371.

Wu H, Ivkovic S, Murray RC, Jaramillo S, Lyons KM, Johnson JE, Calof AL (2003) Autoregulation of neurogenesis by GDF-11. Neuron 37:197–207.

Yamamoto N, Tanigaki K, Tsuji M, Yabe D, Ito J, Honjo T (2006) Inhibition of Notch/RBP-J signaling induces hair cell formation in neonate mouse cochleas. J Mol Med 84:37–45.

Yamashita H, Oesterle EC (1995) Induction of cell proliferation in mammalian inner ear sensory epithelia by transforming growth factor-alpha and epidermal growth factor. Proc Natl Acad Sci USA 92: 3152–3155.

Yang Q, Bermingham NA, Finegold MJ, Zoghbi HY (2001) Requirement of Math1 for secretory cell lineage commitment in the mouse intestine. Science 294:2155–2158.

Zhang N, Martin GV, Kelley MW, Gridley T (2000) A mutation in the Lunatic fringe gene suppresses the effects of a Jagged2 mutation on inner hair cell development in the cochlea. Curr Biol 10:659–662.

Zhang M, Ding D-L, Salvi R (2002) Expression of heregulin and erbB/HER receptors in the adult chinchilla cochlear and vestibular sensory epithelium. Hear Res 169:56–68.

Zheng JL, Gao W-Q (1997) Analysis of rat vestibular hair cell development and regeneration using calretinin as an early marker. J Neurosci 17:8270–8282.

Zheng JL, Gao W-Q (2000) Overexpression of Math1 induces robust production of extra hair cells in postnatal rat inner ears. Nat Neurosci 3:580–586.

Zheng JL, Helbig C, Gao W-Q (1997) Induction of cell proliferation by fibroblast and insulin-like growth factors in pure rat inner ear epithelial cell cultures. J Neurosci 17:216–226.

Zheng JL, Frantz G, Lewis AK, Sliwkowski M, Gao W-Q (1999) Heregulin enhances regenerative proliferation in postnatal rat utricular sensory epithelium after ototoxic damage. J Neurocytol 28:901–912.

Zheng JL, Shou J, Guillemot F, Kageyama R, Gao WQ (2000) HES1 is a negative regulator of inner ear hair cell differentiation. Development 127:4551–4560.

Zine A (2003) Molecular mechanisms that regulate auditory hair-cell differentiation in the mammalian cochlea. Mol Neurobiol 27:223–238.

Zine A, de Ribaupierre F (1998) Replacement of mammalian auditory hair cells. Neuro Report 9:263–268.

Zine A, de Ribaupierre F (2002) Notch/Notch ligands and Math1 expression patterns in the organ of Corti of wild-type and Hes1 and Hes5 mutant mice. Hear Res 170:22–31.

Zine A, Van De Water TR, de Ribaupierre F (2000) Notch signaling regulates the pattern of auditory hair cell differentiation in mammals. Development 127:3373–3383.

Zine A, Aubert A, Qiu J, Therianos S, Guillemot F, Kageyama R, de Ribaupierre F (2001) Hes1 and Hes5 activities are required for the normal development of the hair cells in the mammalian inner ear. J Neurosci 21:4712–4720.

6
Protection and Repair of Inner Ear Sensory Cells

ANDREW FORGE AND THOMAS R. VAN DE WATER

1. Introduction

Sensory hair cells of the inner ear detect mechanical stimuli produced by sound (in the cochlea) or by motion and changes in head position (in the vestibular system) and translate them into neural signals. The information is transferred to the appropriate afferent neuronal complexes of the membranous labyrinth and then to the auditory and vestibular tracts of the central nervous system. This translation and transfer of external sensory and mechanical stimuli is the essential function of the inner ear and its sensory ganglia. Loss of function of the inner ear sensory receptor cells—the hair cells—and of their ganglia results in disorders of hearing and balance. The development of effective therapies to prevent the development of these sensory deficits requires a better understanding of the molecular mechanisms that underlie loss of sensory hair cells and their primary neurons, and of protection and repair mechanisms within the inner ear. This chapter reviews the current understanding of how injury to hair cells and neurons in the inner ear leads to their death and of how their survival after exposures to traumatizing agents might be promoted. The efficacy of oto- and neuro- protective strategies to prevent the loss of injured hair cells and neurons from traumatized inner ears are presented and discussed.

1.1 Mechanisms of Cell Death

The concept of preventing the loss of sensory cells and ganglion neurons arises from the recognition that there are several different ways in which the demise of a cell can be brought about, and an increase in the understanding of fundamental biochemical mechanisms that underlie the initiation of these modes of cell death. Such understanding has had widespread implications across biomedicine in relation to cancer, neurodegenerative disease and ontogenetic development. Broadly, three different routes of cell death are currently recognized— necrosis, apoptosis, and autophagy (Debnath et al. 2005)—each of which can be distinguished by characteristic morphological features. Necrosis encompasses what is generally regarded as acute "passive" cell death. It is characterized by loss of integrity of the plasma membrane; cellular swelling, as a consequence of osmotic imbalance resulting from loss of membrane integrity; swelling of the nucleus and other organelles; rupture of the plasma membrane; and spillage of the necrotic cell's contents into the local microenvironment. The exposure of the cell contents to the extracellular environment provokes an inflammatory response. This in itself may cause further localized tissue disruption as inflammatory cells are recruited to the lesion site. Apoptosis (type 1 cell death) and autophagy (type 2) are different manifestations of "programmed cell death," active processes that involve a cascade of biochemical events triggered within a cell such that the affected cell essentially destroys itself (colloquially known as "cell suicide"). During programmed cell death, the integrity of the plasma membrane is largely retained so that, unlike in necrosis, an inflammatory response is not initiated. However, it is becoming increasingly apparent that a clear distinction between these active cell death processes and a passive one is somewhat blurred as a necrotic-like morphology may be the end stage of a programmed cell death pathway, and cells that initiate the program that normally leads to apoptosis may become necrotic (Danial and Korsmeyer 2004).

Of the defined programmed cell death pathways, autophagy ("self-eating") has been recognized only relatively recently as a distinctive process that can lead to cell death, rather than just a mechanism for turnover and recycling of cell components which is meant to protect the cell rather than eliminate it (Debnath et al. 2005; Meijer and Codogno 2006). It involves the enclosure of cellular components within specialized vacuoles and their delivery to lysosomes where they are destroyed (Reggiori and Klionsky 2005). Under normal circumstances, this serves to recycle cellular material and provide energy sources alternative to those obtained by normal nutrition. The double-membraned autophagic vacuoles are the most characteristic morphological feature (Reggiori and Klionsky 2005). There is no evidence as yet that autophagy is a significant contributing factor in the loss of hair cells or neurons from the inner ear, and therefore it is not considered further in this chapter. However, with the recent indications of the contribution of autophagy to some neurodegenerative diseases and in aging (Reggiori and Klionsky 2005; Rubinzstein et al. 2005; Bergamini 2006; Boland and Nixon 2006) it may be that it does play a role in the removal of cells at a site of damage within the inner ear.

Apoptosis is the end stage of a biochemical pathway that produces an orderly destruction of the cell from within (Danial and Korsmeyer 2004; Green and Kroemer 2004). Activation of the pathway either through external signals (the extrinsic pathway) or by intracellular stress (the intrinsic pathway) leads to distinctive morphological features that include condensation of nuclear chromatin accompanied by shrinkage of the nucleus ("pyknotic" nuclei); nuclear fragmentation; cell shrinkage; membrane blebbing; and breakdown of the cell into fragments, so-called "apoptotic bodies." The apoptotic bodies are phagocytosed by "professional" phagocytic cells such as circulating macrophages and/or by neighboring undamaged cells within the tissue (Thornberry and Lazebnik 1998; Savill and Fadok 2000; Monks et al. 2005). In this way, a dying apoptotic cell is removed without compromising tissue integrity or stimulating an inflammatory process. The biochemical cascade within an apoptotic cell involves the orderly destruction of DNA by endonucleases that results in a distinct "ladder" pattern of different size DNA fragments that are evident on an agarose gel (Fraser et al. 1996). The fragmentation of DNA by those endonucleases activated during apoptotic cell death also leads to the formation of single-strand ends at the sites where the DNA double strand has been cleaved, i.e. the "nick ends." These offer multiple sites for binding a nucleotide that can be labeled during preparation of tissue for examination by microscopy, providing a means to mark nuclei in which the orderly destruction of DNA has occurred. Such Terminal deoxynucleotide–UTP-Nick-End Labeling (TUNEL) provides for the objective identification of cells dying by apoptosis. In necrotic cells, although DNA is also broken down, several different cleavage events may occur. Consequently, as the destruction is more random than with apoptosis, the DNA gel pattern of a cell dying by necrosis consists of a smear of DNA fragments without any definitive size order. Further, although some nick ends may be randomly generated during necrosis, they are usually relatively few so that TUNEL staining, if present at all, is quite weak. Thus, shrunken condensed nuclei and positive TUNEL staining distinguish apoptotic cells from cells undergoing necrosis, which are TUNEL negative (or weakly positive) and which have swollen nuclei (see Fig. 6.1). One link between autophagic cell death and apoptosis in that both of these types of programmed cell death can be inhibited by the anti–cell-death action of Bcl-2 (Miller and Girgenrath 2006). In addition, there is now evidence that calpain-mediated cleavage of the autophagy-related gene-5 product (Atg-5) in a cell dying by the process of autophagy can switch the cell over to apoptosis (Yousefi et al. 2006).

1.2 Apoptotic Pathways

Apoptosis occurs naturally in many tissues to control cell numbers and in the processes that shape tissues and organs during their development. Extracellular signaling molecules can induce cell death on binding to their cognate receptors, thereby activating the extrinsic cell death pathway. Apoptosis may also be triggered when cells are damaged to remove such cells without disrupting tissue integrity. When cells are stressed or when DNA is damaged several biochemical pathways that constitute the "intrinsic" cell death program are activated. These cell death pathways are under tight regulation within each cell. Details of the gene expression patterns and the complex series of interactions that promote or act to inhibit the cell death pathways are described in several reviews (Danial and Korsmeyer 2004; Green and Kroemer 2004), and only a brief description of some features salient to the present discussion is presented here.

The "classical" apoptotic pathway is mediated by caspases (cysteine proteases with aspartate specificity), a family of proteases that have a role specifically in programmed cell death. Caspases are normally present as inactive procaspases within a cell. Triggers of cell death lead to activation of caspases

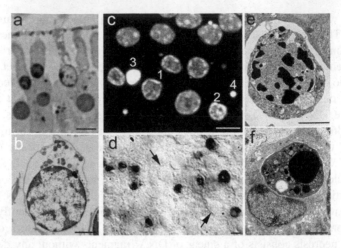

FIGURE 6.1. Morphology of necrosis and apoptosis. (**a, b**) necrosis. (**a**) Toluidine blue–stained section of plastic embedded mouse organ of Corti. Hair cell nucleus is enlarged and less dense in comparison with normal nuclei in neighboring cells. Scale bar $= 10\,\mu m$. (**b**) Transmission electron microscopy (TEM) of thin section of mouse organ of Corti. Outer hair cell with swollen nucleus, disrupted organelles, and loss of cytoplasmic density suggesting ruptured plasma membrane. Scale bar $= 5\,\mu m$. (**c–f**) apoptosis. (**c**) Propidium iodide labeling of nuclei in whole-mount preparation of mouse organ of Corti affected by aminoglycoside (kanamycin). The morphologies of nuclei during progression of apoptosis are revealed: *1*, margination and initial condensation of chromatin, with some shrinkage of nucleus; *2*, more advanced margination and condensation of chromatin; *3*, shrunken nucleus with condensed chromatin; *4*, fragmented nuclei. Scale bar $= 10\,\mu m$. (**d**) TUNEL in whole-mount preparation of guinea pig utricular macula affected by aminoglycoside (gentamicin). The labeled, TUNEL-positive, nuclei have a variety of morphologies similar to those shown in (**c**), and most are shrunken and irregular in shape in comparison with the TUNEL-negative nuclei (arrows) in undamaged, normal cells. Scale bar $= 10\,\mu m$. (**e**) Nucleus in apoptotic hair cell in gentamicin affected utricular macula exhibiting condensation of chromatin. Scale bar $= 5\,\mu m$. (**f**) Apoptotic body of hair cell taken up by and enclosed within a supporting cell. Utricular macula damaged by gentamicin. Scale bar $= 5\,\mu m$.

through a series of regulated interactions that cleave off their pro-domains (see Fig. 6.2a). Activated "initiator" caspases, which include caspases-8, -9 and -10, in combination with other regulators, activate the "effector" caspases. The effector caspases, most notably caspases-3 and/or -7, lead to the activation of proteases that destroy cell proteins and to enzymes that fragment DNA. Different stimulators of apoptosis activate differing sets of caspases and/or different pathways depending on the cellular environment and cell type (Fig. 6.2a). For example, tumor necrosis factor, an extracellular death signaling molecule, activates initiator caspase-8 to directly activate effector caspase-3 (the "extrinsic" pathway). An excess of intracellular free radicals, meanwhile, activate caspase-3 through the activation of initiator caspase-9 (the "intrinsic" pathway)

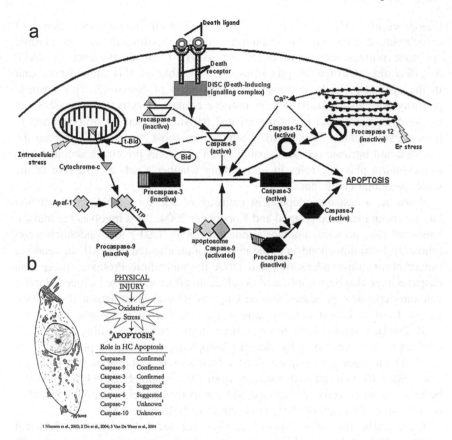

FIGURE 6.2. Caspase-mediated programmed cell death. (a) Diagrammatic representation of major caspase-dependent programmed cell death pathways. On binding of extrinsic cell death factors to the "death receptors," a *d*eath-*i*nducing *s*ignaling *c*omplex" (DISC), composed of several different proteins, is assembled. Different external cell death signals cause the formation of DISCs of differing composition (and some signal/DISC combinations may initiate survival, rather than death, pathways). The assembly of the DISC leads to proteolysis of procaspase-8, releasing initiator capase-8, which catalyzes the proteolysis of procaspase-3 to release activated effector caspase-3. Stress in the endoplasmic reticulum (Er), e.g., from accumulation of damaged or misfolded proteins, leads to release of Ca^{2+} and of procapase-12. Ca^{2+} participates in several cell-death–related events but also in the activation of caspase-12. Caspase-12 activates caspase-3 but also other caspases in the apoptotic cascade. Intracellular stress, in particular accumulation of excessive free radicals, initiates release of cytochrome c (and other proteins) from the mitochondrion. Cytochrome c binds to Apaf1 (apoptosis activating factor-1) to enable Apaf1 to recruit inactive procaspase-9 in the presence of ATP to form the apoptosome complex in which caspase-9 is activated by self-proteolysis and is then able to catalyze the breakdown of procaspase-3 to activated caspase-3. Effector caspase-8 may also act upon cytoplasmic Bid to release truncated (t)-Bid. The activity of t-Bid leads to the release of cytochrome c from mitochondria, and thus the activation of caspase-9 and the ensuing steps in the cell death cascade. Caspase-3, probably in conjunction with activated caspase-9, -7 are the

(Katoh et al. 2004). Activation of caspase-9 itself requires the release of cytochrome c from the mitochondria and then binding to the cytoplasmic apoptotic protease activating factor-one (APAF-1) in the presence of dATP to effect the cleavage of procaspase-9. Assembly of this apoptosome unit in the cytoplasm leads to downstream activation of caspase-3. This intrinsic pathway may also be activated by initiator caspase-8; activated caspase-8 can cleave cytoplasmic Bid to form truncated (t) Bid that acts on mitochondria to release cytochrome c (Fig. 6.2a). There is thus some "cross-talk" between the extrinsic and intrinsic cell death pathways, and various proteins associated with mitochondria play key roles in determining whether a cell will survive or die under particular conditions.

However, a caspase-independent pathway of apoptotic DNA fragmentation has also been identified (Danial and Korsmeyer 2004). This involves the translocation of two proteins, apoptosis-inducing factor (AIF) and endonuclease-G (endo-G), from mitochondria to the nucleus of an affected cell. AIF and endo-G induce chromatin condensation and DNA fragmentation. Proteases other than caspases have also been implicated as cell death effectors. These include cytosolic calcium-dependent proteases termed calpains which, when activated on dissociation from an inhibitor, translocate to the cell membrane (Wang 2000; Goll et al. 2003). Calpains have recently been implicated in switching the process of autophagy to apoptosis by cleaving autophagy-related gene product-5, i.e., atg-5, which is normally required for the formation of autophagosomes (Yousefi et al. 2006). Certain cathepsins that are normally contained inside lysosomes can be released when cells are stressed and act to mediate either apoptotic-like or necrotic-like cell death (Roberg et al. 2002; Jaattela et al. 2003).

The identification of pathways (see Figs. 6.2 and 6.3 and Section 5.1) that lead to cell death has provided potential targets for intervention to rescue cells when one or more of these pathways has been triggered. The identification of which proteins are activated or expressed during cell death allows for the definition and testing of likely effective intervention strategies. Activation of several procaspases has been found to occur during the programmed cell death of hair cells (see Fig. 6.2b).

FIGURE 6.2. activates procaspase-7 to release effector caspase-7. Caspases-3 and major effector caspases. They initiate the cascade of events (involving additional caspases) that result in the controlled proteolysis and DNA fragmentation by which the cell self-destructs. (b) A cartoon depicting the potential role that the different activated caspases play in the apoptosis of an injured hair cell that has been traumatized by an injury (e.g., ototoxicity or sound trauma) that has generated excessive levels of oxidative stress (i.e., production of ROS and RNS) within the hair cells of the inner ear. (Reprinted from Eshraghi and Van De Water. Cochlear implantation trauma and noise induced hearing loss: apoptosis and therapeutic strategies. Anat Rec A Disc Mol Cell Evol Biol 288:473–481, 2006, with permission from John Wiley & Sons, Inc.)

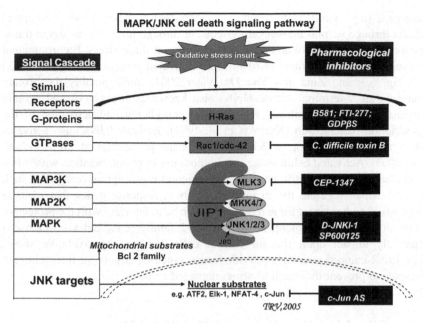

FIGURE 6.3. The MAPK/JNK signaling pathway and the site of action of pharmacological inhibitors. A typical MAPK pathway highlights the similarity in organization shared by these pathways, where an oxidative stress generating insult interacts with a receptor(s) and leads to the activation of small G-proteins (e.g., H-Ras) and GTP-ases (e.g., Cdc-42). G-proteins can be inhibited with Ft-ases (e.g., B-581) and G-proteins with GTPase inhibitors such as the toxin derived from *Clostridium difficile*. The protein kinase cascade that is subsequently activated is composed of up to four tiers of kinase molecules, culminating in activation of a specific MAPK (JNK molecule). In the specific example of the mammalian MAPK/JNK pathway, diversity in signaling is seen with multiple different kinases at the various levels of the pathway. JNKs are the products of three different genes, yielding the protein products: JNK-1, JNK-2, and JNK-3. These protein kinases phosphorylate a variety of target proteins, including nuclear substrates and mitochondrial substrates, and substrates at other locations within the damaged cell. Two small molecule pharmacological inhibitors of this pathway are CEP-1347, which inhibits the signal cascade at the level of MLK3 (i.e., MAPKKK) and D-JNKI-1 and SP 600125, which directly inhibit the signal cascade at the level of the JNKs. The action of the c-Jun, a downstream target of activated JNKs, can be partially blocked by using an antisense oligonucleotide (i.e., *c-jun AS*) to block the amplification and therefore the downstream effects of this transcription factor. (Reprinted from Eshraghi and Van De Water. Cochlear implantation trauma and noise induced hearing loss: apoptosis and therapeutic strategies. Anat Rec A Disc Mol Cell Evol Biol 288:473–481, 2006, with permission from John Wiley & Sons, Inc.)

1.3 Stress-Activated Pathways Leading to Apoptosis

Upstream of programmed cell death pathways are regulated signaling pathways whose activation leads to the initiation of a cell death program. These may be activated through the binding of extrinsic signals at the cell's plasma membrane (e.g., a cell death receptor complex such as TNF-$\tilde{\alpha}$TNF-α receptor) that act through second messenger systems, or through intracellular signal trans- duction pathways that are activated in response to cellular stress. Environmental stresses can activate a number of *m*itogen-*a*ctivated *p*rotein *k*inases (MAPKs) (see Fig. 6.3; and Zine and Van De Water 2004), most notably the protein kinase c-*J*un-*N*-terminal *k*inase (JNK) also known as *s*tress-*a*ctivated *p*rotein *k*inase (SAPK). This pathway leads to phosphorylation and activation of the transcription factor c-Jun (Kyriakis et al. 1994). Activated JNK also activates Elk-1 transcription factor, which upregulates the production of c-Fos (Wang et al. 2003). Activated c-Jun alone as a homodimer or in combination with c-Fos as a heterodimer can contribute to the formation of transcription complexes such as AP-1 which regulate the expression of stress response genes that mediate apoptosis. Another target for activated JNK can be to interfere with the protective action of antiapoptotic members of the Bcl-2 family (e.g., Bcl-xl) which can negatively impact upon the survival of cells damaged by oxidative stress. Such stress-induced pathways provide further potential targets for interventional therapies to prevent the death of stress-damaged cells.

2. Acquired Hearing Loss and Vestibular Dysfunction

Sensorineural deafness or balance dysfunctions are the consequences of hair cell death and accompanying degeneration of afferent innervation. These functional impairments may occur congenitally, as a result of certain deve- lopmental defects or specific genetic conditions, but in the majority of cases hearing loss or balance dysfunction are acquired postnatally. Aging, noise, and certain ototoxic chemicals are the major causes of acquired hearing impairment (Hawkins 1973). Aging and ototoxins also affect the vestibular system. Clini- cally, the most significant ototoxins, and those most widely studied, are the aminoglycoside antibiotics (Forge and Schacht 2000) and the antitumor drug cis-platinum (Rybak and Kelly 2003; Eshraghi et al. 2006a), but several other clinically useful drugs as well as certain organic solvents have been shown to cause hair cell death and hearing loss (Forge and Harpur 2000). Acquired hearing loss is estimated to affect about one in seven people in industri- alized countries, with presbycusis (age-related hearing loss [AHL]) affecting a significantly higher proportion of the elderly population; about 50% of those older than 60 years of age (Royal National Institute for Deaf People website; www.rnid.org.uk/information resources/fact sheets/deaf_awareness/fact sheets_leaflets/deaf_and_hard_of_hearing_people.htm). Vestibular dysfunctions also occur with age and are considered a major underlying factor in falls in the elderly population. Excessive noise has been reported to be at least partially the cause in more than one third of those with hearing impairment in developed countries and represents the biggest occupational hazard for which compensation can be obtained by affected individuals (World Health Organi- zation {WHO} website; www.who.int/whr/2004/en/report04.pdf). The incidence of noise-induced hearing loss (NIHL) is increasing rapidly in developing nations. Aminoglycoside antibiotics are effective antibacterial agents that are inexpensive because they are no longer protected by patent, and although their use in industrialized countries is restricted because of their potential side effects, in developing countries they are often the drugs of first choice for minor infections. Aminoglycosides are also an effective therapy both for tuberculosis and for patients suffering from cystic fibrosis. Consequently, aminoglycoside-induced deafness and balance disorders remain a significant health care problem. Likewise, cis-platinum is an effective antitumor agent, especially for a variety of tumors in children. The potential ototoxic side effects are a major consideration for quality of life of a patient following the use of these ototoxic drugs and can limit their use (Eshraghi et al. 2006a).

In mammals, the functional deficits resulting from hair cell losses caused by these agents are permanent, because unlike the sensory epithelia in the inner ear in birds and other nonmammalian vertebrates (Staecker and Van De Water 1998; Stone et al. 1998; Taylor and Forge 2005), the organ of Corti does not spontaneously regenerate hair cells to replace those lost. In addition, the mammalian vestibular organs possess only a limited capacity to regenerate hair cells (Lopez et al. 1997; Forge et al. 1998; Kopke et al. 2001b). Alternative potential strategies to reduce the incidence of acquired hearing and balance dysfunction or ameliorate their effects on a patient's quality of life are thus to protect hair cells from death and/or to maintain the inner ear's neuronal elements. The latter strategy would enable the most effective use of cochlear implants, especially if bimodal stimulation of hearing via a hybrid device combining both acoustic and electrical stimulation is a desired outcome (Staecker et al. 1996; Van De Water et al. 1996; Marzella and Clark 1999; Shinohara et al. 2002; Gantz and Turner 2004; Eshraghi 2006).

2.1 Patterns of Sensory Cell Damage

In the cochlea, with aging, exposure to ototoxins, and noise, the extent of outer hair cell (OHC) loss is much greater than that of inner hair cells (IHCs) and studies following the progression of hair cell loss over time have shown that after most insults the OHCs die before the IHCs. With aging and exposure to ototoxins, it is generally found that IHCs are absent only in regions where all of the OHCs are lost, and where IHCs are absent, afferent innervation is found to have degenerated. This is manifested as a decrease in the number of neuronal cell bodies within the spiral ganglion. Thus, although spiral ganglion cells as well as hair cells have been reported to be primary targets of some ototoxins such as cis-platinum (van Ruijven et al. 2004), and primary effects on IHCs are seen with carboplatin in chinchillas (Ding et al. 1999), the general pattern of damage within the cochlea is one in which OHCs, then IHCs, and then afferent nerves are successively involved and depleted.

In addition, the effects of aminoglycosides and cis-platinum as well as those of aging are first apparent in the cochlea's basal turn, with damage spreading toward the apex with age and time postexposure (Wright et al. 1987; Hequembourg and Liberman 2001). This pattern is also observed with time and/or increasing dose post-drug exposure (Hawkins 1973; Laurell and Bagger-Sjöbäck 1991; Cardinaal et al. 2000; Forge and Schacht 2000; Minami et al. 2004). This corresponds functionally to hearing impairment that initially affects the high frequencies and then spreads progressively to involve successively lower frequencies. After acoustic overexposure, the loss of hair cells occurs at a location along the organ of Corti that is related to the frequency(ies) of the damaging sound(s) (Saunders et al. 1985; Slepecky 1986; Yoshida et al. 2000), but it also occurs in the basal turn with the extent of this hair cell loss increasing and spreading toward the apex of the cochlea with time postexposure and with increases in sound intensity (Yoshida et al. 2000). These effects in the cochlea's basal coil after acoustic trauma are considered to reflect "metabolic" stresses, including oxidative stress, on hair cells rather than the mechanical damage that occurs at the frequency place (Saunders et al. 1985; Quaranta et al. 1998; Yoshida et al. 2000). Thus, there is a differential vulnerability to metabolic challenges between the base and apex of the cochlea; damage is initiated at the basal end of the cochlea and spreads apically to produce a base to apex gradient of loss among the affected sensory cells.

Differential patterns of damage are also evident in the vestibular sensory epithelia. Type I vestibular hair cells are reported to be more susceptible to aminoglycoside-induced injury than are the type II hair cells (Wersall 1995). Also, damage is first apparent at the crest of the cristae and in the striolar regions of the utricular and saccular maculae, then spreads down the skirts of the cristae and toward the peripheries of the maculae both during aging (Wright 1983; Nakayama et al. 1994), and with time and/or increasing dose post-ototoxin exposure (Lindeman 1969; Forge et al. 1998). In addition, during aging (Wright 1983; Nakayama et al. 1994) or in response to an ototoxic drug (Lindeman 1969; Forge et al. 1998), a larger proportion of the hair cells are lost in the cristae than in the utricular macula which shows proportionally greater hair cell loss than has been observed in the saccular macula. The macula of the saccule may in fact show no obvious hair cell loss when cristae show extensive hair cell losses. As has been demonstrated in the organ of Corti (Staecker et al. 1996), loss of hair cells is followed by loss of neuronal cell bodies from the vestibular (Scarpa's) ganglion (Wang W et al. 2004).

2.2 Repair of Sensory Epithelia by Supporting Cells After Hair Cell Injury

With the exception of very high intensity noise that causes mechanical disruption of the organ of Corti (Henderson and Hamernik 1986), agents that cause hair cell death do not appear to damage the supporting cells. In both the organ of Corti and the vestibular sensory epithelia hair cell loss is accompanied by a repair of the sensory epithelium that is accomplished by expansion of the surrounding supporting cells into the sites that were occupied by the dying hair cells to close the lesion (Forge 1985; Raphael and Altschuler 1991; Meiteles and Raphael 1994; Li et al. 1995). This results in the formation of a characteristic "scar" at the apical surface of the epithelium in the site of a lost hair cell. The scar's morphology is created by newly formed intercellular junctions between the supporting cells within the lesion site. The initial lesion repair by the supporting cells appears to proceed concomitantly with hair cell loss in such a way that there is no breach of the permeability barriers at the apical surface of the epithelium (McDowell et al. 1989; Raphael 2002). The maintenance of these barriers during "wound-healing" prevents the entry of K^+ ions from endolymph into the body of the epithelium, thereby protecting the undamaged regions of the sensory epithelium from the potentially toxic effects of high extracellular K^+ levels and assist in the maintenance of tissue integrity. This is of relevance not only to strategies aimed at inducing hair cell regeneration but also to attempts to prevent hair cell death as it means that if hair cell death is prevented after a challenge by damaging agents it would be expected that the sensory epithelium will survive intact. In the organ of Corti, when all hair cells are lost, further changes among supporting cells and a cellular reorganization occurs such that ultimately the sensory epithelium of the recognizable organ of Corti may be replaced by an unspecialized, simple epithelium that covers the basilar membrane.

2.3 Differential Susceptibility of Cells Within the Sensory Epithelia

The base to apex gradient of hair cell loss that is characteristic of the cochlea exposed to a variety of damaging agents in vivo is retained in organotypic explants of the neonatal organ of Corti from mice or rats exposed directly to aminoglycosides (Richardson and Russell 1991; Kotecha and Richardson 1994) or cis-platinum (Zheng and Gao 1996; Kopke et al. 1997). Only after exposure to the highest concentrations of an ototoxin does hair cell death extend into the most apical regions of the organ of Corti explants. Likewise, in organ cultures of utricular and saccular maculae exposed to aminoglycosides hair cell loss is initiated in the striolar regions, spreading over a wider area toward the peripheries with increasing drug dosage. Moreover, utricular maculae show more hair cell loss than saccular maculae under identical conditions of ototoxin exposure (Li and Forge 1995). Also, in preparations of the isolated organ of Corti maintained in short-term culture the OHCs in the basal coil die more quickly than do those in the apical coil and IHCs survive longer than OHCs, which can remain viable after all of the OHCs in their vicinity have died (Sha et al. 2001a). Supporting cells in these ototoxin-exposed explants survive for prolonged periods. These in vitro observations suggest that differential hair cell losses observed in vivo are the result of inherent characteristics of different hair cells types and of a variation in the resistance to damage by hair cells in different locations within the cochlea and also within the vestibular receptors. This leads to the generalizations that OHCs in the basal coil are inherently more sensitive to damage than are those at the apical coil and IHCs are more resistant to damage than OHCs.

One factor now recognized as underlying the differential vulnerability of OHCs to damage and loss is their susceptibility to free-radical attack. Free radicals, most commonly singlet oxygen (superoxide), nitrogen, and hydroxyl radicals, are redox active molecules that contain an unpaired electron. They are highly unstable and regain stability by losing the unpaired electron to other molecules within a cell that become oxidized. This redox activity is potentially highly disruptive to cellular components, leading to lethal damage, and excessive free-radical production can trigger apoptosis (Evans and Halliwell 1999; Katoh et al. 2004). Although free radicals are produced during normal cellular metabolism, their effects are usually contained by the cell's endogenous free-radical scavenger systems that inactivate them. Glutathione is the most ubiquitous free-radical scavenger in most cells along with the enzyme superoxide dismutase, which catalyzes the inactivation of reactive oxygen species (ROS). Cell damage occurs when free-radical production exceeds the capacity of these endogenous scavenger systems to neutralize them (Evans and Halliwell 1999). In isolated organ of Corti preparations maintained in short-term culture the differential vulnerability of basal coil OHCs relative to those from the apical coil can be overcome by incubating the preparations in the presence of free-radical scavengers (Sha et al. 2001a). This suggests that OHCs are differentially vulnerable to free-radical activity. There is evidence that aminoglycosides may be toxic through a mechanism that generates free radicals (Forge and Schacht 2000), and cis-platinum may inhibit endogenous free-radical scavenging systems and/or cause generation of free radicals themselves (Ravi et al. 1995; Kopke et al. 1997; Minami et al. 2004; Eshraghi et al. 2006a). Excessive noise may also cause free-radical generation from the high levels of oxidative metabolism concomitant with overstimulation of the hair cells (Henderson et al. 1999; Ohlemiller et al. 1999; Yamashita et al. 2004). Accumulation of free radicals is also thought to be associated with the cellular and tissue deterioration that is associated with aging (Evans and Halliwell 1999). This has led to the possibility of preventing OHC loss through preventing the formation of an excess of free radicals by using antioxidants to scavenge them and/or enhancing endogenous scavenger systems. These approaches are discussed in greater detail in Section 5 of this chapter.

The lesser vulnerability of IHCs that is commonly observed in vivo in response to different damaging agents as well as in vitro in isolated organ of Corti preparations suggests that IHCs possess protective mechanisms that render them less susceptible to metabolic insults. There is currently no evidence for this, but identification and characterization of an endogenous survival factor(s) could potentially provide therapies to prevent hair cell loss. However, because the death of IHCs appears to follow the loss of OHCs, the death of an IHC may be a secondary event and the key to their survival would then be the preservation of OHCs.

An alternative or possibly additional explanation proposed to account for the differential vulnerability of hair cells is differences in their physiology, in particular the probability of the open state of the transduction channels at the tips of the stereocilia. OHCs in the basal coil of the cochlea have an open probability of 50% but the open-state probability decreases along the cochlear spiral toward the apex (Russell and Kossl 1992). IHC transduction channels have a 5% probability of being in the open state. It has been suggested that aminoglycosides may enter hair cells through the transduction channels and thus the basal coil OHCs would take up these drugs faster and to a greater extent than do OHCs in the apical coil or than the IHCs (Gale et al. 2001). Further, the resting intracellular potential of an OHC (-70 mV) is much greater than that of an IHC (-40 mV). This fact, taken together with the difference in the probability of transduction channels being open, means that the K^+ current load through basal coil OHCs is greater than that through apical coil OHCs and significantly greater than that through the IHCs. These differences may make basal coil OHCs more vulnerable to metabolic stresses and to disturbances of the homeostasis that maintains fluid composition around the hair cell body. It is of note that a number of different genetic mutations that affect proteins concerned with maintenance of cochlear homeostasis and fluid composition cause progressive hearing impairment and hair cell loss in patterns that resemble those described to occur in acquired hearing loss (Steel and Kros 2001; Boettger et al. 2002; Cohen-Salmon et al. 2002; Rozengurt et al. 2003; Teubner et al. 2003).

Loss of auditory neurons is usually delayed relative to hair cell loss and rarely observed if the IHCs are healthy and intact. This neuronal loss may in many cases be a secondary effect that is consequent to and dependent on the loss of the auditory hair cells. During ontogenetic development, the afferent innervation to the sensory epithelia of the inner ear is dependent on the expression of hair cell (target)–derived growth factors (Van De Water 1988; Rubel and Fritzsch 2002). Neurotrophin-3 (NT-3) and brain-derived neurotrophic factor (BDNF) have been identified as being important to the homeostasis and survival of afferent neurons and the innervation in the inner ear sensory receptors (Ernfors et al. 1995; Fritzsch et al. 2004) (as discussed in Section 6.1). Transgenic animals bearing deletions in genes for one or other or both of these neurotrophin proteins produce hair cells, but neuronal survival is disrupted and afferent innervation of the cochlea in double-knockout animals is completely lacking (Ernfors et al. 1995; Rubel and Fritzsch 2002; Fritzsch et al. 2004). The relationship between NT-3 and BDNF expression patterns within the cochlea and neural innervation of the cochlea is more complex than originally thought, with NT-3 thought to be most important for the survival of basal turn neurons and BDNF for the support of the cochlea's more apical neurons (see Fritzsch et al. 2004). Therefore the absence of neurotrophic factor support that is supplied from the auditory hair cells may be one reason for the degeneration of innervation and spiral ganglion neurons after a loss of auditory hair cells (Ernfors et al. 1996; Staecker et al. 1996; Miller et al. 1997; Shinohara et al. 2002). The loss of an excitatory input to the afferent neurons after the death of the hair cells can also be a cause of neuronal degeneration within the spiral ganglion. Excitatory input is known to be necessary for the maintenance of nerves throughout the nervous system. Thus, the preservation of healthy IHCs would prevent neuronal degeneration within the spiral ganglion and potentially treatment with neurotrophins (e.g., BDNF) and/or provision for the continued stimulation of these neurons after the loss of the hair cells could both act to preserve these neurons (Ernfors et al. 1996; Staecker et al. 1996; Shinohara et al. 2002; Fritzsch et al. 2004).

It has also been suggested that some neuronal losses may be the result of localized excitotoxicity (Pujol and Puel 1999; Rebillard et al. 2003). Excessive release onto afferent terminals of excitatory neurotransmitters, especially glutamate, which is the neurotransmitter at the IHC synapse, can be toxic to the auditory neurons. Binding of glutamate to ionotropic N-methyl-D-aspartate (NMDA) receptors opens ion channels producing a depolarization of the afferent terminal. Excessive glutamate release may therefore cause an excessive number of ion channels to be open, the consequent sustained depolarization creating ionic and osmotic disturbances that lead to swelling and degeneration of the dendrites. Eventually, if neurotransmitter release is sustained at an excessive level it can result in the death of the auditory neurons. Swelling of afferent terminals synapsing with IHCs is seen with acoustic overstimulation and anoxia, and glutamate may be released from damaged or dying hair cells (Pujol and Puel 1999). Swelling of afferent nerve terminals has also been noted in the vestibular system during aminoglycoside-induced hair cell death (Li et al. 1995). There is now evidence that the initial swelling of the nerve terminals may be reversible, and pharmaceutical therapies that will prevent such excitotoxicity damage are being developed (Pujol and Puel 1999; Guitton et al. 2004). In addition, glutamate binds to a cysteine/glutamate antiporter that couples removal of glutamate from the cell to the uptake of cysteine. The maintenance of glutathione in the cell requires the uptake of cysteine because this amino acid is used in the synthesis of glutathione. High levels of extracellular glutamate inhibit this antiporter and sustained inhibition results in a depletion of cellular glutathione. Neurons depleted of glutathione then become sensitive to oxidative stress, leading to cell death initiated by damage generated from an excessive level of oxidative stress (Schubert and Piasecki 2001; Tan et al. 2001).

3. Genes Involved in Hair Cell Maintenance and Survival

The susceptibility of hair cells to damage from environmental factors is influenced by an individual's genetic background. More than 90 different genes that affect inner ear development and/or function have been identified. Mutations in many of these genes can lead to congenital deafness by affecting hair cell differentiation and/or function. There is also the possibility that particular alleles of these genes not directly linked to congenital deafness may contribute to the susceptibility of the inner ear to a delayed onset of hair cell death. It is beyond the scope and intention of this chapter to detail all the genes that have been identified in human studies or in mouse models to cause deafness. The reader can gain access to them elsewhere in the literature (Steel and Kros 2001; Bitner-Glindzicz 2002; Goldfarb and

Avraham 2002; Sabag and Avraham 2005; Petit 2006) and/or via the Internet (e.g., http://webhost.ua.ac.be/hhh/ or www.hpcgg.org/lmm).

However, there are alleles of certain genes that may predispose an individual to an acquired hearing loss or a balance dysfunction. Some genes that are expressed in hair cells play a role in their maintenance. Hair cell survival may also depend on cochlear homeostasis; loss of hair cells appears to be a corollary of mutations in several different genes whose products have a primary role in maintaining the ionic environment of the cochlea (Steel and Kros 2001; Bitner-Glindzicz 2002; Boettger et al. 2002; Cohen-Salmon et al. 2002; Rozengurt et al. 2003; Teubner et al. 2003). A genetic predisposition to hair cell loss will influence the effectiveness of treatments to preserve hair cells, and therefore the characterization of such genes and/or of other genes or gene products with which they interact may provide insights needed for the development of therapeutic intervention(s).

Certain inbred mouse strains show a predisposition to a progressive loss of hearing that has characteristics resembling those of age-related hearing loss (AHL) in humans. Initially AHL characteristics were identified in the C57BL/6J strain of mice (Henry and Chole 1980; Li and Borg 1991; Johnson et al. 1997) which begin to show hearing loss unusually early, i.e., at about 3 months of age. Sensitivity to sounds of high frequency is affected first with progressive involvement of lower frequencies as the mice advance in age. This corresponds to a progressive base to apex loss of hair cells but with more extensive loss of OHCs than IHCs (Spongr et al. 1997; Hequembourg and Liberman 2001). Genetic analysis of strains of mice that exhibit AHL has revealed recessive alleles of two different genes, *Ahl1* and *Ahl2*, which are associated with predisposition to an early onset, progressive type of deafness (Johnson et al. 1997, 2000; Johnson and Zheng 2002). Some mouse strains carrying *Ahl* alleles have been reported also to be more susceptible to noise-induced hearing loss (NIHL) than strains of normal mice (Davis et al. 2001), but this does not seem to be a universal finding. One example of this is the 129/SvEv strain of mice that carries an *Ahl* gene but is particularly insensitive to NIHL (Yoshida et al. 2000). The expression of an *Ahl* gene also does not predispose a strain of mouse to aminoglycoside-induced hair cell loss (Wu et al. 2001). The gene products of *Ahl* genes are not known, but a notable and consistent histological feature of the cochlea in mice with *Ahl*-related hearing impairment is a loss of spiral ligament fibrocytes. In particular, there is a loss type IV fibrocytes in the "basilar crest" region of the spiral ligament where the basilar membrane is anchored to the lateral wall of the cochlea (Hequembourg and Liberman 2001). The possibility is that this gene affects cochlear homeostasis through its effects on the spiral ligament and that the hair cell loss in the cochlea of these animals is a secondary consequence of the damage to cells within the spiral ligament.

A predisposition to aminoglycoside-induced hearing loss has been linked to mutations in a mitochondrial gene. Maternally inherited hypersensitivity to aminoglycoside-induced deafness has been found to be associated with development of hearing loss after exposure to aminoglycoside at doses that do not normally produce ototoxic damage to hearing (Prezant et al. 1993). Affected individuals carry a mutation in the mitochondrial gene for 12S ribosomal RNA, a substitution of guanine for adenosine at position 1555 in the mitochondrial DNA (i.e., the *A1555G* mutation) (Fischel-Ghodsian et al. 1993; Hutchin et al. 1993). Although identified through its inheritance pattern, the mutation has also been found in individuals with no familial history of deafness who have shown hypersensitivity to aminoglycoside antibiotics and thus it may also account for sporadic cases of unexpected aminoglycoside ototoxicity. It has been estimated that between 17% and 33% of patients with aminoglycoside-induced hearing loss may be carriers of this *A1555G* mutation (Fischel-Ghodsian 1999). A second mutation in the same gene, a thymidine (T) deletion at nucleotide 961 with insertion of a varying number of cytosines (*T961Cn*), has subsequently been identified in a family with inherited susceptibility to aminoglycoside ototoxicity (Casano et al. 1999). Some biochemical studies have suggested that the *A1555G* mutation allows interaction of an aminoglycoside with ribosomal RNA, thereby reducing the level of translation of the mitochondrial proteins (Guan et al. 2000). However, the mutation is also associated with nonsyndromic deafness without aminoglycoside intervention (Prezant et al. 1993; Bitner-Glindzicz 2002), and although the mutation would be present in all mitochondria in all tissues, it does not appear to enhance the sensitivity of the hair cells of the vestibular system to aminoglycoside-induced damage. This suggests that this mutation may have quite subtle effects on the auditory receptor that are not yet fully understood.

Two transcription factors, *Brn3c* (Erkman et al. 1996; Xiang et al. 1997, 1998) and *Barhl1* (Li G et al. 2002), have been implicated as necessary for the maintenance of mature hair cells and their continued survival. Both these proteins are expressed in the inner ear exclusively by the hair cells. *Brn3c* and *Barhl1* gene expression first appear after the terminal mitosis of as yet undifferentiated hair cell precursors in the presumptive sensory patches and continues throughout adulthood in both cochlear and vestibular hair cells. A mutation in the human *Brn3c* gene (also known as *POU4f3,* formerly *Brn3.1*), i.e., an 8 base pair deletion, has been identified as the cause of an autosomal dominant, nonsyndromic deafness in humans (Vahava et al. 1998). This deafness is late in onset, beginning at about 18 years of age and subsequently progressing. Mice homozygous for targeted deletion of the *Brn3c* gene are completely deaf and also show balance dysfunction. In *Brn3c* knockout mice, hair cells are initially generated, differentiate, and then die postnatally. Afferent neurons also die in these *Brn3c* mutant mice but the loss of neurons and innervation is delayed relative to hair cell death and thus neuronal loss is considered to be secondary to the loss of hair cells, most probably due to a lack of trophic support (Xiang et al. 1998). No human hearing impairment has yet been associated with *Barhl1*, but targeted disruption of this gene in mice leads to progressive hair cell death and corresponding hearing loss that is first apparent postnatally. The OHCs in the apical coil of *Barhl1*$^{-/-}$ null animals are affected first, with the death of these sensory cells spreading successively to more basal regions of the cochlea (Li S et al. 2002). IHC death in *Barhl1*$^{-/-}$ mice is initiated only after all OHCs have died, and loss of IHCs begins in the basal coils and then progresses to the apical turn. Whether this IHC death is a secondary effect after loss of OHCs or a direct consequence of the absence of the Barhl1 protein is not known. Exactly how these transcription factors are involved in hair cell maintenance is also currently not known because the downstream genes that they regulate have not yet been identified. Nevertheless, *Brn3c* and *Barhl1* are prime candidates for genes that can predispose the cochlea to late-onset deafness, and identification of the downstream genes they regulate could potentially provide therapeutic targets for the treatment of some forms of progressive deafness.

Progressive loss of hair cells has also been noted in association with mutations in genes whose products are involved in maintaining the ionic environment in the cochlea. The K–Cl cotransporter Kcc4 is localized in the cochlea to the supporting cells of the organ of Corti (Boettger et al. 2002) and is thought to be necessary for the uptake of K$^+$ to maintain a low K$^+$ ion concentration around the cell bodies of the hair cells. Disruption of this gene in mice leads to a rapidly progressive loss of hair cells, initiated first among the OHCs within the basal coil (Boettger et al. 2002). Likewise a progressive degeneration of hair cells accompanies disruption of the gene for the K$^+$-channel protein KCNJ10 (*Kir4.1*), which is present in the cochlea within the intermediate cells of the stria vascularis and is required for the generation of the endocochlear potential (EP; Marcus et al. 2002; Rozengurt et al. 2003). Deafness is also associated with mutations in the genes for at least three different connexins, the proteins that form gap junction channels. Gap junctions are present between supporting cells in the organ of Corti and vestibular sensory epithelia and between fibro-cytes within the spiral ligament; they are also associated with basal cells of the stria vascularis (Kikuchi et al. 1994, 2000; Forge et al. 2003). The pathways of intercellular communication created by these gap junctions are thought to be involved in recycling of K$^+$ from the organ of Corti via supporting cells back to endolymph and in maintaining the EP (Wangemann 2002). Connexin (cx) 26 and cx30 are localized in gap junctions in the organ of Corti and vestibular sensory epithelia, the basal cell region of the stria vascularis, and the fibro-cytes of the spiral ligament that are below the stria (Lautermann et al. 1998; Kikuchi et al. 2000; Forge et al. 2003). Cx31 localizes to type 2 fibrocytes below the spiral prominence and in the region of the ligament where Reissner's membrane joins this complex (Xia et al. 2000). Mutations in the gene that encodes cx26 are the most common cause of hereditary deafness, and it is estimated that approximately 4% of the general population are carriers of *cx26* mutations (Bitner-Glindzicz 2002). Although the deafness associated with a *cx26* mutation is often congenital and profound, late-onset progressive deafness has also been observed in some cases with high-frequency hearing preferentially affected (Bitner-Glindzicz 2002). In transgenic mice with a targeted knockout of *cx26* from the supporting cells of the cochlea, there is a progressive death of the hair cells (Cohen-Salmon et al. 2002). Likewise, a knockout of the *cx30* gene causes a high–mid-frequency pattern of hearing loss (Grifa et al. 1999) and progressive hair cell death as well as an inhibition of the EP (Teubner et al. 2003). In humans, mutations in the gene that encodes cx31 cause a late-onset progressive pattern of deafness that affects the high frequencies first (Xia et al. 1998). The observation that hair cell death results from defects in genes whose products are essential for maintenance of cochlear homeostasis indicates the high degree of sensitivity of hair cells, particularly of OHCs, to their local ionic environment. The fact that high frequencies are preferentially affected may be a reflection of the previously discussed inherent differential sensitivity of auditory hair cells along the length of the organ of Corti in a base to apex pattern of sensitivity.

Hair cell survival may also be influenced by the levels of expression of certain genes that encode proteins involved in the inactivation of free radicals. Superoxide dismutases (SODs) are among the first lines of cellular defence against free radical damage. They catalyze the conversion of the superoxide radical to hydrogen peroxide, which is then broken down by peroxide-removing enzymes. Two major forms of SOD are expressed by almost all animal cells: one, Cu/Zn SOD (SOD1), is localized in the cytoplasm; the other, MnSOD (SOD2), is localized specifically to mitochondria. In mice, manipulation of gene expression has implicated the gene that encodes SOD1 in protecting cochlear hair cells from the effects of noise trauma, aminoglycoside ototoxicity, and aging (although whether it plays a role in survival of vestibular hair cells has not been determined). Ablation of *SOD1* in mice (*SOD1*-null mutants or *SOD1*-knockout [KO] animals) results in increased susceptibility to noise-induced hearing loss (Ohlemiller et al. 1999), and to both a faster and a greater extent of auditory hair cell loss with aging in comparison with wild-type littermates (McFadden et al. 1999a, b; Keithley et al. 2005). Conversely, in transgenic mice in which *SOD1* is overexpressed, hair cells are protected from aminoglycoside-induced hair cell death (Sha et al. 2001). However, these transgenic mice that overexpress *SOD1* do not show any increases in the survival of auditory hair cells with aging (Coling et al. 2003; Keithley et al. 2005). Nevertheless, the consequences on the extent of hair cell survival under stress resulting from manipulation of *SOD1* expression provide further support for the hypothesis that excessive free radical production is a major factor involved in initiating the hair cell loss and hearing impairment that results from noise trauma, ototoxic drug challenges, and aging. These results have also led to investigation of possible means for therapeutic intervention to prevent hair cell loss, either through the use of drugs that mimic the effects of SOD (McFadden et al. 2003), or by transfer of genes that encode proteins involved in free radical scavenging (Kawamoto et al. 2004). However, it should be noted that as yet there is no evidence that susceptibility to acquired hearing loss in humans is linked to mutations or variations in genes involved in protection from free radical damage. One analysis of genetic variability within several different human "oxidative stress"–related genes failed to identify any correlation with variability in susceptibility of individuals to noise-induced hearing loss (Carlsson et al. 2005).

4. Self-Repair of Hair Cells

Hair cells appear to possess a limited potential to repair themselves after a sublethal injury. Explants of the immature mouse organ of Corti damaged by scratching with a glass micropipette or with a laser microbeam (Sobkowicz et al. 1992, 1996, 1997), and explants of early postnatal rat utricles (Zheng et al. 1999) and bullfrog saccules (Baird et al. 2000; Gale et al. 2002) exposed to an aminoglycoside have all shown that hair cells can respond to these injuries by losing their apical structures (the cuticular plate and stereociliary bundle) and can then repair and regrow these structures. In some cases, notably the aminoglycoside-exposed rat utricles, only the stereociliary bundles were lost; the hair cells retained their cuticular plates and were maintained at the apical surface of the epithelium. Stereocilia then were observed to reform on these injured hair cells within seven days after the initial trauma. In the bull frog saccules (Gale et al. 2002) and in damaged, immature organ of Corti explants (Sobkowicz et al. 1997), hair cells not only lost their stereociliary bundles and often their cuticular plates but they also were observed to withdraw from contact with the apical surface of the sensory epithelium and to become entirely enclosed within this epithelium, surrounded completely by their adjacent supporting cells which had closed the space at the apical surface of the sensory epithelium previously occupied by the cuticular plate and hair bundle of the injured cell. The bodies of these hair cells remained within the sensory epithelium for some time in what appeared to be a partially dedifferentiated state (Baird et al. 2000; Gale et al. 2002). These injured hair cells then reemerged at the apical surface of the sensory epithelium and reformed cuticular plates and hair bundles by 4–7 days after the initial injury (Gale et al. 2002). Hair cells that are lethally injured in the mammalian utricle and the mature organ of Corti die within the sensory epithelium. The cell ruptures just below the cuticular plate and the surrounding adjacent supporting cells enclose the fragments of the cell body whose nucleus shows morphological signs of apoptosis (Forge 1985; Li et al. 1995; Forge and Li 2000). Thus, the loss of the apical structures of an injured hair cell and withdrawal of its cell body into the sensory epithelium may be a common response of mammalian hair cells whether they are lethally or sublethally damaged (although in birds injured hair cells are extruded in their entirety from the apical surface of the sensory epithelium [Cotanche et al. 1987; Cotanche and Dopyera 1990; Janas et al. 1995]). The cellular and biochemical mechanisms underlying the self-repair of injured hair cells are at present not fully understood. Further, it is not clear whether such self-repair of hair cells after a sublethal injury in explant cultures occurs in the mammalian inner ear in vivo. Immature hair bundles that indicate regrowth and repair have not been observed in the mammalian organ of Corti in regions where sublethal damage might be expected, such as at the leading edge of a progressive hair cell loss between the area where hair cells are lost and where hair cells are still intact. Some evidence does in fact suggest that after an aminoglycoside exposure, hair cells in the mammalian inner ear either die or show no obvious signs of damage and remain intact (Forge and Li 2000). This observation of damage or no damage does not exclude the possability that hair cells affected by aminoglycosides can suffer nonlethal injuries that are not recognizable at the cellular level but that do adversely affect function from which they recover and return to function after internal repair.

Recovery of auditory sensitivity after a transiently induced sensorineural hearing loss, i.e., temporary threshold shift (TTS), does occur spontaneously especially after exposure to noise at moderate intensity levels. Immediately after this noise exposure, the hair bundles become disarrayed and stereocilia appear bent and irregular, suggesting a loss of stiffness, probably as a result of the disorganization of the actin bundling within these affected stereocilia (Saunders et al. 1985; Slepecky 1986). Some studies (reviewed in Saunders et al. 1985) have suggested that such damage is repairable because an organ of Corti exposed to conditions that induce a TTS and stereociliary damage immediately after exposure does not show any anomalies when examined at later postexposure times when hearing thresholds have recovered. There is evidence that actin within the stereocilia is in a state of constant turnover. Transfection of organ of Corti explants with a β-actin-GFP (*G*reen *F*luorescent *P*rotein) c-DNA construct that when expressed allows visualization of the dynamics of actin filament assembly has demonstrated constant delivery of GFP-tagged β-actin monomers to the distal tips of the stereociliary actin filaments and passage of the tagged actin down the stereocilia from tip to base with time (Schneider et al. 2002; Rzadzinska et al. 2004). The continual renewal of actin in stereocilia implied by these β-actin turnover experiments could provide a mechanism for the repair of damaged stereocilia. The turnover time of stereociliary actin filaments was estimated to be about 48 hours, purportedly approximating the time for functional recovery of hair cells after a TTS in some studies (Schneider et al. 2002). However, there is some dispute over whether the damage to stereocilia seen immediately after noise exposure is actually recoverable. As discussed by Saunders and colleagues (Saunders et al. 1985), it is difficult to know whether the normal stereocilia seen after recovery of the hair cells from TTS were actually the ones that were damaged in the first place or are replacements. Stereocilia appear to be unaffected after exposures to noise conditions that consistently produce TTS whereas they appear to be damaged only in conditions that produce permanent hearing loss (Wang et al. 2002). Further, "floppy" bent, and disarrayed stereocilia persist for prolonged periods in the organ of Corti in regions outside those where there is hair cell loss when there is a permanent hearing deficit (Thorne and Gavin 1985). Thus it is not clear that a potential for the repair of excessively damaged stereocilia actually exists in animals exposed to sound trauma.

However, it may be possible that more subtle damage to a hair cell's stereocilia can be repaired. Moderate mechanical disarray of stereocilia might damage the tip-links that gate the transduction channels resulting in loss of the ability to transduce auditory signals. Sound conditions that produce TTS can break tip links (Pickles et al. 1987) and inhibit transduction (Patuzzi et al. 1989). Tip-links have been shown to regenerate spontaneously following their disruption

(Zhao et al. 1996). Exposure of chick basilar papilla explants to the calcium chelator BAPTA (1,2-bis(o-aminophenoxy)ethane-N,N,N',N'-tetraacetic acid) caused loss of the tip-links and inhibition of transduction but within 24 hours these links reform and transduction recovers. Recovery during some forms of TTS may therefore be related to reformation of broken tip-links, although the mechanism by which this extracellular structure is repaired is currently not known. It has been reported that proteins isolated from sea anemones (Watson et al. 1998) that possess mechanosensitive cells with hair bundle-like structures can enhance recovery of tip-links in the hair cells of fish neuromasts that have been damaged by exposure to low-calcium conditions (Berg and Watson 2002). Behavioral recovery from the effects of the calcium depletion occurred spontaneously over 9 days but in the presence of the "tip-link repair proteins" the recovery was initiated within hours. It has been suggested that these "repair proteins" may have a therapeutic potential for the treatment of noise damaged inner ears.

5. Agents and Procedures that Protect Hair Cells from Lethal Damage and Loss

The identification of putative triggers, most notably excess free-radical production, and characterization of the cell death pathways subsequently activated, offers potential opportunities for the development of therapeutic interventions to prevent hair cell death. Agents that inhibit the cell death pathway or which suppress the generation of or enhance the scavenging of free radicals have been demonstrated to protect hair cells from the lethal consequences of exposure to ototoxins or noise and to preserve hearing and/or vestibular function (Cheng et al. 2005; Lynch and Kil 2005; Rybak and Whitworth 2005). Protection against noise-induced hair cell death and hearing loss has also been shown to be afforded by preexposure to a controlled nondamaging sound, so-called "sound conditioning" (Canlon et al. 1988; McFadden et al. 1997; Yoshida and Liberman 2000) although there is some evidence that at the cellular level the effectiveness of this sound conditioning may be due to enhancement of protection against free radical attack (Wang and Liberman 2002). The receptors at afferent terminals beneath IHCs have also been implicated as a means to suppress excitotoxicity responses and preserve neurons (Puel et al. 1994; Pujol and Puel 1999).

5.1 Inhibition of Programmed Cell Death

Inhibition of caspases is one target for intervention in the prevention of caspase-dependent programmed cell death (PCD) (Van De Water et al. 2004). Tri- and tetrapeptides whose amino acid sequences correspond to the specific substrate recognition sites of the caspases act as competitive inhibitors. Addition of fluoro- or chloromethyl ketones (i.e., -cmk or -fmk, respectively) to the peptide structure produces irreversible inhibition as the ketone groups interact with the active site to inactivate the enzyme (Ekert et al. 1999). Further addition of a benzyloxycarbonyl group (BOC- or a z-) enhances cell permeability so that these inhibitors potentially can be used in vivo as well as in vitro. The amino acid sequence in the peptide, usually a tetrapeptide, can be tailored to the unique active site of each particular caspase to achieve nearly specific inhibition, or a shorter "consensus" sequence can be used to broaden the scope of inhibitory action thereby, inhibiting almost all caspases. For example, Boc-aspartate-fmk (BAF) and z-valine-alanine-aspartate-fmk (z-VAD-fmk) are broad-spectrum, pan-caspase inhibitors while z-DEVD-fmk (z-asp-glu-val- asp-fmk) inhibits mainly caspase-3 and z-LEHD-fmk (z-leu-glu-his-asp-fmk) inhibits predominantly caspase-9. Animal experiments have shown the potential of such caspase inhibitors as therapeutic agents for conditions arising from inappropriate triggering of apoptosis, such as neurodegenerative diseases (Onteniente 2004).

Caspase inhibition can rescue hair cells from the lethal effects of aminoglycosides or cis-platinum. In organotypic cultures of mature mammalian utricular maculae (Forge and Li 2000) and of the utricular maculae and basilar papillae of birds (Cunningham et al. 2002; Matsui et al. 2002; Cheng et al. 2003), the presence of pan-caspase inhibitors BAF and z-VAD-fmk or of a specific inhibitor of caspase-9 at a concentration of approximately $100\,\mu M$ for 2 hours before and during incubation with an aminoglycoside almost completely prevented aminoglycoside-induced hair cell loss. Caspase inhibitors preserved 80% of the auditory hair cells from the lethal effects of cis-platinum in neonatal rat organ of Corti explants (Liu et al. 1998). These studies confirmed apoptosis as a major and universal mechanism of hair cell death following an ototoxin challenge. They also identify the specific pathways activated and indicated possibilities for therapeutic intervention to prevent hair cell death. Subsequent in vivo studies have established the potential of caspase inhibition as a means to preserve hair cell function. In guinea pigs treated with cis-platinum, continuous perfusion of the scala tympani with specific caspase inhibitors via an implanted osmotic pump during the period of exposure resulted in significant preservation of auditory CAP thresholds as well as hair cell integrity (Wang et al. J 2004). Likewise, direct perfusion of the inner ear in birds with a pan-caspase inhibitor delivered from an implanted osmotic pump protected vestibular organs from aminoglycoside-induced hair cell loss (Matsui et al. 2003). It was also observed in birds that effective protection of hair cells could be obtained by systemic treatment with caspase inhibitors delivered at a daily dose of 1.5 mg/kg (Matsui et al. 2003). The extent of hair cell survival in vestibular organs exposed to streptomycin in animals receiving z-VAD-fmk systemically was less than when the inhibitor was delivered via direct perfusion into the perilymphatic compartment of the inner ear, which may reflect the difficulty of a relatively large inhibitor molecule gaining entry to the inner ear fluids from the bloodstream. Nevertheless, with the systemic dosing of caspase inhibitors given concomitantly with aminoglycoside treatment, the vestibular–ocular reflexes of the treated animal were significantly preserved, indicating preservation of vestibular function (Matsui et al. 2003). Activation of caspases occurs downstream of the initial effects of aminoglycoside and cis-platinum; therefore, the preservation of hearing and of vestibular function when caspases are inhibited indicates that the initial effects of these ototoxins may not of themselves significantly affect hair cell function. However, the preservation of hair cells in the long term after caspase inhibition requires further investigation.

Although there is little evidence for widespread side effects of systemic application of caspase inhibitors (Onteniente 2004), one potential difficulty is the possibility of inappropriate inhibition of apoptosis in tissues where it is occurring normally as part of cell turnover and tissue maintenance. This side effect of systemic treatment with caspase inhibitors could potentially have serious consequences. Blocking of apoptosis may have a significant impact on the body's ability to prevent tumor development. Unwanted side effects are less likely when inhibitors are perfused into inner ear fluids (i.e., local therapy), but the requirement for a local delivery device and the necessity for a surgical procedure to position it may restrict clinical application of caspase inhibition for preservation of hearing after a physical trauma (e.g. acoustic overexposure) that generates a high level of oxidative stress and the subsequent activation of initiator and effector caspases (Nicotera et al. 2003; Do et al. 2004; Hu et al. 2006) . An alternative means to inhibit a cell death pathway is to target the biochemical pathway(s) upstream of the caspase cascade, such as the biochemical pathways activated in response to the stress induced by a damaging agent that subsequently triggers apoptosis. Inhibition of such stress-activated pathways has the advantage that it should be relatively specific as it will target only cells which have suffered stress through exposure to a damaging agent or trauma.

The c-Jun stress-activated pathway may be of special importance as a potential target in this regard. Several small, cell-permeable molecules that interfere with activation of the JNK cascade have been developed and can inhibit programmed cell death (Bonny et al. 2005; Kuan and Burke 2005). Inhibition of the MAPK/JNK cell death signal pathway has been shown to enhance survival of hair cells after aminoglycoside- and noise-induced injury (Pirvola et al. 2000; Ylikoski et al. 2002; Wang et al. 2003; Matsui et al. 2004). However, blocking JNK signaling does not protect against cis-platinum ototoxicity (Wang et al. 2004), suggesting that the pathways leading to programmed cell death in hair cells after a cis-platinum challenge are different from those activated by either aminoglycosides or noise. Hair cell death is almost completely prevented in neonatal mouse organ of Corti explants exposed to an ototoxic level of neomycin by treatment of the explants with CEP 1347 which is a synthetic inhibitor of the MAPK/JNK signal pathway at the level of the mixed lineage kinases (i.e., MLK-3; see Fig. 6.3) (Pirvola et al. 2000). Hair cells in utricular macula explants from chickens also survive after aminoglycoside-induced damage when cultured in the presence of a similar MAPK/JNK pathway inhibitor, CEP 11004 (Matsui et al. 2004). Continuous perfusion of a JNK-specific inhibitor, D-JNKI-1 peptide, delivered from an osmotic pump into the cochlea, protected against the ototoxic side effects of a course of a systemically administered aminoglycoside and from the damaging effects of exposure to a traumatizing level of noise (Wang et al. 2003) (see Figs. 6.4 and 6.5, respectively). A recent development with the ability of D-JNKI-1 inhibitor to prevent sound induced hearing loss is that this inhibitor has now been demonstrated to be effective when delivered via the round window membrane either as a liquid for 30 minutes or as a bioreleased compound delivered in a hyaluronan gel (Wang et al. 2007). This is an important advance because this mode of drug delivery can and has been used clinically

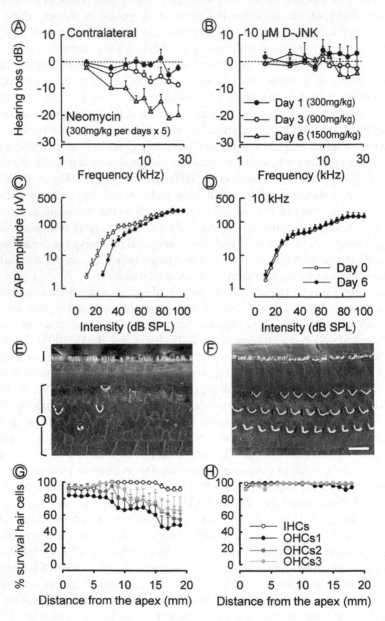

FIGURE 6.4. Local delivery of D-JNKI-1 into the cochlea strongly protected against neomycin-induced hearing and hair cell losses. (**A, B**) A comparison of hearing threshold

to deliver steroid therapy and therefore the therapeutic efficacy of D-JNKI-1 peptide for the treatment of inner ear disorders that involve the mitochondrial cell death pathway and a stress-activated protein kinase (i.e., activated JNK) can be performed. Similar to caspase inhibition, the maintenance of auditory function when the stress-activated cell death pathway was inhibited indicates that the initial stresses caused by the damaging agents may not of themselves affect hair cell function permanently. This is also true in an animal model of cochlear implantation trauma where there is both an immediate and a progressive loss of hearing threshold after exposure of the guinea pig to electrode insertion trauma

◄───

FIGURE 6.4. shifts measured with the compound action potential (CAP) of contralateral unperfused left cochleae (**A**) to those of the D-JNKI-1 perfused right cochleae (**B**) from the same neomycin-treated animals ($n = 7$). Hearing loss was calculated as the difference in decibels between auditory thresholds before neomycin treatment and after 1 day (black circles), 3 days (white circles), and 6 days (white triangles). Changes in hearing thresholds are expressed as mean values ± SEM. Note that the neomycin treatment (3 mg/kg day during 5 days) induced dose-dependent cumulative hearing lossaes in the contralateral unperfused cochleae (**A**) but there were no significant hearing losses in the neomycin-exposed, D-JNKI-1–perfused inner ears (**B**). (**C, D**) Protective effect of a 10 μM solution of D-JNKI-1 in AP against neomycin ototoxicity on amplitude-intensity function of the CAP evoked by stimulation with 8-kHz tone bursts. Shown are the results obtained before neomycin treatment (white circles) and at 6 days after neomycin (black circles). In the contralateral, unperfused cochleae, neomycin treatment caused a reduction in the CAP amplitudes, predominantly in the lower portion of the amplitude-intensity function (**C**). In contrast, there was no reduction in the CAP amplitude seen in the cochleae perfused with a 10 μM solution of D-JNKI-1 (**D**). (**A–D**) All points represent mean ± SEM values calculated from seven animals. (**E, F**) Scanning electron micrographs of the same area of the basal turns of cochleae from the same neomycin-exposed animal. Note the extensive loss of OHCs from the organ of Corti of the basal turn of the contralateral unperfused, neomycin-exposed left cochlea. Only four OHCs remain in the area viewed in e as apposed to the 29 OHCs present in the image presented in (**F**), in which perfusion of a 10 μM solution of D-JNKI-1 prevented loss of OHCs, O, Area of all three rows of OHCs; I, single row of IHCs. Scale bar = 15 μm. (**G, H**) Cytocochleograms obtained from contralateral unperfused left cochleae (**G**; $n = 3$) and D-JNKI-1–perfused right cochleae (**H**; $n = 3$) of the same neomycin-treated animals 10 days after the start of neomycin injections. Cytocochleograms show the percentage of surviving IHCs (white circles) and OHCs from the first (black circles), second (dark gray circles), and third (light gray circles) rows as a function of the distance from the cochlear apex (millimeters). Note the extensive loss of OHCs in the basal turns of the contralateral unperfused cochleae. An average of only 58.2% of the OHCs remained intact in the damaged area of the basal turns of the contralateral unperfused cochleae (**G**). In contrast, the D-JNKI-1–perfused cochleae show only minimal losses of OHCs (i.e., 4%), which occur predominantly in the basal turns (**H**). (Reprinted from Wang, Van De Water, Bonny, de Ribaupierre, Puel and Zine. A peptide inhibitor of c-Jun N-terminal kinase protects against both aminoglycoside and acoustic trauma-induced auditory hair cell death and hearing loss. J Neurosci 23:8596-8607, 2003, with permission.)

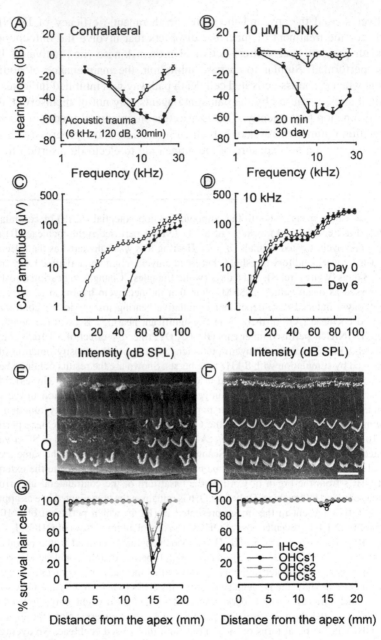

FIGURE 6.5. Perfusion of D-JNKI-1 into the scala tympani protected against acoustic trauma-induced hair cell loss and development of a permanent hearing loss. (**A, B**) Hearing thresholds (mean values ± SEM) from contralateral noise exposed, unperfused left cochleae (A; n = 7) and the noise exposed right cochleae perfused with a 10 μM solution of D-JNKI-1 (**B**; $n = 7$) from the same animals. Hearing loss was calculated as the difference in decibels between auditory thresholds before acoustic trauma, 20 minutes (filled circles) and 30 days (empty circles) after noise exposure. Note that the maximum

(Eshraghi et al. 2006b). There is apoptosis of auditory hair cells at sites within the cochlea that are distal to the site of immediate trauma but if the animal is treated via D-JNKI-1 peptide perfusion of the scala tympani immediately after the insertional trauma and for the following 7 days then there is a significant level of protection against electrode insertion-induced hearing loss for up to 2 months after the initial trauma (Eshraghi et al. 2006b).

Significant in terms of therapeutic strategies is the finding that systemic delivery of CEP 1347 via subcutaneous injections, although not providing complete preservation, can afford some protection from hair cell death and auditory threshold shifts (Pirvola et al. 2000; Ylikoski et al. 2002). Guinea pigs

---

FIGURE 6.5. hearing loss of 40–60 dB when measured 20 minutes after acoustic trauma and the spontaneous, but incomplete recovery of thresholds in the contralateral, unperfused cochleae (A). Protection against a permanent hearing loss was clearly observed for the 10 μM D-JNKI-1–treated cochleae, with an initial hearing loss that was similar to the contralateral unperfused cochleae at 20 minutes but with a near complete recovery of hearing function by 30 days postexposure (B). (C, D) Protective effect of 10 μM D-JNKI-1 against acoustic trauma on amplitude-intensity function of the CAP evoked by stimulation with 8-kHz tone bursts. Shown are the results obtained before (white circles) and 6 days after exposure to the acoustic trauma paradigm (black circles). Acoustic trauma induced a drastic decrease in the CAP amplitude for all intensity levels of sound stimulation (8 kHz) in the contralateral noise-exposed unperfused left cochleae by day 6 (C). Note a near complete recovery of the amplitude-intensity function by 6 days after exposure in cochleae perfused with 10 mM D-JNKI-1 (D). In (C) and (D) all points represent mean ± SEM values calculated from six animals. (E, F) Scanning electron micrographs of the area of acoustic trauma damage in cochleae from the same noise-exposed animal. In the damaged area of the contralateral unperfused cochleae, the most severe damage was observed in the row of IHCs (I) and the first row of OHCs (O) with a gradation of damage in the second and the third rows of OHCs (E). Note that direct delivery of 10 μM of D-JNKI-1 into the scala tympani of the cochlea effectively prevented acoustic trauma induced hair cell loss (F). Scale bar in F = 15 μm for (E, F). (G, H) Quantitative analysis of hair cell damage consisted of counting all hair cells along the entire length of the cochlear ducts. Cochleograms represent the mean survival of hair cells as the function of the distance from the apex (in mm) in contralateral unperfused cochleae (G; n = 3) and in the 10 μM D-JNKI-1–perfused cochleae (H; n = 3) of the same animals. Noise exposure caused a narrow band of hair cell trauma in the cochlea located 14–16 mm from the apex of the cochlea. Ninety one percent of the IHCs (thick black line) and 43% of the OHCs were lost from this area by 30 days after the initial acoustic trauma in the unprotected cochleae. Note the typical gradient of loss from the first row (thin black line) to the third row (thin gray line) of OHCs. In contrast, only 6% OHCs and 11.9% IHCs were lost as a consequence of acoustic trauma if cochleae that were treated with local application of a 10 μM solution of D-JNKI1. (Modified from Wang,Van De Water, Bonny, de Ribaupierre, Puel and Zine. A peptide inhibitor of c-Jun N-terminal kinase protects against both aminoglycoside and acoustic trauma-induced auditory hair cell death and hearing loss, J Neurosci, 23:8596-8607, 2003 with permission from Soc for Neurosci.)

injected systemically with 1 mg/kg of CEP 1347 daily for 2 days before exposure to damaging noise and daily for 2 weeks thereafter showed significantly less hair cell loss and reduced threshold shifts at 2 weeks postexposure than untreated noise-exposed animals (Pirvola et al. 2000). Animals that were chronically treated with gentamicin by daily subcutaneous injections for 2 weeks and with daily injections of CEP 1347 beginning 1 day before and continuing concurrently with and for 2 weeks after gentamicin treatment all showed enhanced auditory and vestibular hair cell survival and protection of hearing in comparison with animals that received only gentamicin (Ylikoski et al. 2002). However, systemic CEP 1347 treatment prevented only about 30% of the auditory hair cell loss that occurred with gentamicin only. Further, while the permanent threshold shifts at 4 and 8 kHz were reduced by about 50% there was still quite a significant elevation of thresholds, and at 32 kHz, the CEP 1347 treatment did not provide any protection against the gentamicin-induced functional impairment. These observations should be compared with effects in aminoglycoside-challenged organ of Corti explants where CEP 1347 effectively prevented all hair cell loss in vitro. This discrepancy between in vitro and in vivo protective action of this molecule could result from a relatively poor uptake of the CEP 1347 inhibitor into the cochlear fluids from the bloodstream after systemic injection although the degree of hair cell survival in vestibular organs appeared to be greater than that in the cochlea, suggesting this inhibitor gained reasonable access to the inner ear. Alternatively, there may be major differences in the effects of CEP 1347 (and caspase inhibition) on the acute hair cell loss occurring in aminoglycoside-challenged immature organ of Corti explants from those occurring during progressive hair cell loss after chronic, systemic treatment of mature animals which is a more clinically relevant condition. In mature animals, there may be multiple forms of hair cell death induced by prolonged exposure to an aminoglycoside with only a proportion of these programmed cell deaths involving MAPK/JNK activation of caspase-mediated apoptosis (Jiang et al. 2006). Multiple forms of hair cell death have also been described during the progression of hair cell loss after exposure to damaging noise which includes both apoptosis and necrosis of affected hair cells within the same lesion site (see Fig. 6.6; Yang et al. 2004). There may also be differences in the patterns of cell death activated by the different aminoglycoside antibiotics because D-JNKI-1 treatment of neomycin-exposed guinea pigs afforded near complete protection of both hair cells and hearing from ototoxicity (Wang et al. 2003; Eshraghi et al. 2006c). Consequently, targeting particular components of stress activated pathways and/or the downstream caspase cascades may in some cases only partially prevent hair cell death and protect hearing against a stress-induced loss.

5.2 Inhibition of Free Radical Action

As discussed previously (see Section 2.3), a major stress-induced factor activating the cell death programs within injured hair cells appears to be the generation of an excess of free radicals. Procedures and agents that scavenge reactive oxygen

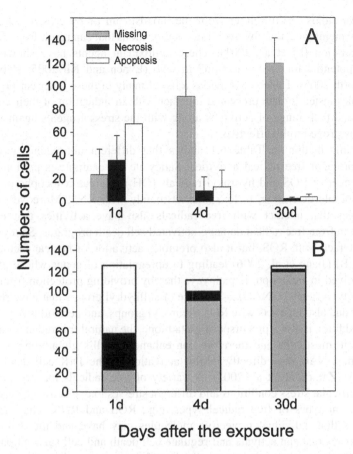

FIGURE 6.6. (**A**) Comparison of the average numbers of apoptotic, necrotic, and missing OHCs across the three time points after exposure to a 104-dB noise. The number of apoptotic cells is significantly greater than the number of necrotic cells at 1 day after the noise exposure. The number of missing OHCs gradually increases and the numbers of necrotic and apoptotic OHCs decrease with time after the noise exposure. (**B**) Comparison of the sum of the numbers of apoptotic, necrotic, and missing OHCs across the three time points after the noise exposure. Notice that from day 1 to day 30 after the noise exposure, there is no significant increase in the total number of damaged OHCs. (Reprinted from Yang, Henderson, Hu and Nicotera. Quantitative analysis of apoptotic and necrotic outer hair cells after exposure to different levels of continuous noise. Hear Res 196:69–76, Copyright 2004, with permission from Elsevier.)

species (ROS) or reactive nitrogen species (RNS) or enhance the production of natural free radical scavenger systems within a cell have been shown to enhance hair cell survival after exposure to noise or ototoxins (Campbell et al. 1996; Henderson et al. 1999; Reser et al. 1999; Sha and Schacht 1999; Kopke et al. 2000). Many molecules with such antioxidant activity, as well as being essentially nontoxic, are relatively small and able to gain access to the inner ear fluids readily either from the bloodstream or by crossing the round window membrane, thereby avoiding complications that can result from systemic administration (Li et al. 2001). These antioxidant agents have shown therapeutic potential for "otoprotection" in vivo (Lynch and Kil 2005; Rybak and Whitworth 2005). Table 6.1 provides a list of many of these inner ear protective molecules tested to date presented together with an indication of their chemical nature and their range of activities along with the stressor agents against which each has proved to be effective.

The information in Table 6.1 shows that different antioxidant molecules have particular free radical activities. Salicylate for example is predominantly a scavenger of ROS and hydroxyl radicals (OH·) whereas α-tocopherol and its water-soluble derivative, trolox, interact mainly with RNS. Moreover, several molecules that interact with free radicals also have activities that enhance protection from free radical damage-activated cell death programs. Salicylate not only interacts with ROS, but it also promotes activation of nuclear factor kappa B (NFκB) (Jiang et al. 2006) leading to upregulation of genes whose products are involved in antiapoptotic pathways thereby providing protection from stress. L-N-acetylcysteine (L-NAC) is a source of sulfhydryl groups that have reducing activity and also interacts with ROS, hydroxyl groups, and peroxides (e.g., H_2O_2) but in addition is also a precursor of glutathione, the natural free radical scavenger present in most cells and therefore can enhance a cell's glutathione levels. In addition, L-NAC may directly inhibit activation of the JNK cell death signal pathway (Zafarullah et al. 2003). A variety of free radicals may be produced from particular stress conditions and different stresses may lead to the production of different types of free radical types, e.g., ROS and RNS. The variety of activities that individual protective molecules may have and the diversity of mechanisms that can activate and regulate cell death and cell survival pathways means that particular antioxidants may be effective against a specific damaging agents whereas for other stress-causing conditions significant survival of hair cells may only be achieved with a combination of antioxidants that possess differing specificities of action (see Table 6.1).

To appreciate the significance of this point consider that salicylate on its own can protect against both aminoglycoside and cis-platinum ototoxicity when coadministered with either of these ototoxins (Sha and Schacht 1999; Li G et al. 2002) but is effective against noise-induced hearing loss only when used in combination with another antioxidant, e.g., L-NAC. Treatment of animals with salicylate and L-NAC, administered together systemically 1 hour before noise exposure and then twice daily afterwards, reduced the degree of ABR threshold shifts that occurred immediately after exposure (temporary threshold shifts [TTS]) as well as permanent threshold shifts (PTS) and hair cell loss (Kopke et al. 2000). However, if the first antioxidant treatment was delayed for only 1 hour after noise exposure then there was little reduction in the extent of hair cell loss in these noise-exposed animals. It has been found that after noise exposure an initial production of ROS is followed by the generation of RNS and the continuation of damage for several days after noise exposure may be

TABLE 6.1. Agents that act to enhance detoxification of free radicals, their actions and the damaging agents against which they have been successfully tested.

Protective agent	Relevant chemistry	Free radicals scavenged	Other activities	Stresses effective against[a]
Salicylate[b] (2-hydroxybenzoate)		Hydroxyl radical ($^{\bullet}$OH)	NFκB activation (antiapoptosis regulation)	Aminoglycoside, cis-platinum noise (in combination)
L-NAC (N-L-acetyl cysteine)	Thiol	$^{\bullet}$OH, H_2O_2	Cysteine donor; enhances glutathione	Noise
Ebselen (2-phenyl-1,2-benzisoselenazol-3[2H]-one)	Selenium derivative	Peroxynitrite (ONOO$^-$) Organic hydroperoxides	Glutathione peroxidase (GPx) mimetic and inducer of GPx; catalyzes reduction of H_2O_2 (and other peroxides) by glutathione	Noise, aminoglycoside, cis-platinum
D,L-Methionine	Thiol	Nitric oxide (NO)	Precursor for glutathione	Cis-platinum, noise
Trolox		ONOO$^-$	Prevents nitrosylation of proteins and DNA	Noise (in combination), cis-platinum
α-Tocopherol		ONOO$^-$	Inhibits lipid peroxidation	Noise, cis-platinum, aminoglycoside
Thiosulfate	Thiol		Precursor for glutathione	Cis-platinum
Glutathione monoethyl ester			Maintains glutathione levels by intracellular conversion to active glutathione	Noise, cis-platinum, 4-hydroxy-2,3-nonenal
ALCAR (acetyl-n-camitine)			Protects mitochondria; ROS production reduced	Noise (in combination)
R-PIA (R – N^6-phenylisopropyl adenosine)			Increases levels of antioxidant enzymes and glutathione	Noise
Ginkgo biloba extract	Antioxidant	?		Aminoglycoside
Danshen	Antioxidant	?		Aminoglycoside
M40403			Superoxide dismutase mimetic $O_2^{-\bullet} \rightarrow H_2O_2 + O_2$	Aminoglycoside but not cis-platinum (in vitro only)

Except for M40403, all tests performed in vivo and show a greater or lesser degree of protection against hair cell loss, and preservation of auditory thresholds. (Compiled from Lynch and Kil 2005; Rybak and Whitworth 2005, and articles cited in text.)

[a]The degree of effectiveness, i.e., extent of hair cell protection, varies. Assessments of effectiveness included in Rybak and Whitworth (2005).

[b]On interaction with $^{\bullet}$OH, salicylate (2-hydroxybenzoate) is coverted to 2,3- or 2,5-dihyroxybenzoate, which chelates iron, thereby suppressing continuing $^{\bullet}$OH production amplified by Fe^{2+}, as well as acting directly as a free radical scavenger. Dihydoxybenzoate has been shown to be effective in vivo in preventing aminoglycoside-induced hair cell loss and hearing impairment (Song et al. 1997).

because of this delayed secondary generation of RNS (Henderson et al. 1999; Yamashita et al. 2004). Salicylate does not interact with RNS but used in combination with trolox, a known scavenger of RNS, signifcant attenuation of noise-induced threshold shifts and hair cell loss could be achieved even with a delay in administration of the combination of up to 3 days after the noise exposure (Yamashita et al. 2005).

It should also be noted that a number of the antioxidants shown in Table 6.1, in particular L-NAC, D-methionine, and ebselen (2-phenyl-1,2-benzisoselenazol-3[2H]-one), are involved in enhancing the intracellular level or metabolism of glutathione. Glutathione (GSH, sometimes incorrectly called "reduced glutathione") is a tripepetide (γ-L-glutamyl-L-cysteinyl-glycine) containing cysteine that is the principal natural free radical scavenger in most cells of the body. Oxidation of the sulfydryl group by ROS, in particular peroxides through the action of glutathione peroxidase (GPX), generates glutathione disulfide (GSSG, "oxidized glutathione") through the formation of sulfur–sulfur bonds between the cysteines of two GSH molecules. This acts to "scavenge" the free radicals. Both L-NAC and D-methionine contain active thiol groups and can act as precursors of glutathione through being a cysteine donor. Ebselen is a mimetic for glutathione peroxidase (GPx) and also acts to induce the production of GPx when acting to protect against noise-induced hearing loss (Kil et al. 2007). Ebselen when used in combination with allopurinol has also been shown both to protect against cis-platinum–induced hearing loss and to enhance the antitumor action of this chemotherapeutic agent in models of breast and ovarian cancer (Lynch et al. 2005). Glutathione itself enhances hair cell survival in vitro (Sha et al. 2001a; Feghali et al. 2001; Ruiz et al. 2006) and increasing the cellular levels of glutathione in vivo with esterified analogs provides protection against noise-induced hearing loss (Hight et al. 2003). As glutathione is synthesized naturally from its constituent amino acids, the maintenance of glutathione levels is dependent n an oorganism's proper protein nutrition. Animals maintained on a low-protein diet have greater susceptibility to aminoglycoside and cis-platinum–induced hearing impairment and hair cell loss but dietary supplementation with glutathione will reduce the extent of the ototoxic damage (Garetz et al. 1994; Lautermann et al. 1995a, b). This emphasizes the need for proper nutrition in maintaining auditory health and resistance to stress-induced hearing loss and that a poor diet may be a contributing factor to susceptibility to acquired hearing loss.

5.3 Sound Conditioning (Toughening)

The degree of hair cell loss and permanent threshold shifts (PTS) that normally develop after exposure to a loud, traumatizing noise has been found to be signif-icantly reduced in cochleae that have been exposed previously to a moderate level of nondamaging sound. This phenomenon is known as sound conditioning (Canlon 1996, 1997). This was initially observed in guinea pigs that had been exposed first to a 1-kHz tone at 81 decibels (dB) sound pressure level (SPL) continuously for 24 days, a condition that did not cause damage to the cochlea, then subsequently to the same tone at 105 dB SPL for 72 hours. This latter noise condition produced a significant PTS measured at 8 weeks after exposure in animals not previously exposed to the low-level noise, but in the "sound conditioned" animals there was essentially no PTS (Canlon et al. 1988). Protection from the damage normally caused by loud noise, usually in the order of a 30- to 40-dB reduction in thresholds, has subsequently been obtained with a variety of different nondamaging preexposure sound conditions varying from 81 dB to 95 dB SPL for exposure periods of several days to 15 minutes (Canlon 1997; McFadden et al. 1997; Kujawa and Liberman 1999; Yoshida and Liberman 2000) and has been observed in several different species including guinea pigs, chinchillas, gerbils, rabbits, and mice (Yoshida and Liberman 2000). The effects have been found to last for up to at least 60 days after the end of the conditioning period before exposure to the traumatizing noise (McFadden et al. 1997). Such protection is not dependent on middle ear muscle reflexes because paralyzing the muscles does not prevent it (Dagli and Canlon 1995), nor is it dependent on the cochlea's efferent neural innervation acting on the outer hair cells (Yamasoba and Dolan 1998; Kujawa and Liberman 1999). Significant protection occurs where the efferent effects on OHCs are minimal (Kujawa and Liberman 1999) and when efferent activity has been inhibited with strychnine (Yamasoba and Dolan 1998). Further, the sound conditioning appears to be operating locally at the level of the cochlea; in animals unilaterally deafened by an ear plug during the conditioning period only the unblocked ear was protected when both ears were subsequently exposed to a traumatizing level of noise, suggesting that systemic effects during conditioning do not play a major role in this phenomenon (Yamasoba et al. 1999).

While the actual basis of the protective effect underlying sound conditioning is not yet clear, it is likely that it results from the local enhancement of stress-activated protective mechanisms such as NFκB and glucocorticoid receptors (Tahera et al. 2006; Canlon et al. 2007 ; Tahera et al. 2007). Reduction in PTS from a traumatizing noise exposure can also be achieved by prior application of other physical stresses. Imposing a mild physical restraint on an animal creates a stress condition and when applied up to 2 hours before the period of noise exposure it reduces the PTS in comparison to animals that are not physically prestressed prior to noise exposure (Wang and Liberman 2002). Restraint stress induces a rise in an animal's level of circulating glucocorticoid (Wang and Liberman 2002). Increases in glucocorticoid levels have also been observed in animals after exposure to noise of a moderate intensity level (Rarey et al. 1995). The binding of glucocorticoids to their cognate receptors regulates expression of genes that are involved in inflammatory responses, and cell survival pathways including those that enhance free radical scavenging (Wang and Liberman 2002) and glucocorticoid receptors have been localized in the cochlea (Rarey et al. 1995; Rarey et al. 1995; Terunuma et al. 2003). Thus, systemic stress may lead to tissue specific responses in the inner ear that provide protection from the effects of noise-induced free radical generation. Daily intraperitoneal injections of methyl prednisolone (a potent glucocorticoid) after moderate sound exposure can significantly reduce the noise-induced threshold shifts in comparison with untreated animals (Takahashi et al. 1996).

Corticosteroids have also been shown to afford some protection against hearing loss following local application via middle ear injection and absorption through the round window membrane (Kopke et al. 2001a; Plontke et al. 2005). This supports the concept of a local corticosteroid-mediated protective effect against noise exposure in the cochlea and suggests that there is therapeutic potential in the activation of this pathway (Tahera et al. 2006). Direct infusion of dexamethasone into the cochlea of a guinea pig has also been shown to partially protect hearing against loss induced by aminoglycoside ototoxicity (Himeno et al. 2002; Takemura et al. 2004). Corticosteroid treatment may also be effective in enhancing hair cell survival after implantation of a cochlear prosthesis. An animal study using a guinea pig model of electrode insertion trauma-induced hearing loss (Eshraghi et al. 2007b) has shown that local treatment of the cochlea with high-dose dexamethasone immediately after electrode insertion and continued for 1 week results in a significantly conservation of hearing against trauma-induced hearing loss and that this protection of hearing is stable at 1 month posttrauma (Eshraghi et al. 2007a). In human cochlear implant patients, the corticosteroid triamcinolone acetonide was applied in conjunction with electrode array insertion and resulted in improved impedances from the cochlear implant electrode arrays (De Ceulaer et al. 2003). This same steroid preparation has been used in conjunction with a soft surgery approach when inserting the electrode array for the conservation of a patient's low-frequency hearing (Kiefer et al. 2004). Direct evidence that a corticosteroid may be otoprotective has also come from the result of an in vitro study. Triamcinolone acetonide protected the hair cells of neonatal rat cochlear explants from the ototoxic effect of 4-hydroxy-2,3-nonenal (HNE), which is an ototoxic by-product of peroxidation of cell membrane lipids by free radicals (Guzman et al. 2006). The benefit of using a corticosteroid as a protective molecule in the inner ear is that it has already been demonstrated to be an effective treatment against sudden hearing loss (Parnes et al. 1999; Kopke et al. 2001a; Plontke et al. 2005) and this class of drugs have both anti-inflammatory (De Ceulaer et al. 2003) and otoprotective qualities (Guzman et al. 2006).

5.4 Hyperthermia and Hypothermia

A second physical stress that can provide protection is moderate but extended exposure to hyperthermia. Warming animals to about 41°C for 6–24 hours before exposure to a traumatizing level of noise (Yoshida et al. 1999; Altschuler et al. 2002) reduces the level of the PTS that develops post-noise exposure. The cochlea expresses certain heat shock proteins (HSPs), which are proteins that assist in survival of cells when they are stressed. They act as molecular chaperones protecting damaged proteins from destruction and assisting in their repair (Jakob and Buchner 1994; Winter and Jakob 2004). Heat shock proteins may also exhibit free radical scavenging and antiapoptotic activities. The HSPs expressed in the cochlea include HSP 27, 32, and 70 (Lim et al. 1993; Yoshida et al. 1999; Altschuler et al. 2002; Leonova et al. 2002; Fairfield et al. 2004). The expression levels of HSP 70 (Yoshida et al. 1999) and HSP 32 (Fairfield et al. 2004) are upregulated in the cochlea after an animal has been subjected to hyperthermia and are subsequently downregulated over a time course that mirrors the period during which heat stress "preconditioning" confers protection against noise trauma. "Heat shock" conditioning also leads to upregulation of heat shock factor-1 (HSF-1), which is a stress-activated transcription factor that regulates the expression of *HSP* genes (Fairfield et al. 2004). HSP 70 is upregulated in the cochlea after exposure to damaging sound (Lim et al. 1993) and transgenic mice in which *HSF-1* has been ablated suffer a much greater PTS and hair cell loss after a noise exposure than their wild-type littermates. This suggests a role for HSPs in the protection of the cochlea from noise-induced injury. Upregulation of HSPs may be another factor underlying the protective effect of sound conditioning (Fairfield et al. 2005). As some HSPs, including HSP32, act as free radical scavengers (Winter and Jakob 2004), sound conditioning may enhance the levels of free radical scavengers available in the cochlea at the time of the traumatizing noise exposure, thereby conferring protection against this noise exposure. The antiapoptotic activity of HSPs is also likely to be significant in this context. These effects may also have wider significance. HSP 70 has been shown to be upregulated following cis-platinum–induced injury to the cochlea (Oh et al. 2000) and exposure of explants of vestibular sensory tissues to the agent geranylgeranylacetone, which induces HSP 70 expression, can protect against gentamicin-induced hair cell loss (Takumida and Anniko 2005). Thus, manipulating intracellular levels of HSPs in the inner ear may potentially be a future therapeutic strategy for preventing or ameliorating the extent of acquired hearing loss.

Hypothermia may also be effective in protecting hair cells from damage. Moderate hypothermia has been used clinically to lessen the effects of traumatic brain injury (Marion et al. 1997) and the neurological damage that can occur as a result of cardiac arrest (Bernard et al. 2002). Systemic hypothermia has been demonstrated to cause a temperature-dependent loss of the cochlea's sensitivity to auditory stimuli (Drescher 1974) and to reduce noise-induced elevation of the cochlear microphonic response (Drescher 1976). In human patients that were undergoing cooling during cardiovascular-bypass surgery, the depth of a patient's hypothermia was observed to correlate with a reduction in a patient's transient evoked otoacoustic emissions (TEOAEs), suggesting that the motile activity of the outer hair cells is temperature sensitive in humans (Veuillet et al. 1997). This effect of hypothermia on TEOAEs in patients may be explained by the laboratory finding that hypothermia decreases glutamate efflux in the perilymph of guinea pigs after a transient ischemia of the cochlea (Hyodo et al. 2001). Moderate hypothermia (i.e., 32^0C) has been reported to prevent hearing loss and the progression of hair cell loss in gerbils that occurs after an ischemia–reperfusion injury (Watanabe et al. 2001). In addition, induction of systemic hypothermia in mice has been demonstrated to be protective against noise-induced hearing loss

TABLE 6.2. Hypothermia protection against hearing loss induced by electrode insertion trauma; total changes in auditory brainstem response (ABR) thresholds and distortion product ototacoustic emission (DPOAE) amplitudes.

	37°C	34°C	p Value
ABR threshold	15.9[a]	4.2	<0.0001
DPOAE amplitude	5.3	1.0	<0.001

[a] Mean, dB.
Modified from Laryngoscope 115:1543–1547, 2005, with permission.

(Henry and Chole 1984; Henry 2003). In contrast, induction of hyperthermia during sound exposure, rather than before sound exposure, may exacerbate the amount of hearing loss induced by a defined noise exposure (Henry 2003). Systemic mild hypothermia of the laboratory rat (i.e., 34°C) can also protect against the hearing loss induced by the trauma of electrode insertion (Balkany et al. 2005; see Table 6.2). Taken together these results strongly suggest that mild hypothermia may have clinical application in the prevention of hearing loss initiated by the vibration trauma of a surgical drill and by the physical trauma that is created during the insertion of an electrode array into the cochlea of a cochlear implant patient with residual hearing (Balkany et al. 2005; Eshraghi 2006; Eshraghi and Van De Water 2006).

6. Sensory Neurons of the Inner Ear

The neurons of the inner ear's VIIIth nerve complex are derived from the otic placode, from which all of the hair cells of the inner ear also arise. The support cells and ensheathing cells of the neuronal complex are all derived from contributions by the migrating cephalic neural crest cells of the head region (D'Amico-Martel and Noden 1983).

6.1 Genes Involved in Neuronal Survival and Homeostasis

The control of the neural fate of cells within these otic anlagen before the outward migration of neuroblasts that form the Scarpas' (vestibular) and spiral (auditory) ganglia of the VIIIth nerve is determined by the activity of *Tbx1* in suppressing the proneural gene neurogenin-1 (*ngn1*). *Tbx1*$^{-/-}$ losses of function (knockout [KO]) mutant mice have rudimentary sensory receptors and a partial duplication of the VIIIth nerve ganglion complex (Raft et al. 2004). Normal expression of the *Tbx1* gene is critical to both normal ear development (i.e., specification of sensory cell fate) and specification of neural cell fate for the formation of the inner ear ganglia during the placode stage of mammalian inner ear development (Raft et al. 2004). The proneural gene *ngn1* is essential for the development of all of the inner ear sensory neurons as shown by the analysis of the inner ears

of $ngn1^{-/-}$ gene KO mice (Ma et al. 1998). The inner ears from these animals
had neither a Scarpa's ganglion (no vestibular neurons) nor a spiral ganglion
(no auditory neurons) and yet they possessed morphologically normal hair cells
in the appropriate locations within the patches of inner ear sensory receptor
epithelia that are completely devoid of innervation (Ma et al. 2000). *NeuroD*,
another proneural gene, is a basic helix–loop–helix transcription factor that acts
downstream of *ngn1*, is also important in the early survival of all of the inner
ear sensory afferent neurons. *NeuroD* null ("knockout" [KO]) mutants do not
have auditory evoked potentials and histological examination of their inner ears
shows that their sensory ganglia have died during development. These developing neurons never express either TrkB or TrkC, which are the high-affinity
neurotrophin receptors for BDNF and NT-3, respectively (Kim et al. 2001). The
importance of TrkB and TrkC and their respective BDNF and NT-3 neurotrophin
ligands in the inner ear was suggested by in situ hybridization that localized
these receptors and their respective neurotrophin ligands in developing and
adult inner ears of mammals (Pirvola et al. 1992, 1994; Ylikoski et al. 1993).
Analysis of the inner ears of $bdnf^{-/-}$ and of $nt3^{-/-}$ null mutant mice and of the
$bdnf^{-/-} + nt3^{-/-}$ double null mutant mouse (Ernfors et al. 1995) indicated that
a lack of neurotrophin gene expression within the inner ear has adverse effects
on the survival of both the auditory and vestibular ganglion neurons. The results
obtained from these mutants demonstrate that NT-3 and BDNF are in general
responsible for the support of, respectively, the auditory and vestibular ganglion
neurons with the inner ears of the double mutant devoid of inner ear afferent
sensory neurons. However, this relationship of inner ear neuronal cell type to
neurotrophic factor is not as simple as originally described (Ernfors et al. 1995).
A much more complex set of relationships has been defined through the analyses
of the inner ears of a series of gene knockout and gene knockout/knockin mice
(for comprehensive reviews see Pirvola and Ylikoski 2003; Fritzsch et al. 2004).
The effects of hair cells on neuron survival have been studied in mice that lack
expression of the *atoh1* gene and that never develop differentiated hair cells
(Bermingham et al. 2001). Despite the absence of hair cells, afferent nerve fibers
grow into the areas where the hair cells would have differentiated. Moreover,
if there is local production of BDNF by support cells in these areas that are
devoid of hair cells then some of the neuronal processes survive and are retained
(Fritzsch et al. 2005). It appears therefore that the presence of hair cells is not
required for the initial stages of afferent innervation of the inner ear to take place
(Fritzsch et al. 2005). In mice deficient in the *brn3c* gene ($brn3c^{-/-}$), where hair
cells initially differentiate but are then lost soon after birth, an almost normal
pattern of innervation is generated, and even long after the loss of their sensory
hair cells some innervation of both the auditory receptor (apical area) and the
vestibular epithelium persists (Xiang et al. 2003). This observation suggests that
there are several different factors necessary for neuronal homeostasis, in addition
to neurotrophins, that are involved in the initial pattern of neuritic in-growth, in
retention of nerve fibers and in support of the peripheral innervation of inner ear
sensory epithelium (Fritzsch et al. 2003).

6.2 Protection and Rescue Therapies for the Support and Homeostasis of Inner Ear Sensory Neurons

In an animal model of complete hair cell loss (i.e., treatment with an amino-glycoside antibiotic immediately followed by intravenous administration of a loop diuretic) the perfusion of either BDNF or NT-3 into the scala tympani sustained the overall survival of the spiral ganglion neurons by replacing the trophic support that had been supplied by an intact peripheral sensory epithelium (Staecker et al. 1996). Similarly, after chronic aminoglycoside treatment that induced hair cell loss the delivery of NT-3 into the perilymph of the scala tympani also maintained auditory neurons that had lost their peripheral target supply of neurotrophins (Ernfors et al. 1996). In addition, the treatment of deafened guinea pigs with local neurotrophin therapy perfused into the scala tympani after loss of all of the auditory hair cells has been shown to improve electrical stimulus evoked auditory brainstem responses (eABR responses) (Shinohara et al. 2002). However, as a cautionary note, the cessation of locally delivered neurotrophin therapy in deafened animals has been demonstrated to result in an accelerated loss of spiral ganglion neurons that are without a peripheral target tissue to supply trophic support (Gillespie et al. 2003). However, in a clinical setting neurotrophin therapy is likely to be used in conjunction with a cochlear implant. It is therefore important to determine the combined effect of electrical stimulation of auditory neurons with neurotrophin replacement therapy and whether electrical stimulation could sustain neuronal survival after the cessation of neurotrophin therapy. Addressing these questions via both in vitro (Hegarty et al. 1997) and in vivo approaches (Kanzaki et al. 2002; Shepherd et al. 2005) has suggested that electrical stimulation in combination with neurotrophin therapy enhances neuronal survival above that obtained with either depolarization alone or neurotrophin alone. However, a further consideration for clinical application of neurotrophins is that specific neurotrophic factors (e.g., NT-3 and BDNF) can have quite different effects on the firing patterns of spiral ganglion neurons in a base to apex pattern. This then poses the problem of how to mimic to some degree the gradients of NT-3 and BDNF present within the uninjured cochlea if these trophic factors are to be used as an effective therapy to support both neuronal survival and function within the spiral ganglion (Adamson et al. 2002; Davis 2003; Zhou et al. 2005).

Viral vectors have also been used to deliver genes encoding neurotrophic factors to cochleae in which hair cells have been lost after exposure to an ototoxin. Both the herpes amplicon gene therapy vectors (i.e., HSV_{bdnf} and HSV_{nt-3}) and an adenoviral gene therapy vector (i.e., Ad_{gdnf}) proved to provide successful maintenance of auditory neuron viability after total loss of the auditory hair cells (Staecker et al. 1998; Yagi et al. 2000; Bowers et al. 2002). Neurotrophin therapy can also be delivered into a damaged cochlea by encapsulation of NT-3 in biodegradable alginate beads via either direct injection of the beads into the scala tympani or via placement of the beads on the round window membrane

(Noushi et al. 2005). An important question to answer is whether delayed delivery of neurotrophic factor therapy is of any use to prevent further degeneration of spiral ganglion neurons once a pattern of trophic factor deprivation-induced cell death has already begun in the auditory neurons of the spiral ganglion. Studies addressing this question have demonstrated that delayed therapy with neurotrophins can be effective in preventing the ongoing apoptosis of the auditory neurons and in the prevention of any additional neuronal losses (Gillespie et al. 2004; Yamagata et al. 2004). In addition, there is some evidence that those auditory neurons rescued by the delayed neurotrophin therapy have an enhanced responsiveness to electrical stimulation (i.e., lowered eABR thresholds; Yamagata et al. 2004). A recent histological study that evaluated the survival of human spiral ganglion neurons in temporal bone specimens has not established a direct correlation between the number of surviving auditory neurons and (1) loss of hair cells, (2) loss of supporting cells, and (3) duration of deafness (Teufert et al. 2006) as has been well established in numerous animal studies (Staecker et al. 1996; Shinohara et al. 2002).

In neomycin-deafened rats in vivo following the death of the hair cells there is an upregulation of transcripts that encode for *bdnf, TGFβ1 and TGFβR1* (TGFβreceptor-1 in the auditory nerve and a downregulation of *trkB*, which is the high-affinity receptor for BDNF. The most dramatic up- and downregulation occurs in the BDNF ligand high-affinity receptor system (Wissel et al. 2006). Immunostaining also suggested an increased presence of both BDNF and TGFβ1/2 proteins in the auditory nerve (Wissel et al. 2006). TGFβ1 has also been found to be upregulated in auditory neurons after transection of the auditory nerve. These results suggest that upregulation of *TGFβ1* gene expression and of its receptor (*TGFβR1*) is an early autocrine signal in response to either injury (Lefebvre et al. 1992) or loss of trophic support (Wissel et al. 2006). This concept is consistent with the observation that TGFβ1 does not in itself directly support neuronal survival of adult rat auditory neurons in culture but does increase the neuron's responsiveness to the trophic action of other growth factors such as basic fibroblast growth factor (bFGF; Lefebvre et al. 1991) and BDNF (Van De Water, unpublished data). Neurotrophic factors alone and in combination with cytokines (Staecker et al. 1995; Hartnick et al. 1996) are powerful inhibitors of programmed cell death of auditory neurons that have been deprived of trophic support (Van De Water et al. 1996; Gillespie and Shepherd 2005). However, there are other factors such as substance P and activation of protein kinase C beta one (PKCβ1) signaling that can also act to prevent apoptosis of neurotrophin-deprived auditory neurons in vitro. This type of approach may hold greater promise because of their stability and the smaller size of these neurotrophic molecules (Lallemend et al. 2003, 2005). This area of spiral ganglion therapy with small stimulatory molecules (Scarpidis et al. 2003) is an important area of future therapy and needs further exploration, as do other alternatives to neurotrophin therapy within in vivo models of spiral ganglion cell death initiated by deprivation of hair cell derived trophic factors.

7. Summary and Conclusions

A factor common to the initiation of hair cell death caused by a variety of conditions is the creation of oxidative stress within the injured sensory cell. Lethal damage ensues when the level of ROS, RNS, and other free radicals that are generated exceeds the capacity of natural cellular free radical scavenging systems to neutralize them. Caspase-dependent, or caspase-independent programmed cell death, or cell death via necrosis can then be triggered within the affected inner ear sensory cell. The loss of neurons from inner ear ganglia is not as well defined as the loss of hair cells, although it is now clear that apoptosis is a major contributing factor in neuronal loss when auditory or vestibular ganglion neurons are deprived of trophic factor support. At present, it is not known how well the animal studies documenting neuronal degeneration in the inner ear ganglia following a loss of hair cells translates to what occurs in humans after a similar loss of hair cells. It is also apparent that mechanisms of hair cell death identified in vitro, where immature tissues from neonatal animals are generally used to examine cell death occurring acutely, may not always reflect those activated in vivo in more mature animals in which hair cell loss may occur over a prolonged period after the initial insult. This possible disparity between in vitro and in vivo results does not negate the fact that in vitro studies are valuable for the identification and evaluation of agents for their potential to protect hair cells from the lethal effects of stress-induced injury. However, confirmation through experiments performed with live laboratory animals is essential to validate agents for possible clinical application. To date effective otoprotective therapies identified from in vivo studies in animals involve (1) inhibiting caspases and, thus, apoptosis; or (2) blocking a major cell death signaling pathway upstream of the apoptotic cascade (e.g., D-JNKI-1 peptide binding of activated JNK molecules); or (3) inhibiting the buildup of ROS or RNS within injured cells by administration of "antioxidants," e.g., salicylate and L-NAC. Because in some cases both ROS and RNS are generated in response to an insult, often a combinatorial approach to therapy (e.g., L-NAC + salicylate) has proved to be more effective in preventing auditory hair cell death than individual agents administered alone. The clinical effectiveness of an antioxidant approach to maintaining auditory function under conditions where hair cells are likely to be susceptible to damage has recently been demonstrated. Salicylate, which protects laboratory animals against the hearing losses induced both by aminoglycoside (Sha and Schacht 1999) and by cis-platinum (Li G et al. 2002), has been tested in a clinical trial in China. It was found that salicylate, when coadministered with gentamicin, afforded significant protection against aminoglycoside-induced hearing loss in patients receiving this antibiotic for common infections, without compromising the efficacy of the antibiotic activity of gentamicin (Sha et al. 2006).

Future insights into the mechanisms of cell death and of the repair of sublethally damaged hair cells, as well as of the long-term survival of hair cells rescued by otoprotective agents are essential for the continued development and improvement of therapeutic interventions that will aid restoration of normal function within a damaged but healing inner ear. For such therapies to be effective, however, they must be present in the right place at the right time; that is, the most appropriate protective agent must be available to injured hair cells at the time when the cascade of events that leads to cell death is initiated in individual, injured sensory cells. Currently it appears that effective protection can be achieved when agents are administered before the initiation of a trauma, be it an ototoxin challenge or noise exposure, but protection of hair cells posttrauma has so far proved elusive. This may be because the cell death mechanisms operating and/or the specific initiator of the cell death may change as hair cell death progresses with time after the original damaging event. This could, for the present, limit the use of therapies that prevent hair cell death to situations where damage might be predicted—when drugs with potential ototoxic side effects are to be used, or prior to exposure to known noisy environments—and to relatively short time windows of opportunity postexposure when intervention is possible. This emphasizes the importance of gaining further understanding of how and why hair cell death in many situations continues for quite a prolonged period after the initial damaging event(s) has passed in order to identify further targets for posttrauma intervention. It would also seem likely that therapeutic dosing may need to be continuous for some extended period. Certainly, if it is the case that the major initiating factor in age-related hair cell loss is accumulation of free radicals and subsequent apoptotic cell death, then presumably any preventative therapy aimed at ameliorating presbyacusis that targets those events would need to be administered over an extended period of time. This raises the question of how such therapies are to be delivered. In the case of protecting hair cells from trauma-induced damage during the insertion of cochlear implants, it may be possible to administer the therapeutic agent directly into the scala tympani of the cochlea during the procedure in which the electrode array is inserted into the cochlea. This would appear to be one area in which hair cell protection strategies may be successful. However, in many experimental studies in animals where there is documented protection of hair cell against noise or ototoxins the therapeutic agents have been delivered either (1) through implanted osmotic pumps directly into perilymph, or (2) from the middle space perhaps using hyaluronan gel or a similar delivery preparation placed on the round window membrane, or (3) by systemic treatment through repeated injections. It seems questionable that physical interventional strategies such as implantable delivery systems or even surgical opening of the middle ear cavity to locate a therapeutic agent or delivery device with a therapetic agent close to the round window membrane niche will be considered clinically acceptable from either a risk–benefit or a cost–benefit analysis at the present time. The need for repeated injections of an otoprotective drug may also not be acceptable to many patients. Ideally, therapeutic agents could be administered orally. This requires some knowledge of the pharmacokinetics of the potential otoprotectant drug, i.e., how well it is taken up into the blood stream and how easily it can gain entry to the fluid spaces and then to the targeted tissues of the inner ear. There is thought to be a "blood–perilymph barrier" that limits uptake of chemicals into the inner ear fluids from the blood (Wangemann and Schacht, 1996). It is well known that salicylate can cross this barrier and gain rapid entry to perilymph, which is clearly one reason why its ROS scavenging properties can be exploited for otoprotection. Other antioxidants that show promise as hair cell protectant molecules—NAC, ebselen, and D-methionine, for example—are also relatively small molecules that might be expected to enter perilymph, although validation of this concept requires further investigation. But it may be that larger bioactive molecules such as those involved in inhibiting cell death pathways may not enter the inner ear so readily after systemic application. Relatively little is known about the pharmacokinetics of drug entry into and distribution within the fluids of the inner ear, and initial studies have shown some of the complexities that need to be addressed (Salt and Plontke 2005). Further understanding of these phenomena would aid identification and design of candidate therapeutic agents able to gain access to the fluid spaces and therby access to the tissues of the inner ear.

Nevertheless, in the light of the proof of principle derived from the clinical trial cited (Sha et al. 2006) and the fact that a number of antioxidants are being considered for or are entering clinical trails (Lynch and Kil 2005; Rybak and Whitworth 2005), it seems reasonable to predict that pharmaceutics that can prevent hair cell death may become available for routine clinical use in the not too distant future. At present the prospect of hair cell regeneration in response to gene therapy (Izumikawa et al. 2005) is very encouraging but not yet ready for a clinical application, and acting on the precept that prevention is better than cure, such a therapeutic approach could lead to a significant reduction in the incidence of acquired hearing loss with major health care and economic impacts.

References

Adamson CL, Reid MA, David RL (2002) Opposite actions of brain-derived neurotrophic factor and neurotrophin-3 on firing features and ion channel composition of murine spiral ganglion neurons. J Neurosci 22:1385–1396.

Altschuler RA, Fairfield D, Cho Y, Leonova E, Benjamin IJ, Miller JM, Lomax MI (2002) Stress pathways in the rat cochlea and potential for protection from acquired deafness. Audiol Neurootol 7:152–156.

Baird RA, Burton MD, Fashena DS, Naeger RA (2000) Hair cell recovery in mitotically blocked cultures of the bullfrog saccule. Proc Natl Acad Sci USA 97:11722–11729.

Balkany TJ, Eshraghi AA, He J, Polak M, Mou CH, Dietrich D, Van De Water TR (2005) Mild hypothermia protects auditory function during cochlear implant surgery. Laryngoscope 115:1543–1547.

Berg A, Watson GM (2002) Rapid recovery of sensory function in blind cave fish treated with anemone repair proteins. Hear Res 174:296–304.

Bergamini E (2006) Autophagy: a cell repair mechanism that retards ageing and age-associated diseases and can be intensified pharmacologically. Mol Asp Med 27:403–410.

Bermingham NA, Hassan BA, Wang VY, Fernandez M, Banfi S, Bellen HJ, Fritzsch B Zoghbi HY (2001) Proprioceptor pathway development is dependent on Math1. Neuron 30, 411–422.

Bernard SA, Gray TW, Buist MD, Jones BM, Silvester W, Gutteridge G, Smith K (2002) Treatment of comatose survivors of out-of-hospital cardiac arrest with induced hypothermia. N Engl J Med 346:557–563.

Bitner-Glindzicz M (2002) Hereditary deafness and phenotyping in humans. Br Med Bull 63:73–94.

Boettger T, Hubner CA, Maier H, Rust MB, Beck FX, Jentsch TJ (2002) Deafness and renal tubular acidosis in mice lacking the K-Cl co-transporter Kcc4. Nature 416:874–878.

Boland B, Nixon RA (2006) Neuronal macroautophagy: from development to degeneration. Mol Asp Med 27: 503–519.

Bonny C, Borsello T, Zine A (2005) Targeting the JNK pathway as a therapeutic protective strategy for nervous system diseases. Rev Neurosci 16:57–67.

Bowers WJ, Chen X, Guo H, Frisina DR, Federoff HJ, Frisina RD (2002) Neurotrophin-3 transduction attenuates cisplatin spiral ganglion neuron ototoxicity in the cochlea. Mol Ther 6:12–18.

Campbell KC, Rybak LP, Meech RP, Hughes L (1996) D-methionine provides excellent protection from cisplatin ototoxicity in the rat. Hear Res 102:90–98.

Canlon B (1996) The effects of sound conditioning on the cochlea. In Salvi R, Henderson D (eds) Auditory System Plasticity and Regeneration. New York: Thieme, pp 118–127.

Canlon B (1997) Protection against noise trauma by sound conditioning. Ear Nose Throat J 76:248–250, 253–255.

Canlon B, Borg E, Flock A (1988) Protection against noise trauma by pre-exposure to a low level acoustic stimulus. Hear Res 34:197–200.

Canlon B, Meltser I, Johansson P, Tahera Y (2007) Glucocorticoid receptors modulate auditory sensitivity to acoustic trauma. Hear Res 226:61–69.

Cardinaal RM, de Groot JC, Huizing EH, Veldman JE, Smoorenburg GF (2000) Dose-dependent effect of 8–day cisplatin administration upon the morphology of the albino guinea pig cochlea. Hear Res 144:135–146.

Carlsson PI, Van Laer L, Borg E, Bondeson ML, Thys M, Fransen E, Van Camp G (2005) The influence of genetic variation in oxidative stress genes on human noise susceptibility. Hear Res 202:87–96.

Casano RA, Johnson DF, Bykhovskaya Y, Torricelli F, Bigozzi M, Fischel-Ghodsian N (1999) Inherited susceptibility to aminoglycoside ototoxicity: genetic heterogeneity and clinical implications. Am J Otolaryngol 20:151–156.

Cheng AG, Cunningham LL, Rubel EW (2003) Hair cell death in the avian basilar papilla: characterization of the in vitro model and caspase activation. J Assoc Res Otolaryngol 4:91–105.

Cheng AG, Cunningham LL, Rubel EW (2005) Mechanisms of hair cell death and protection. Curr Opin Otolaryngol HNS 13:343–348.

Cohen-Salmon M, Ott T, Michel V, Hardelin JP, Perfettini I, Eybalin M, Wu T, Marcus DC, Wangemann P, Willecke K, Petit C (2002) Targeted ablation of connexin26 in the inner ear epithelial gap junction network causes hearing impairment and cell death. Curr Biol 12:1106–1111.

Coling DE, Yu KC, Somand D, Satar B, Bai U, Huang TT, Seidman MD, Epstein CJ, Mhatre AN, Lalwani AK (2003) Effect of SOD1 overexpression on age- and noise-related hearing loss. Free Radic Biol Med 34:873–880.

Cotanche DA, Dopyera CE (1990) Hair cell and supporting cell response to acoustic trauma in the chick cochlea. Hear Res 46:29–40.

Cotanche DA, Saunders JC, Tilney LG (1987) Hair cell damage produced by acoustic trauma in the chick cochlea. Hear Res 25:267–286.

Cunningham LL, Cheng AG, Rubel EW (2002) Caspase activation in hair cells of the mouse utricle exposed to neomycin. J Neurosci 22:8532–8540.

D'Amico-Martel A, Noden DM (1983) Contributions of placodal and neural crests cells to avian cranial peripheral ganglia. Am J Anat 166:445–468.

Dagli S, Canlon B (1995) Protection against noise trauma by sound conditioning in the guinea pig appears not to be mediated by the middle ear muscles. Neurosci Lett 194:57–60.

Danial NN, Korsmeyer SJ (2004) Cell death: critical control points. Cell 116:205–219.

Davis RL (2003) Gradients of neurotrophins, ion channels, and tuning in the cochlea. Neuroscientist 9:311–316.

Davis RR, Newlander JK, Ling X, Cortopassi GA, Krieg EF, Erway LC (2001) Genetic basis for susceptibility to noise-induced hearing loss in mice. Hear Res 155:82–90.

Debnath J, Baehrecke EH, Kroemer G (2005) Does Autophagy contribute to cell death? Autophagy 1: 66–74.

De Ceulaer G, Johnson S, Yperman M, Daemers K, Offeciers FE, O'Donoghue GM, Govaerts PJ (2003) Long-term evaluation of the effect of intracochlear steroid deposition on electrode impedance in cochlear implant patients. Otol Neurotol 24:769–774.

Ding DL, Wang J, Salvi R, Henderson D, Hu BH, McFadden SL, Mueller M (1999) Selective loss of inner hair cells and type-I ganglion neurons in carboplatin-treated chinchillas. In Mechanisms of damage and protection. Ann NY Acad Sci 884:152–170.

Do K, Baker K, Praetorius M, Staecker H (2004) A mouse model of implantation trauma. Int Cong Ser 1273:167–170.

Drescher DG (1974) Noise-induced reduction of inner-ear microphonic response: dependence on body temperature. Science 185:273-274.

Drescher DG (1976) Effect of temperature on cochlear responses during and after exposure to noise. J Acoust Soc Amer 59:401-407.

Ekert PG, Silke J, Vaux DL (1999) Caspase inhibitors. Cell Death Differ 6:1081–1086.

Erkman L, McEvilly RJ, Luo L, Ryan AK, Hooshmand F, O'Connell SM, Keithley EM, Rapaport DH, Ryan AF, Rosenfeld MG (1996) Role of transcription factors Brn-3.1 and Brn-3.2 in auditory and visual system development. Nature 381:603–606.

Ernfors P, Van De Water T, Loring J, Jaenisch R (1995) Complementary roles of BDNF and NT-3 in vestibular and auditory development. Neuron 14:1153–1164.

Ernfors P, Duan ML, El Shamy WM, Canlon B (1996) Protection of auditory neurons from aminoglycoside toxicity by neurotrophin-3. Nat Med 2:463–467.

Eshraghi AA (2006) Prevention of cochlear implant electrode damage. Curr Opin Otolaryngol HNS 14:323–328.

Eshraghi AA, Van De Water TR (2006) Cochlear implantation trauma and noise-induced hearing loss: apoptosis and therapeutic strategies. Anat Rec A Discov Mol Cell Evol Biol 288:473–481.

Eshraghi AA, Bublik M, Van De Water TR (2006a) Mechanisms of chemotherapeutic-induced hearing loss and otoprotection. Drug Discov Today Dis Mech 3:125–130.

Eshraghi AA, He J, Mou CH, Polak M, Zine A, Bonny C, Balkany TJ, Van De Water TR (2006b) D-JNKI-1 treatment prevents the progression of hearing loss in a model of cochlear implantation trauma. Otol Neurotol 27:504–511.

Eshraghi AA, Adil E, He J, Graves R, Balkany TJ, Van De Water TR (2007a) Local dexamethasone therapy conserves hearing in an animal model of electrode insertion trauma. Otol Neurotol 28:842–849.

Eshraghi AA, Wang J, Adil E, He J, Zine A, Bublick M, Bonny C, Puel JL, Balkany TJ, Van De Water TR (2007b) Blocking c-Jun-N-terminal kinase signaling can prevent hearing loss induced by both electrode insertion trauma and neomycin ototoxicity. Hear Res 226:168–177.

Evans P, Halliwell B (1999) Free radicals and hearing. Cause, consequence, and criteria. Ann NY Acad Sci 884:19–40.

Fairfield DA, Kanicki AC, Lomax MI, Altschuler RA (2004) Induction of heat shock protein 32 (Hsp32) in the rat cochlea following hyperthermia. Hear Res 188:1–11.

Fairfield DA, Lomax MI, Dootz GA, Chen S, Galecki AT, Benjamin IJ, Dolan DF, Altschuler RA (2005) Heat shock factor 1–deficient mice exhibit decreased recovery of hearing following noise overstimulation. J Neurosci Res 81:589–596.

Feghali J, Liu W, Van De Water TR (2001) L-N-Acetyl-cysteine protection against cisplatin-induced auditory neuronal and hair cell toxicity. Laryngoscope 111:1147–1155.

Fischel-Ghodsian N (1999) Genetic factors in aminoglycoside toxicity. Ann NY Acad Sci 884:99–109.

Fischel-Ghodsian N, Prezant TR, Bu X, Oztas S (1993) Mitochondrial ribosomal RNA gene mutation in a patient with sporadic aminoglycoside ototoxicity. Am J Otolaryngol 14:399–403.

Forge A (1985) Outer hair cell loss and supporting cell expansion following chronic gentamicin treatment. Hear Res 19:171–182.

Forge A, Harpur ES (2000) Ototoxicity. In Ballantyne B, Marrs T, Syversen T (eds) General and Applied Toxicology. Basingstoke: Macmillan, pp 775–801.

Forge A, Li L (2000) Apoptotic death of hair cells in mammalian vestibular sensory epithelia. Hear Res 139:97–115.

Forge A, Schacht J (2000) Aminoglycoside antibiotics. Audiol Neurootol 5:3–22.

Forge A, Li L, Nevill G (1998) Hair cell recovery in the vestibular sensory epithelia of mature guinea pigs. J Comp Neurol 397:69–88.

Forge A, Becker D, Casalotti S, Edwards J, Marziano N, Nevill G (2003) Gap junctions in the inner ear: comparison of distribution patterns in different vertebrates and assessement of connexin composition in mammals. J Comp Neurol 467:207–231.

Fraser A, McCarthy N, Evans GI (1996) Biochemistry of cell death. Curr Opin Neurobiol 6:71–80.

Fritzsch B (2003) Development of inner ear afferent connections: forming primary neurons and connecting them to the developing sensory epithelia. Brain Res Bull 60:423–433.

Fritzsch B, Tessarollo L, Coppola E, Reichardt LF (2004) Neurotrophins in the ear: their roles in sensory neuron survival and fiber guidance. Prog Brain Res 146:265–278.

Fritzsch B, Matei VA, Nichols DH, Bermingham N, Jones K, Beisel KW, Wang VY (2005) Atoh1 null mice show directed afferent fiber growth to undifferentiated ear sensory epithelia followed by incomplete fiber retention. Dev Dyn 233:570–583.

Gale JE, Marcotti W, Kennedy HJ, Kros CJ, Richardson GP (2001) FM1-43 dye behaves as a permeant blocker of the hair-cell mechanotransducer channel. J Neurosci 21:7013–7025.

Gale JE, Meyers JR, Periasamy A, Corwin JT (2002) Survival of bundleless hair cells and subsequent bundle replacement in the bullfrog's saccule. J Neurobiol 50:81–92.

Gantz, BJ Turner C (2004) Combining acoustic and electrical speech processing: Iowa/nucleus hybrid implant. Acta Otolaryngol 124:344–347.

Garetz SL, Altschuler RA, Schacht J (1994) Attenuation of gentamicin ototoxicity by glutathione in the guinea pig in vivo. Hear Res 77:81–87.

Gillespie LN, Shepherd RK (2005) Clinical application of neurotrophic factors: the potential for primary auditory neuron protection. Eur J Neurosci 22:2123–2133.

Gillespie LN, Clark GM, Bartlett PF, Marzella PL (2003) BDNF-induced survival of auditory neurons in vivo: Cessation of treatment leads to accelerated loss of survival effects. J Neurosci Res 71:785–790.

Gilliespie LN, Clark GM, Marzella PL (2004) Delayed neurotrophin treatment supports auditory neuron survival in deaf guinea pigs. NeuroReport 15:1121–1125.

Goldfarb A, Avraham KB (2002) Genetics of deafness:recent advances and clinical implications. J Basic Clin Physiol Pharmacol 13:75–88.

Goll DE, Thompson VF, Li H, Wei W, Cong J (2003) The calpain system. Physiol Rev 83:731–801.

Green DR, Kroemer G (2004) The pathophysiology of mitochondrial cell death. Science 305:626–629.

Grifa A, Wagner CA, D'Ambrosio L, Melchionda S, Bernardi F, Lopez-Bigas N, Rabionet R, Arbones M, Monica MD, Estivill X, Zelante L, Lang F, Gasparini P (1999) Mutations in GJB6 cause nonsyndromic autosomal dominant deafness at DFNA3 locus. Nat Genet 23:16–18.

Guan MX, Fischel-Ghodsian N, Attardi G (2000) A biochemical basis for the inherited susceptibility to aminoglycoside ototoxicity. Hum Mol Genet 9:1787–1793.

Guitton MJ, Wang J, Puel JL (2004) New pharmacological strategies to restore hearing and treat tinnitus. Acta Otolaryngol 124:411–415.

Guzman J, Ruiz J, Eshraghi AA, Polak M, Garnham C, Balkany TJ, Van De Water TR (2006) Triamcinolone acetonide protects auditory hair cells from 4–hyrdoxy-2, 3–nonenal (HNE) ototoxicity in vitro. Acta Otolaryngol 162:685–690.

Hartnick CJ, Staecker H, Malgrange B, Lefebvre PP, Liu W, Moonen G, Van De Water TR (1996) Neurotrophic effect of BDNF and CNTF, alone and in combination, on postnatal day 5 rat acoustic ganglion neurons. J Neurobiol 30:246–254.

Hawkins JE Jr (1973) Comparative otopathology: aging, noise, and ototoxic drugs. Adv Otorhinolaryngol 20:125–141.

Hegarty JL, Kay AR, Green SH (1997) Tropic support of cultured spiral ganglion neurons by depolarization exceeds and is additive with that by neurotrophins or cAMP and requires elevation of $[Ca^{2+}]i$ within a set range. J Neurosci 17:1959–1970.

Henderson D, Hamernik RP (1986) Impulse noise: critical review. J Acoust Soc Am 80:569–584.

Henderson D, McFadden SL, Liu CC, Hight N, Zheng XY (1999) The role of antioxidants in protection from impulse noise. Ann NY Acad Sci 884:368–380.

Henry KR (2003) Hyperthermia exacerbates and hypothermia protects from noise-induced threshold elevation of the cochlear nerve envelope response in the C57BL/6J mouse. Hear Res 179:88–96.

Henry KR, Chole RA (1980) Genotypic differences in behavioral, physiological and anatomical expressions of age-related hearing loss in the laboratory mouse. Audiology 19:369–383.

Henry KR, Chole RA (1984) Hypothermia protects the cochlea from noise damage. Hear Res 16:225–230.

Hequembourg S, Liberman MC (2001) Spiral ligament pathology: a major aspect of age-related cochlear degeneration in C57BL/6 mice. J Assoc Res Otolaryngol 2:118–129.

Hight NG, McFadden SL, Henderson D, Burkard RF, Nicotera T (2003) Noise-induced hearing loss in chinchillas pre-treated with glutathione monoethylester and R-PIA. Hear Res 179:21–32.

Himeno C, Komeda M, Izumikawa M, Takemura K, Yagi M, Weiping Y, Doi T, Kuriyama H, Miller JM, Yamashita T (2002) Intra-cochlear administration of dexamethasone attenuates aminoglycoside ototoxicity in the guinea pig. Hear Res 167:61–70.

Hu BH, Henderson D, Nicotera TM (2006) Extremely rapid induction of outer hair cell apoptosis in the chinchilla cochlea following exposure to impulse noise. Hear Res 211:16–25.

Hutchin T, Haworth I, Higashi K, Fischel-Ghodsian N, Stoneking M, Saha N, Arnos C, Cortopassi G (1993) A molecular basis for human hypersensitivity to aminoglycoside antibiotics. Nucleic Acids Res 21:4174–4179.

Hyodo J, Hakuba N, Koga K, Watanabe F, Shuduo M, Taniguchi M, Gyo K (2001) Hypothermia reduces glutamate efflux in perilymph following transient ischemia. NeuroReport 12:1983–1987.

Izumikawa M, Minoda R, Kawamoto K, Abrashkin KA, Swiderski DL, Dolan DF, Brough DE, Raphael Y (2005) Auditory hair cell replacement and hearing improvement by *Atoh1* gene therapy in deaf animals. Nat Med 11:240–250.

Jaattela M, Tschopp J (2003) Caspase-independent cell death in T lymphocytes. Nat Immunol 4:416–23.

Jakob U, Buchner J (1994) Assisting spontaneity: the role of Hsp90 and small Hsps as molecular chaperones. Trends Biochem Sci 19:205–211.

Janas JD, Cotanche DA, Rubel EW (1995) Avian cochlear hair cell regeneration: stereological analyses of damage and recovery from a single high dose of gentamicin. Hear Res 92:17–29.

Jiang H, Sha SH, Forge A, Schacht J (2006) Caspase-independent pathways of hair cell death induced by kanamycin in vivo. Cell Death Differ 13:20–30.

Johnson KR, Zheng QY (2002) Ahl2, a second locus affecting age-related hearing loss in mice. Genomics 80:461–464.

Johnson KR, Erway LC, Cook SA, Willott JF, Zheng QY (1997) A major gene affecting age-related hearing loss in C57BL/6J mice. Hear Res 114:83–92.

Johnson KR, Zheng QY, Erway LC (2000) A major gene affecting age-related hearing loss is common to at least ten inbred strains of mice. Genomics 70:171–180.

Kanzaki S, Stover T, Kawamoto K, Prieskorn DM, Altschuler RA, Miller JM, Raphael Y (2002) Glial cell line-derived neurotrophic factor and chronic electrical stimulation prevent VIII cranial nerve degeneration following denervation. J Comp Neurol 454:350–360.

Katoh I, Tomimori Y, Ikawa Y, Kurata S (2004) Dimerization and processing of procaspase-9 by redox stress in mitochondria. J Biol Chem 279:15515–15523.

Kawamoto K, Sha SH, Minoda R, Izumikawa M, Kuriyama H, Schacht J, Raphael Y (2004) Antioxidant gene therapy can protect hearing and hair cells from ototoxicity. Mol Ther 9:173–181.

Keithley EM, Canto C, Zheng QY, Wang X, Fischel-Ghodsian N, Johnson KR (2005) Cu/Zn superoxide dismutase and age-related hearing loss. Hear Res 209:76–85.

Kiefer J, Gstoettner W, Baumgartner W, Pok SM, Tillein J, Te Q, von Ilberg C (2004) Conservation of low-frequency hearing in cochlear implantation. Acta Otolaryngol 124:272-280.

Kikuchi T, Adams JC, Paul DL, Kimura RS (1994) Gap junction systems in the rat vestibular labyrinth: immunohistochemical and ultrastructural analysis. Acta Otolaryngol 114:520–528.

Kikuchi T, Kimura RS, Paul DL, Takasaka T, Adams JC (2000) Gap junction systems in the mammalian cochlea. Brain Res Brain Res Rev 32:163–166.

Kil J, Pierce C, Tran H, Gu R, Lynch ED (2007) Ebselen treatment reduces noise induced hearing loss via the mimicry and induction of glutathione peroxidase. Hear Res 226:44–51.

Kim WY, Fritzsch B, Serls A, Bakel LA, Huang EJ, Reichardt LF, Barth DS, Lee JE (2001) Neuro D-null mice are deaf due to a severe loss of the inner ear sensory neurons during development. Development 128:417–426.

Kopke RD, Liu W, Gabaizadeh R, Jacono A, Feghali J, Spray D, Garcia P, Steinman H, Malgrange B, Ruben RJ, Rybak L, Van De Water TR (1997) Use of organotypic cultures of Corti's organ to study the protective effects of antioxidant molecules on cisplatin-induced damage of auditory hair cells. Am J Otol 18:559–571.

Kopke RD, Weisskopf PA, Boone JL, Jackson RL, Wester DC, Hoffer ME, Lambert DC, Charon CC, Ding DL, McBride D (2000) Reduction of noise-induced hearing loss using L-NAC and salicylate in the chinchilla. Hear Res 149:138–146.

Kopke RD, Hoffer ME, Webster D, O'Leray MJ, Jackson RL (2001a) Targeted topical steroid therapy in sudden sensorineural haering loss. Otol Neurotol 22: 475–479.

Kopke RD, Jackson RL, Li G, Rasmussen MD, Hoffer ME, Frenz DA, Costello M, Schultheiss P, Van De Water TR (2001b) Growth factor treatment enhances vestibular hair cell renewal and results in improved vestibular function. Proc Natl Acad Sci USA 98:5886–5891.

Kotecha B, Richardson GP (1994) Ototoxicity in vitro: effects of neomycin, gentamicin, dihydrostreptomycin, amikacin, spectinomycin, neamine, spermine and poly-L-lysine. Hear Res 73:173–184.

Kuan CY, Burke RE (2005) Targeting the JNK signaling pathway for stroke and Parkinson's diseases therapy. Curr Drug Targets CNS Neurol Disord 4:63–67.

Kujawa SG, Liberman MC (1999) Long-term sound conditioning enhances cochlear sensitivity. J Neurophysiol 82:863–873.

Kyriakis LM, Banerjee P, Nikolakaki E, Dai T, Rubie EA, Ahmad MF, Avruch J, Woodjett JR (1994) The stress-activated protein kinase subfamily of c-Jun kinases. Nature 369:156-160.

Lallemend F, Lefebvre PP, Hans G, Rigo JM, Van De Water TR, Moonen G, Malgrange B (2003) Substance P protects spiral ganglion neurons from apoptosis via PKC-Ca^{2+}-MAPK/ERK pathways. Neurochem 87:508–521.

Lallemend F, Hadjab S, Hans G, Moonen G, Lefebvre PP, Malgrange B (2005) Activation of protein kinase Cbeta1 constitutes a new neurotrophic pathway for deafferented spiral ganglion neurons. J Cell Sci 118:4511–4525.

Laurell G, Bagger-Sjöbäck D (1991) Dose-dependent inner ear changes after i.v. administration of cisplatin. J Otolaryngol 20:158–167.

Lautermann J, McLaren J, Schacht J (1995a) Glutathione protection against gentamicin ototoxicity depends on nutritional status. Hear Res 86:15–24.

Lautermann J, Song B, McLaren J, Schacht J (1995b) Diet is a risk factor in cisplatin ototoxicity. Hear Res 88:47–53.

Lautermann J, ten Cate WJ, Altenhoff P, Grummer R, Traub O, Frank H, Jahnke K, Winterhager E (1998) Expression of the gap-junction connexins 26 and 30 in the rat cochlea. Cell Tissue Res 294:415–420.

Lefebvre PP, Van De Water TR, Weber T, Rogister B, Moonen G (1991) Growth factor interactions in cultures of dissociated adult acoustic ganglia: neurotrophic effects. Brn Res 567:306–312.

Lefebvre PP, Martin D, Staecker H, Weber T, Moonen G, Van De Water TR (1992) TGF β1 expression is initiated in adult auditory neurons by sectioning of the auditory nerve. NeuroReport 3:295–298.

Leonova EV, Fairfield DA, Lomax MI, Altschuler RA (2002) Constitutive expression of Hsp27 in the rat cochlea. Hear Res 163:61–70.

Li G, Frenz DA, Brahmblatt S, Feghali JG, Ruben RJ, Berggren D, Arezzo J, Van De Water TR (2001) Round window membrane delivery of L-methionine provides protection from cisplatin ototoxicity without compromising chemotherapeutic efficacy. Neurotoxicology 22:163–176.

Li G, Sha SH, Zotova E, Arezzo J, Van De Water T, Schacht J (2002) Salicylate protects hearing and kidney function from cisplatin toxicity without compromising its oncolytic action. Lab Invest 82:585–596.

Li HS, Borg E (1991) Age-related loss of auditory sensitivity in two mouse genotypes. Acta Otolaryngol 111:827–834.

Li L, Forge A (1995) Cultured explants of the vestibular sensory epithelia from adult guinea pigs and effects of gentamicin: a model for examination of hair cell loss and epithelial repair mechanisms. Audit Neurosci 1:111–125.

Li L, Nevill G, Forge A (1995) Two modes of hair cell loss from the vestibular sensory epithelia of the guinea pig inner ear. J Comp Neurol 355:405–417.

Li S, Price SM, Cahill H, Ryugo DK, Shen MM, Xiang M (2002) Hearing loss caused by progressive degeneration of cochlear hair cells in mice deficient for the Barhl1 homeobox gene. Development 129:3523–3532.

Lim HH, Jenkins OH, Myers MW, Miller JM, Altschuler RA (1993) Detection of HSP 72 synthesis after acoustic overstimulation in rat cochlea. Hear Res 69:146–150.

Lindeman HH (1969) Regional differences in sensitivity of the vestibular sensory epithelia to ototoxic antibiotics. Acta Otolaryngol 67:177–189.

Liu W, Staecker H, Stupak H, Malgrange B, Lefebvre P, Van De Water TR (1998) Caspase inhibitors prevent cisplatin-induced apoptosis of auditory sensory cells. NeuroReport 9:2609–2614.

Lopez I, Honrubia V, Lee SC, Schoeman G, Beykirch K (1997) Quantification of the process of hair cell loss and recovery in the chinchilla crista ampullaris after gentamicin treatment. Int J Dev Neurosci 15:447–461.

Lynch ED, Kil J (2005) Compounds for the prevention and treatment of noise-induced hearing loss. Drug Discov Today 10:1291–1298.

Lynch ED, Gu R, Pierce C, Kil J (2005) Combined oral delivery of ebselen and allopurinol reduces multiple cisplatin toxicities in rat breast and ovarian cancer models while enhancing anti-tumor activity. Anticancer Drugs 16:569–579.

Ma Q, Chen Z, del Barco Barrantes, de la Pompa JL, Anderson DJ (1998) Neurogenin-1 is essential for the determination of neuronal precursors for proximal cranial sensory anlagen Neuron 20:469–482.

Ma Q, Anderson DJ, Fritzsch B (2000) Neurogenin 1 null mutant ears develop fewer, morphologically normal hair cells in smaller sensory epithelia devoid of innervation. J Assoc Res Otolaryngol 1:129-143.

Marcus DC, Wu T, Wangemann P, Kofuji P (2002) KCNJ10 (Kir4.1) potassium channel knockout abolishes endocochlear potential. Am J Physiol Cell Physiol 282:C403–407.

Marion DW, Penrod LE, Kelsey SF, Obrist WD, Kochanek PM, Palmer AM, Wisnewski SR, DeKosy ST (1997) Treatment of traumatic brain injury with moderate hypothermia. N Engl J Med 336:540–546.

Marzella PL, Clark GM (1999) Growth factors, auditory neurones and cochlear implants: a review. Acta Otolaryngol 119:407–412.

Matsui JI, Ogilvie JM, Warchol ME (2002) Inhibition of caspases prevents ototoxic and ongoing hair cell death. J Neurosci 22:1218–1227.

Matsui JI, Haque A, Huss D, Messana EP, Alosi JA, Roberson DW, Cotanche DA, Dickman JD, Warchol ME (2003) Caspase inhibitors promote vestibular hair cell survival and function after aminoglycoside treatment in vivo. J Neurosci 23:6111–6122.

Matsui JI, Gale JE, Warchol ME (2004) Critical signaling events during the aminoglycoside-induced death of sensory hair cells in vitro. J Neurobiol 61:250–266.

McDowell B, Davies S, Forge A (1989) The effect of gentamicin-induced hair cell loss on the tight junctions of the reticular lamina. Hear Res 40:221–232.

McFadden SL, Henderson D, Shen YH (1997) Low-frequency 'conditioning' provides long-term protection from noise-induced threshold shifts in chinchillas. Hear Res 103:142–150.

McFadden SL, Ding D, Burkard RF, Jiang H, Reaume AG, Flood DG, Salvi RJ (1999a) Cu/Zn SOD deficiency potentiates hearing loss and cochlear pathology in aged 129, CD-1 mice. J Comp Neurol 413:101–112.

McFadden SL, Ding D, Reaume AG, Flood DG, Salvi RJ (1999b) Age-related cochlear hair cell loss is enhanced in mice lacking copper/zinc superoxide dismutase. Neurobiol Aging 20:1–8.

McFadden SL, Ding D, Salvemini D, Salvi RJ (2003) M40403, a superoxide dismutase mimetic, protects cochlear hair cells from gentamicin, but not cisplatin toxicity. Toxicol Appl Pharmacol 186:46–54.

Meijer AJ, Codogno P (2006) Signalling and autophagy regulation in health, aging and disease. Mol Asp Med 27: 411–425

Meiteles LZ, Raphael Y (1994) Scar formation in the vestibular sensory epithelium after aminoglycoside toxicity. Hear Res 79:26–38.

Miller JB, Girgenrath M (2006) The role of apoptosis in neuromuscular diseases and prospectus for anti-apoptosis therapy. Trends Mol Med 12:279–285.

Miller JM, Chi DH, O'Keeffe LJ, Kruszka P, Raphael Y, Altschuler RA (1997) Neurotrophins can enhance spiral ganglion cell survival after inner hair cell loss. Int J Dev Neurosci 15:631–643.

Minami SB, Sha SH, Schacht J (2004) Antioxidant protection in a new animal model of cisplatin-induced ototoxicity. Hear Res 198:137–143.

Monks J, Rosner D, Geske FJ, Lehman L, Hanson L, Neville MC, Fadok VA (2005) Epithelial cells as phagocytes: apoptotic epithelial cells are engulfed by mammary alveolar epithelial cells and repress inflammatory mediator release. Cell Death Differ 12:107–114.

Nakayama M, Helfert RH, Konrad HR, Caspary DM (1994) Scanning electron microscopic evaluation of age-related changes in the rat vestibular epithelium. Otolaryngol HNS 111:799–806.

Nicotera TM, Hu BH, Henderson D (2003) The caspase pathway in noise-induced apoptosis of the chinchilla cochlea. J Asoc Res Otolaryngol 4:466–477.

Noushi F, Richardson RT, Hardman J, Clark G, O'Leary S (2005) Delivery of neurotrophin-3 to the cochlea using alginate beads. Otol Neurotol 26:528–533.

Oh SH, Yu WS, Song BH, Lim D, Koo JW, Chang SO, Kim CS (2000) Expression of heat shock protein 72 in rat cochlea with cisplatin-induced acute ototoxicity. Acta Otolaryngol 120:146–150.

Ohlemiller KK, McFadden SL, Ding DL, Flood DG, Reaume AG, Hoffman EK, Scott RW, Wright JS, Putcha GV, Salvi RJ (1999) Targeted deletion of the cytosolic Cu/Zn-superoxide dismutase gene (*Sod1*) increases susceptibility to noise-induced hearing loss. Audiol Neurootol 4:237–246.

Onteniente B (2004) Natural and synthetic inhibitors of caspases: targets for novel drugs. Curr Drug Targets CNS Neurol Disord 3:333–340.

Parnes LS, Sun AH, Freeman DJ (1999) Corticosteroid pharmacokinetics in the iner ear fluids: an animal study followed by clinical application. Otol Neurotol 109:1–17.

Patuzzi RB, Yates GK, Johnstone BM (1989) Changes in cochlear microphonic and neural sensitivity produced by acoustic trauma. Hear Res 39:189–202.

Petit C (2006) From deafness genes to hearing mechanisms: harmony and counterpoint. Trends Mol Med 12:57–64.

Pickles JO, Osborne MP, Comis SD (1987) Vulnerability of tip links between stereocilia to acoustic trauma in the guinea pig. Hear Res 25:173–183.

Pirvola U, Ylikoski J (2003) Neurotrophic factors during inner ear development. Curr Top Dev Biol. 51:207–223.

Pirvola U, Ylikoski J, Palgi J, Lehtonen E, Arumae U, Saarma M (1992) Brain-derived neurotrophic factor and neurotrophin 3 mRNAs in the peripheral target fields of developing inner ear ganglia. Proc Natl Acad Sci USA 89:9915–9919.

Pirvola U, Arumae U, Moshnyakov M, Palgi J, Saarma M, Ylikoski J (1994) Coordinated expression and function of neurotrophins and their receptors in the rat inner ear during target innervation. Hear Res 75:131–144.

Pirvola U, Xing-Qun L, Virkkala J, Saarma M, Murakata C, Camoratto AM, Walton KM, Ylikoski J (2000) Rescue of hearing, auditory hair cells, and neurons by CEP-1347/KT7515, an inhibitor of c-Jun N-terminal kinase activation. J Neurosci 20:43–50.

Plontke S, Lowenheim H, Preyer S, Liens P, Dietz K, Koitschev A (2005) Outcomes research analysis of continuous intratympanic glucocorticoid delivered in patients with acute severe to profound hearing loss: basis for randomized controlled trials. Acta Otolaryngol 125:830–839.

Prezant TR, Agapian JV, Bohlman MC, Bu X, Oztas S, Qiu WQ, Arnos KS, Cortopassi GA, Jaber L, Rotter JI, et al. (1993) Mitochondrial ribosomal RNA mutation associated with both antibiotic-induced and non-syndromic deafness. Nat Genet 4:289–294.

Puel JL, Pujol R, Tribillac F, Ladrech S, Eybalin M (1994) Excitatory amino acid antagonists protect cochlear auditory neurons from excitotoxicity. J Comp Neurol 341:241–256.

Pujol R, Puel JL (1999) Excitotoxicity, synaptic repair, and functional recovery in the mammalian cochlea: a review of recent findings. Ann NY Acad Sci 884:249–254.

Quaranta A, Portalatini P, Henderson D (1998) Temporary and permanent threshold shift: an overview. Scand Audiol Suppl 48:75–86.

Raft S, Nowotschin S, Liao J, Morrow BE (2004) Suppression of neural fate and control of inner ear morphogenesis by *Tbx1*. Development 131:1801–1812.

Raphael Y (2002) Cochlear pathology, sensory cell death and regeneration. Br Med Bull 63:25–38.

Raphael Y, Altschuler RA (1991) Reorganization of cytoskeletal and junctional proteins during cochlear hair cell degeneration. Cell Motil Cytoskel 18:215–227.

Rarey KE, Curtis LM (1996) Receptors for glutocorticoids in the human inner ear. Otolaryngol HNS 115:38–41.

Rarey KE, Gerhardt KJ, Curtis LM, ten Cate WJ (1995) Effect of stress on cochlear glucocorticoid protein: acoustic stress. Hear Res 82:135–138.

Ravi R, Somani SM, Rybak LP (1995) Mechanism of cisplatin ototoxicity: antioxidant system. Pharmacol Toxicol 76:386–394.

Rebillard G, Ruel J, Nouvian R, Saleh H, Pujol R, Dehnes Y, Raymond J, Puel JL, Devau G (2003) Glutamate transporters in the guinea-pig cochlea: partial mRNA sequences, cellular expression and functional implications. Eur J Neurosci 17:83–92.

Reggiori F, Klionsky DJ (2005) Autophagosomes: biogenesis from scratch? Curr Opin Cell Biol 17:415–422.

Reser D, Rho M, Dewan D, Herbst L, Li G, Stupak H, Zur K, Romaine J, Goldbloom L, Kopke R, Frenz D, Arezzo J, Van De Water, TR (1999) L- and D-methionine provide equivalent long term protection against CDDP-induced ototoxicity in vivo, with partial in vitro and in vivo retention of antineoplastic activity. Neurotoxicology 20:731–748.

Richardson GP, Russell IJ (1991) Cochlear cultures as a model system for studying aminoglycoside induced ototoxicity. Hear Res 53:293–311.

Roberg K, Kagedal K, Ollinger K (2002) Microinjection of cathepsin d induces caspase-dependent apoptosis in fibroblasts. Am J Pathol 161:89–96.

Rozengurt N, Lopez I, Chiu CS, Kofuji P, Lester HA, Neusch C (2003) Time course of inner ear degeneration and deafness in mice lacking the Kir4.1 potassium channel subunit. Hear Res 177:71–80.

Rubel EW, Fritzsch B (2002) Auditory system development: primary auditory neurons and their targets. Annu Rev Neurosci 25:51–101.

Rubinzstein DC, DiFaglia M, Heintz N, Nixon RA, Qin Z-H, Ravikumar B, Stefanis L, Tolkovsky A (2005) Autophagy and its possible role in nervous system diseases, damage and repair. Autophagy 1:11–22.

Ruiz J, Guzman J, Polak M, Eshraghi AA, Balkany TJ, Van De Water TR (2006) Glutathione ester protects against hydroxynonenal-induced loss of auditory hair cells. Otolaryngol Head Neck Surg 135:792–797.

Russell IJ, Kossl M (1992) Sensory transduction and frequency selectivity in the basal turn of the guinea-pig cochlea. Philos Trans Royal Soc Lond B Biol Sci 336:317–324.

Rybak LP, Kelly T (2003) Ototoxicity: bioprotective mechanisms. Curr Opin Otolaryngol HNS 11:328–333.

Rybak LP, Whitworth CA (2005) Ototoxicity: therapeutic opportunities. Drug Discov Today 10:1313–1321.

Rzadzinska AK, Schneider ME, Davies C, Riordan GP, Kachar B (2004) An actin molecular treadmill and myosins maintain stereocilia functional architecture and self-renewal. J Cell Biol 164:887–897.

Sabag AD, Avraham KB (2005) Connexins in hearing loss: a comprehensive overview. J Basic Clin Physiol Pharmacol 16:101–116.

Salt AN, Plontke SK (2005) Local inner-ear drug delivery and pharmacokinetics. Drug Discov Today 10:1299–1306.

Saunders JC, Dear SP, Schneider ME (1985) The anatomical consequences of acoustic injury: a review and tutorial. J Acoust Soc Am 78:833–860.

Savill J, Fadok V (2000) Corpse clearance defines the meaning of cell death. Nature 407:784–788.

Scarpidis U, Mandani D, Shoemaker C, Fletcher CH, Kojima K, Eshraghi AA, Staecker H, Lefebvre PP, Malgrange B, Balkany TJ, Van De Water TR (2003) Arrest of apoptosis in auditory neurons: implications for sensorineural preservation in cochlear implants. Otol Neurotol 24:409–417.

Schneider ME, Belyantseva IA, Azevedo RB, Kachar B (2002) Rapid renewal of auditory hair bundles. Nature 418:837–838.

Schubert D, Piasecki D (2001) Oxidative glutamate toxicity can be a component of the excitotoxicity cascade. J Neurosci 21:7455–7462.

Sha SH, Schacht J (1999) Salicylate attenuates gentamicin-induced ototoxicity. Lab Invest 79:807–813.

Sha SH, Taylor R, Forge A, Schacht J (2001a) Differential vulnerability of basal and apical hair cells is based on intrinsic susceptibility to free radicals. Hear Res 155:1–8.

Sha SH, Zajic G, Epstein CJ, Schacht J (2001b) Overexpression of copper/zinc-superoxide dismutase protects from kanamycin-induced hearing loss. Audiol Neurootol 6:117–123.

Sha SH, Qui JH, Schacht J (2006) Aspirin to prevent gentamicin-induced hearing loss. N Engl J Med 354:1856–1857.

Shepherd RK, Coco A, Epp SB, Crook JM (2005) Chronic depolarization enhances the trophic effects of brain-derived neurotrophic factor in rescuing auditory neurons following a sensorineural hearing loss. J Comp Neurol 486:145–158.

Shinohara T, Bredberg G, Ulfendahl M, Pyykko I, Olivius NP, Kaksonen R, Lindstrom B, Altschuler R, Miller JM (2002) Neurotrophic factor intervention restores auditory function in deafened animals. Proc Natl Acad Sci USA 99:1657–1660.

Slepecky N (1986) Overview of mechanical damage to the inner ear: noise as a tool to probe cochlear function. Hear Res 22:307–321.

Sobkowicz HM, August BK, Slapnick SM (1992) Epithelial repair following mechanical injury of the developing organ of Corti in culture: an electron microscopic and autoradiographic study. Exp Neurol 115:44–49.

Sobkowicz HM, August BK, Slapnick SM (1996) Post-traumatic survival and recovery of the auditory sensory cells in culture. Acta Otolaryngol 116:257–262.

Sobkowicz HM, August BK, Slapnick SM (1997) Cellular interactions as a response to injury in the organ of Corti in culture. Int J Dev Neurosci 15:463–485.

Song BB, Anderson DJ, Schacht J (1997) Protection from gentamicin ototoxicity by iron chelators in guinea pig in vivo. J Pharmacol Exp Ther 282:369–377.

Spongr VP, Flood DG, Frisina RD, Salvi RJ (1997) Quantitative measures of hair cell loss in CBA and C57BL/6 mice throughout their life spans. J Acoust Soc Am 101:3546–3553.

Staecker H, Van De Water TR (1998) Factors controlling hair-cell regeneration/repair in the inner ear. Curr Opin Neurobiol 8:480–487.

Staecker H, Liu W, Hartnick C, Lefebvre PP, Malgrange B, Moonen G, Van De Water TR (1995) NT-3 combined with CNTF promotes survival of neurons in modiolus-spiral ganglion explants. NeuroReport. 6:1533–1537.

Staecker H, Kopke R, Malgrange B, Lefebvre P, Van De Water TR (1996) NT-3 and/or BDNF therapy prevents loss of auditory neurons following loss of hair cells. NeuroReport 7:89–94.

Staecker H, Gabaizadeh R, Federoff H, Van De Water TR (1998) Brain-derived neurotrophic factor gene therapy prevents spiral ganglion degeneration after hair cell loss. Otolaryngol HNS 119:7–13.

Steel KP, Kros CJ (2001) A genetic approach to understanding auditory function. Nat Genet 27:143–149.

Stone JS, Oesterle EC, Rubel EW (1998) Recent insights into regeneration of auditory and vestibular hair cells. Curr Opin Neurol 11:17–24.

Tahera Y, Meltser I, Johansson P, Bian Z, Stierna P, Hansson AC, Canlon B (2006) NF-κB mediated glucortoicosteroid response in the inner ear after acoustic trauma. J Neurosci Res 83:1066–1076.

Tahera Y, Meltser I, Johansson P, Salman H, Canlon B (2007) Sound conditioning protects hearing by activating the hypothalamic-pituitary-adrenal axis. Neurobiol Dis 25:189–197.

Takahashi K, Kusakari J, Kimura S, Wada T, Hara A (1996) The effect of methylprednisolone on acoustic trauma. Acta Otolaryngol 116:209–212.

Takemura K, Komeda M, Yagi M, Himeno C, Izumikawa M, Doi T, Kuriyama H, Miller JM, Yamashita T (2004) Direct inner ear infusion of dexamethasone attenuates noise-induced trauma to the guinea pig. Hear Res 196:58–68.

Takumida M, Anniko M (2005) Heat shock protein 70 delays gentamicin-induced vestibular hair cell death. Acta Otolaryngol 125:23–28.

Tan S, Schubert D, Maher P (2001) Oxytosis: a novel form of programmed cell death. Curr Top Med Chem 1:497–506.

Taylor R, Forge A (2005) Hair cell regeneration in sensory epithelia from the inner ear of a Urodele amphibian. J Comp Neurol, 484, 105–120

Terunuma T, Kawauchi S, Kajihara M, Takahashi S, Hara A (2003) Effect of acoustic stress on glucocorticoid receptor mRNA in the cochlea of the guinea pig. Brain Res Mol Brain Res 120:65–72.

Teubner B, Michel V, Pesch J, Lautermann J, Cohen-Salmon M, Sohl G, Jahnke K, Winterhager E, Herberhold C, Hardelin JP, Petit C, Willecke K (2003) Connexin30 (Gjb6)-deficiency causes severe hearing impairment and lack of endocochlear potential. Hum Mol Genet 12:13–21.

Teufert KB, Linthicum FH, Connell SS (2006) The effect of organ of Corti loss on ganglion cell survival in humans. Otol Neurotol 27:1146–1151.

Thornberry NA, Lazebnik Y (1998) Caspases: enemies within. Science 281:1213–1216.

Thorne PR, Gavin JB (1985) Changing relationships between structure and function in the cochlea during recovery from intense sound exposure. Ann Otol Rhinol Laryngol 94:81–86.

Vahava O, Morell R, Lynch ED, Weiss S, Kagan ME, Ahituv N, Morrow JE, Lee MK, Skvorak AB, Morton CC, Blumenfeld A, Frydman M, Friedman TB, King MC, Avraham KB (1998) Mutation in transcription factor POU4F3 associated with inherited progressive hearing loss in humans. Science 279:1950–1954.

Van De Water TR (1988) Tissue interactions and cell differentiation: neurone-sensory cell interaction during otic development. Development 103 (Suppl):185–193.

Van De Water TR, Staecker H, Ernfors P, Moonen G, Lefebvre PP (1996) Neurotrophic Factors as pharmacological agents for the treatment of injured auditory neurons. Ciba Found Symp 196:149–162.

Van De Water TR, Lallemend F, Eshraghi AA, Ahsan S, He J, Guzman J, Polak M, Malgrange B, Lefebvre PP, Staecker H, Balkany TJ (2004) Caspases, the enemy within, and their role in oxidative stress-induced apoptosis of inner ear sensory cells. Otol Neurotol 25:627–632.

van Ruijven MW, de Groot JC, Smoorenburg GF (2004) Time sequence of degeneration pattern in the guinea pig cochlea during cisplatin administration. A quantitative histological study. Hear Res 197:44–54.

Veuillet E, Gartner M, Champsaur G, Neidecker J, Collet L (1997) Effects of hypothermia on cochlear micromechanical properties in humans. J Neurol Sci 145:69–76.

Wang KK (2000) Calpain and caspase: can you tell the difference? Trends Neurosci 23:20–26.

Wang J, Van De Water TR, Bonny C, de Ribaupierre F, Puel JL, Zine A (2003) A peptide inhibitor of c-Jun N-terminal kinase protects against both aminoglycoside and acoustic trauma-induced auditory hair cell death and hearing loss. J Neurosci 23: 8596–8607.

Wang J, Ladrech S, Pujol R, Brabet P, Van De Water TR, Puel JL (2004) Caspase inhibitors, but not c-Jun NH2–terminal kinase inhibitor treatment, prevent cisplatin-induced hearing loss. Cancer Res 64:9217–9224.

Wang J, Ruel J, Ladrech S, Bonny C, Van De Water TR, Puel JL (2007) Inhibition of the JNK-mediated mitochondrial cell death pathway restores auditory function in sound exposed animals. Mol Pharmacol 71:654–666.

Wang, W, Grimmer, F, Van De Water, TR, Lufkin, T (2004) *Hmx2* and *Hmx3* homeobox genes direct development of the murine inner ear and hypothalamus and can be functionally replaced by *Drosophila Hmx*. Dev Cell 7:430–453.

Wang Y, Liberman MC (2002) Restraint stress and protection from acoustic injury in mice. Hear Res 165:96–102.

Wang Y, Hirose K, Liberman MC (2002) Dynamics of noise-induced cellular injury and repair in the mouse cochlea. J Assoc Res Otolaryngol 3:248–268.

Wangemann P (2002) K(+) cycling and the endocochlear potential. Hear Res 165:1–9.

Wangemann P, Schacht J (1996) Homeostatic Mechanisms in the Cochlea. In Dallos P, Popper AN, Fay RR (eds) The Cochlea. New York: Springer.

Watanabe F, Koga K, Hakuba N, Gyo K (2001) Hypothermia prevents hearing loss and progressive hair cell loss after transient cochlear ischemia in gerbils. Neuroscience 102:639–645.

Watson GM, Mire P, Hudson RR (1998) Repair of hair bundles in sea anemones by secreted proteins. Hear Res 115:119–128.

Wersall J (1995) Ototoxic antibiotics: a review. Acta Otolaryngol Suppl 519:26–29.

Winter J, Jakob U (2004) Beyond transcription—new mechanisms for the regulation of molecular chaperones. Crit Rev Biochem Mol Biol 39:297–317.

Wissel K, Wefstaedt P, Miller J, Lenarz T, Stover T (2006) Differential brain-derived neurotrophic factor and transforming growth factor-beta expression in the rat cochlea following deafness. NeuroReport 17:1297–1301.

Wright A (1983) The surface structures of the human vestibular apparatus. Clin Otolaryngol Allied Sci 8:53–63.

Wright A, Davis A, Bredberg G, Ulehlova L, Spencer H (1987) Hair cell distributions in the normal human cochlea. Acta Otolaryngol Suppl 444:1–48.

Wu WJ, Sha SH, McLaren JD, Kawamoto K, Raphael Y, Schacht J (2001) Aminoglycoside ototoxicity in adult CBA, C57BL and BALB mice and the Sprague-Dawley rat. Hear Res 158:165–178.

Xia AP, Ikeda K, Katori Y, Oshima T, Kikuchi T, Takasaka T (2000) Expression of connexin 31 in the developing mouse cochlea. NeuroReport 11:2449–2453.

Xia JH, Liu CY, Tang BS, Pan Q, Huang L, Dai HP, Zhang BR, Xie W, Hu DX, Zheng D, Shi XL, Wang DA, Xia K, Yu KP, Liao XD, Feng Y, Yang YF, Xiao JY, Xie DH, Huang JZ (1998) Mutations in the gene encoding gap junction protein beta-3 associated with autosomal dominant hearing impairment. Nat Genet 20:370–373.

Xiang M, Gan L, Li D, Chen ZY, Zhou L, O'Malley BW Jr, Klein W, Nathans J (1997) Essential role of POU-domain factor Brn-3c in auditory and vestibular hair cell development. Proc Natl Acad Sci USA 94:9445–9450.

Xiang M, Gao WQ, Hasson T, Shin JJ (1998) Requirement for Brn-3c in maturation and survival, but not in fate determination of inner ear hair cells. Development 125:3935–3946.

Xiang M, Maklad A, Pirvola U, Fritzsch B (2003) Brn3c null mutant mice show long term, incomplete retention of some afferent inner ear innervation. BMC Neurosci 4:2.

Yagi M, Kanzaki S, Kawamoto K, Shin B, Shah PP, Magal E, Sheng J, Raphael Y (2000) Spiral ganglion neurons are protected from degeneration by GDNF gene therapy. J Assoc Res Otolaryngol 1:315–325.

Yamagata T, Miller JM, Ulfendahl M, Olivius NP, Altschuler RA, Pyykko I, Bredberg G (2004) Delayed neurotrophic treatment preserves nerve survival and electrophysiological responsiveness in neomycin-deafened guinea pigs. J Neurosci Res 78:75–86.

Yamashita D, Jiang HY, Schacht J, Miller JM (2004) Delayed production of free radicals following noise exposure. Brain Res 1019:201–209.

Yamashita D, Jiang HY, Le Prell CG, Schacht J, Miller JM (2005) Post-exposure treatment attenuates noise-induced hearing loss. Neurosci 134:633–642.

Yamasoba T, Dolan DF (1998) The medial cochlear efferent system does not appear to contribute to the development of acquired resistance to acoustic trauma. Hear Res 120:143–151.

Yamasoba T, Dolan DF, Miller JM (1999) Acquired resistance to acoustic trauma by sound conditioning is primarily mediated by changes restricted to the cochlea, not by systemic responses. Hear Res 127:31–40.

Yang WP, Henderson D, Hu BH, Nicotera TM (2004) Quantitative analysis of apoptotic and necrotic outer hair cells after exposure to different levels of continuous noise. Hear Res 196:69–76.

Ylikoski J, Pirvola U, Moshnyakov M, Palgi J, Arumae U, Saarma M (1993) Expresión patterns of neurotrophin and their receptor mRNAs in the rat inner ear. Hear Res 65:69–78.

Ylikoski J, Xing-Qun L, Virkkala J, Pirvola U (2002) Blockade of c-Jun N-terminal kinase pathway attenuates gentamicin-induced cochlear and vestibular hair cell death. Hear Res 166:33–43.

Yoshida N, Liberman MC (2000) Sound conditioning reduces noise-induced permanent threshold shift in mice. Hear Res 148:213–219.

Yoshida N, Kristiansen A, Liberman MC (1999) Heat stress and protection from permanent acoustic injury in mice. J Neurosci 19:10116–10124.

Yoshida N, Hequembourg SJ, Atencio CA, Rosowski JJ, Liberman MC (2000) Acoustic injury in mice: 129/SvEv is exceptionally resistant to noise-induced hearing loss. Hear Res 141:97–106.

Yousefi S, Perozzo R, Schmidt I, Ziemiecki A, Schaffner T, Scapozza L, Brunner T, Simon H-U (2006) Calpain-mediated cleavage of Atg-5 switches autophagy to apoptosis. Nat Cell Biol 8:1124–1132.

Zafarullah M, Li WQ, Sylvester J, Ahmad M (2003) Molecular mechanisms of N-acetylcysteine actions. Cell Mol Life Sci 60:6–20.

Zhao Y, Yamoah EN, Gillespie PG (1996) Regeneration of broken tip links and restoration of mechanical transduction in hair cells. Proc Natl Acad Sci USA 93:15469–15474.

Zheng JL, Gao WQ (1996) Differential damage to auditory neurons and hair cells by ototoxins and neuroprotection by specific neurotrophins in rat cochlear organotypic cultures. Eur J Neurosci 8:1897–1905.

Zheng JL, Keller G, Gao WQ (1999) Immunocytochemical and morphological evidence for intracellular self-repair as an important contributor to mammalian hair cell recovery. J Neurosci 19:2161–2170.

Zhou Z, Lie Q, Davis RL (2005) Complex regulation of spiral ganglion neuron firing patterns by neurotrophin-3. J Neurosci 25:7558–7566.

Zine A, Van De Water TR (2004) The MAPL/JNK signaling pathway offers potential therapeutic targets for the prevention of acquired deafness. Curr Drug Targets CNS Neurol Disord 3:325–332.

Zuo J, Curtis LM (1995) Glucocorticoid expression in the post natal rat cochlea. Hear Res 87:220–227

7
Gene Arrays, Cell Lines, Stem Cells, and Sensory Regeneration in Mammalian Ears

MARCELO N. RIVOLTA AND MATTHEW C. HOLLEY

1. Introduction

Sensory regeneration in the mammalian inner ear involves replacement of lost hair cells and spiral ganglion neurons (SGNs). The irreversible loss of these cells is associated with most forms of hearing loss (Davis 1998), even if the primary cause may frequently originate in other cell types within or around the sensory epithelium (Hirose and Liberman 2003). Replacement of sensory hair cells is a natural property of mechanosensory epithelia in nonmammals (Corwin and Oberholtzer 1997; Stone and Rubel 2000). Although it is not naturally observed in the mammalian organ of Corti (Forge et al. 1998), functional hair cells can be derived from supporting cells in adult guinea pigs by transfection with the basic helix–loop–helix (bHLH) transcription factor *Atoh1* (Izumikawa et al. 2005). Further, hair cell replacement does occur naturally in mammalian vestibular epithelia, even if to a very limited extent (Forge et al. 1993). This suggests that hair cells could be replaced therapeutically if we can uncover the appropriate targets in terms of cells and signaling molecules. Supporting cells within the sensory epithelia are the most realistic cellular targets because they share a common precursor cell during development (Fekete et al. 1998; Lawlor et al. 1999; Fekete and Wu 2002), they are the source of new hair cells in nonmammals (Jones and Corwin 1996; Stone and Rubel 1999), and they are located appropriately for functional integration. Studies on transgenic and mutant mice (Fekete 1999; Anagnostopoulos 2002; Fekete and Wu 2002) have revealed numerous mammalian genes that regulate hair cell proliferation (Chen et al. 2003; Mantela et al. 2005; Sage et al. 2005), differentiation, and survival (Wallis et al. 2003) during development. The same applies to SGNs (Alsina et al. 2003; Fritzsch 2003). Thus the stage is set to explore the genetic programs and signaling pathways that must be targeted to stimulate sensory regeneration. Gene arrays, cell lines, and stem cells provide important opportunities toward this goal.

There is a large potential market for treating hearing loss (Holley 2005). About 1 in 1000 newborn children suffer profound hearing loss (Morton 1991) and more than one third of those older than 60 years of age suffer significant hearing loss (Davis 1995). The World Health Organization estimates that 250 million people worldwide suffer disabling hearing loss (www.who.int/pbd/deafness/en). There is also a large market for treating balance disorders, which are related to problems of mobility in elderly people. The potential biological therapies broadly include drug therapy, gene therapy, and cell transplantation. Successful drugs would provide the more attractive long-term solution because the technical challenges of delivering genes or cells and the associated need for invasive therapy are likely to be more expensive and labor intensive. Gene arrays, cell lines, and stem cells are valuable tools in the search for drug-based, thera-peutic solutions and they provide complementary experimental approaches for existing technologies such as ex vivo organ cultures and genetically modified animals.

Mammals do not present the most convenient experimental preparations. Most hair cells and associated neurons are born during embryonic development, are few in number, and become surrounded by several layers of bone. Much of what we know about their development has been primed or guided by studies on more accessible species of birds, amphibia, insects, and worms. These animal models have been remarkably powerful in the search for the basic programs underlying sensory development, which usually revolve around transcription factors whose functions have been well conserved through evolution (Bertrand et al. 2002). However, transcription factors do not make good drug targets. From this viewpoint we are more interested in components of the signaling mechanisms that activate them. These components may include membrane receptors, ion channels, G-protein–coupled receptors, proteases, and protein kinases (Vandenberg and Lummis 2000; Dahl et al. 2001; Cohen 2002; Neubig and Siderovski 2002). At this level there are significant mechanistic differences, not only between species but also between auditory and vestibular systems within the same species. For example, growth factors produce different responses in sensory epithelial cultures from chick and mouse (Kuntz and Oesterle 1998), and the fibroblast growth factor receptor 1 (FGFR1) is essential for proliferation and differentiation of auditory but not vestibular sensory cells in the mouse (Pirvola et al. 2002). Thus we need effective experimental preparations to search for drug targets in the relevant mammalian signaling pathways.

The process of drug discovery, screening, and development toward clinical trials requires a "staircase" of experimental tests from cells to whole animals. Although a great deal of fundamental knowledge will continue to emerge from studies on nonmammals, this chapter focuses more on recent advances with mammalian experimental systems. Analysis of natural and induced mutations in the mouse has provided valuable insights into various aspects of development, physiology, and disease (Steel and Kros 2001). At the next level down, organ-otypic cultures (Sobkowicz et al. 1993; Saffer et al. 1996) provide equally powerful in vitro tools, which have additional value for studies concerning ototoxicity and sensory regeneration (Warchol et al. 1993; Ding et al. 2002). Although simplified, these preparations are still complex in terms of cell composition and they are quantitatively limited. Thus it is more difficult to use them for studies on specific signaling pathways or for deeper analysis of the activation and targets of specific regulatory genes. With more than one cell type, the larger scale screening technologies such as gene arrays and proteomics are harder to interpret, especially if different cells respond in a reciprocal manner to the same stimulus. Cell lines have proved to be invaluable in other research fields but they have only recently emerged in hearing research (Rivolta and Holley 2002b). They are particularly useful in conjunction with gene arrays and proteomics. In this chapter we explore this relationship in the context of sensory repair and regeneration.

The application of new experimental approaches demands careful experimental design, which should be guided by the functional context of the question. There are relatively few studies of the inner ear based on cell lines, gene arrays, and stem cells but numerous studies in other research fields illustrate the relative power of different experimental designs. To make these comparisons we place our discussions in the context of fundamental questions concerning cell proliferation, differentiation, and cell death, processes that lie at the heart of normal developmental biology (Fig. 7.1a) and regeneration (Fig. 7.1b). We also discuss the biology and application of stem cells. Stem cells can be employed to effect functional repair in the central nervous system (CNS; Lovell-Badge 2001). They might be transplanted to damaged tissues or it may be possible to awaken regenerative responses from endogenous, tissue-specific stem cells (Nakatomi et al. 2002). We discuss questions of whether potential sensory cell precursors exist in the adult inner ear, whether we can stimulate existing cells to become precursors for sensory cells, and whether it is possible to replace sensory cells or their precursors by cell transplantation.

2. Application of Gene Arrays

The discovery of specific genes involved in inner ear function has provided fundamental knowledge to the field and is likely to continue at an increasing rate (Van Camp and Smith 2007; Morton 2002; Atar and Avraham 2005). Further understanding will demand functional studies and the ability to explore molecular interactions, signaling pathways, and gene networks. Gene arrays allow us to profile the expression of thousands of genes simultaneously. This opens up the opportunity to study complex biological processes more comprehensively. There are numerous levels at which gene arrays can be applied, corresponding to different organizational levels in gene network models (Ruan et al. 2004; Schlitt and Brazma 2005). The most elementary aim is simply to produce a list of genes expressed in cells, organs, or tissues at a given time. In modeling terms this would be equivalent to a parts list, possibly with information on absolute expression levels. The value of this kind of study is highly dependent on the

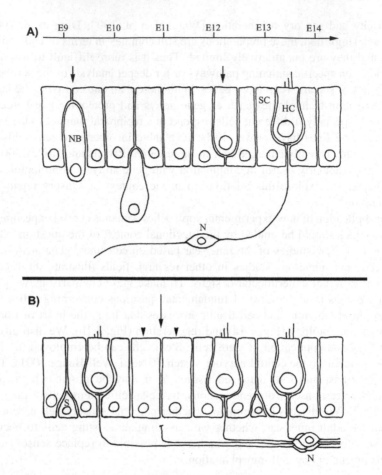

FIGURE 7.1. (**A**) This diagram represents events during the development of the sensory epithelia along a time line from E9 to E14, the 9–14th days of embryonic development in the mouse. After E9, neuroblasts (NB) are selected from the ventral otic epithelium, delaminate and form primary sensory neurons (N) in the cochlear or vestibular ganglia. Cell selection (arrow) probably involves the membrane receptor notch1. At E12, a similar selection process (arrow) leads to the differentiation of hair cells (HC) and supporting cells (SC). Cell selection occurs within a field of equivalent cells. Differentiated hair cells do not contact the basement membrane so that sensory epithelia can be classified as stratified epithelia. (**B**) This diagram represents a regenerative response. Here the sensory epithelium is differentiated and a lost hair cell (arrow) must be replaced by existing, differentiated supporting cells (arrowheads) that contact the basement membrane. There is currently no evidence for undifferentiated stem cells (S) in the epithelium. The lack of regeneration in mammals could be due to the inability of epithelial cells to proliferate and differentiate or to some inhibitory property of existing cells. The context for cell differentiation is quite different from that experienced during development. SGNs (N) occupy a completely different environment from that of the early neuroblasts. We know little of their potential to proliferate or for any associated cells to differentiate into neurons within the adult ganglion.

characterization of the source tissue. At the next level is the topology model, equivalent to a wiring diagram, which provides an indication of direct or indirect functional links between different genes within the list. These models are most effectively derived from gene deletion experiments and they have been exploited extensively in model organisms such as yeast and *Caenorhabitis elegans*, in which it is possible to delete many different genes individually in single cell types. Control logic models incorporate specific information concerning the effects of regulatory molecules. These include regulatory gene network models that describe transcription factors and associated components of cell signaling pathways (Levine and Davidson 2005). Developmental biologists commonly construct small transcriptional networks from gene deletions in vertebrates to describe the molecular regulation of tissue morphogenesis and cell differentiation. While such networks contain information about transcriptional programs, they often contain very little information about cell function, or how individual transcription factors influence cell behavior. Discovery of the links between individual transcription factors and specific signaling pathways is a crucial challenge for our understanding of development and for the identification of drug targets to stimulate regeneration. Dynamic network models aim to simulate and predict gene network behavior in real time following activation by intrinsic or extrinsic factors. The current experimental data available to build such models are extremely limited.

Construction of these different types of gene network model is not always the primary aim of a gene array experiment. For example, array experiments may be designed to identify differences between sensory epithelia before and after induced sensory damage. These kinds of experiment are highly complex and consideration of the different types of information required for the network models described in the preceding text provides a valuable insight into the kind of information that might be obtained. In reality, many researchers have used gene array data to identify and conduct further work on just a few genes linked to their specific field of interest. In this section we review the different types of gene array as they have been applied to research on the inner ear in the context of development and regeneration.

Gene arrays are a technological achievement that evolved from the decades-old dot blot technique. They are based on the same principle, hybridizing labeled RNA or DNA in solution to target DNA fragments immobilized to a solid support. The initial "macroarrays" consisted of DNA fragments spotted onto a nitrocellulose or nylon membrane and then probed with a sample labeled with a radioactive group. The spotted materials were often cDNA fragments derived from libraries. The advantage of this method is that it involves conventional isotopic hybridization techniques, without the need for expensive, sophisticated equipment, thus putting it within reach of most molecular biology laboratories. The approach is still applied today and has been used for several different studies of inner ear tissues but it is applicable to parallel screening of a relatively limited number of genes. The miniaturization introduced with microarrays now allows us to scan many thousands of genes in a single experiment. In a typical microarray many thousands of genes can be arranged in an area as small as $1\,cm^2$. There are two main types, the spotted cDNA arrays and the high-density, oligonucleotide arrays (Blackshaw and Livesey 2002; Ruan et al. 2004).

2.1 cDNA Arrays

cDNA arrays consist of a series of spots of cDNAs, normally obtained via polymerase chain reaction (PCR) amplification from isolated cDNA clones and arranged on a glass coverslip by a robot (Fig. 7.2). These arrays are hybridized with a dual sample. A cDNA pool obtained from one experimental condition and another obtained from a reference sample are labeled with two different fluorochromes, Cy3 (green) and Cy5 (red), and hybridized simultaneously to the array. After washing, the array is scanned at two different wavelengths, and the

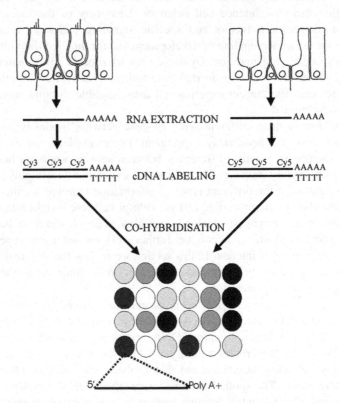

FIGURE 7.2. Representation of a hypothetical gene array experiment using cDNA microarrays. In this example, a preparation of sensory epithelia depleted of hair cells by aminoglycoside treatment is compared to an intact tissue. Samples of RNA from each preparation are reversed-transcribed into cDNA, incorporating two different fluorescent labels (typically Cy3 and Cy5). These labeled cDNAs are cohybridized to the same chip. The process is therefore competitive and ratiometric. In these arrays, each gene is represented by a single spot of cloned cDNA. Relative labels of expression are obtained by scanning the array at the two different wavelengths and determining their ratios.

ratio of the fluorescence intensities relates to the relative expression levels of given genes in either sample.

This approach has strength in that it allows the representation of cDNAs from tissue-specific libraries. Thus it is possible to screen experimental samples against large numbers of genes from more directly relevant inner ear cDNA libraries such as those produced for mouse and human ears (Heller 2002; Beisel et al. 2004; Pompeia et al. 2004; Powles et al. 2004). Powles et al. prepared a library, designed to identify genes expressed during early development of the inner ear, by subtracting cDNA from 800 mouse otic vesicles at embryonic day (E) 10.5 from cDNA derived from adult liver. Eighty genes from a total of 280 derived from the screen were followed up via in situ hybridization, providing a tailored resource for the production of cDNA arrays focused on this developmental period. Pompeia et al. derived expressed sequence tags (ESTs) from organs of Corti dissected from BALB/cJ 5- to 13-day-old mice, which covers the later stages of structural and functional maturation of hair cells and supporting cells. Beisel et al. generated a full-length, small-scale subtracted, nonamplified cDNA library of adult C57BL/6J 5- to 7-week-old mice. It is important to bear in mind that these two commonly used mouse strains (BALB/C and C57) develop early-onset hair cell loss. C57 show signs of hair cell loss by 8 weeks of age in the base of the cochlea. CBA mice, in contrast, retain most of their hair cells to 2 years of age.

2.1.1 Molecular Anatomy

In defining the molecular anatomy of tissues and cells the precision, characterization, and reproducibility of the tissue dissection and cell isolation are critical. Inner ear tissues are complex and the organ of Corti contains many different cell types. Some cell types, such as hair cells, exist in very low numbers and many of the potentially interesting genes are expressed at low levels. Thus the value of any gene array screen and an understanding of its limitations depend heavily on knowledge of the relative proportions of different cell types in the initial preparation. It is technically challenging to derive firm expression data from microarrays and it is necessary to confirm results with methods such as reverse transcriptase (RT)-PCR, in situ hybridization, or immunolabeling. These techniques can provide confirmation of gene expression at cellular resolution, but when using array data as a reference it is essential to check the exact nature of the tissue that has been analyzed.

One relatively early descriptive study applied the Atlas[TM] cDNA expression array from Clontech (Palo Alto, CA), which included 588 rat genes, each represented by a 200- to 800-base-pair PCR product immobilized on a nylon membrane (Cho et al. 2002). Gene expression profiles were obtained for the cochlea, cochlear nucleus, and inferior colliculus in male 8- to 12-week-old Sprague-Dawley rats. In each tissue some 20%–30% of the arrayed genes were judged to be above background and there were clear differences between tissues. The results provided a snapshot of the expression of a selection of genes, thus offering elementary data to inform future studies. Similar arrays were used to study gene expression in the inner ears of rats of the same age (Lin et al. 2003).

The results excluded the vestibular part of the ear so the list of genes should be related primarily to cochlear tissue. A much more focused screen was carried out for the expression of connexin genes in CD-1 mice (Ahmad et al. 2003). Total RNA was extracted from cochlear tissues at postnatal days P10, P12, and P18, and from adults. Some 15 connexin genes were screened and the results were supported by immunofluorescence and immunoprecipitation. Although only a small number of genes were studied, the combination of immunolabeling and the developmental trends observed added an important functional dimension to the screen. In a slightly different study the relative expression of about 100 genes from cDNA libraries of kidney and ear were compared (Liu et al. 2004). This interesting idea was based on the clinical similarities between the two organs, possibly linked to common elements involved in fluid regulation and ion transport. The screen was designed to identify possible common molecular targets for some therapeutic drugs.

The inner ear cDNA library produced by Beisel et al. (2004) was used to construct microarrays that included some 2000 cDNAs that were apparently unique to the inner ear (Morris et al. 2005). The arrays were then used to obtain gene expression profiles for dissected tissue including the stria vascularis, organ of Corti, and spiral ganglion. The results provide an inventory of genes for the complex combinations of different cell types within the dissected samples, many of which are consistent with published data. Although the insight into gene function is limited, the approach helps to identify novel genes in various parts of the mammalian inner ear. Several important genes known to be expressed in the organ of Corti, for example, *prestin* and the α9 subunit of the acetylcholine receptor, were not detected in the organ of Corti sample. Many such genes are expressed at low levels, in relatively few cell types, which mean that they are less likely to be found in complex tissues. False-negative results are often more common than false positives in array studies due to the inability to distinguish expression of rare transcripts from the background signal. Nevertheless, microarrays can be used to identify single genes that are relevant to hearing loss. For example, total RNA isolated from inner ear tissue derived from human patients undergoing labyrinthectomy was screened against 23,040 cDNAs selected from the UniGene database (Abe et al. 2003). This revealed the relatively abundant expression of *CRYM*, which encodes μ-crystallin, a protein that may be associated with nonsyndromic deafness. In another study, DRASIC was identified from analysis of cDNA from the Soares NMIE mouse inner ear library as a candidate for the mechanoelectrical transducer channel in mammalian hair cells (Hildebrand et al. 2004). DRASIC is an amiloride-sensitive ion channel that shares characteristics of the degenerin subfamily from *C. elegans* and vertebrate epithelial sodium channels. Subsequent studies on the DRASIC null mice showed that it was unlikely to be the transduction channel, but it was interesting to note that the mice suffered late-onset hearing loss after 6 months. Although this study did not conclude with the expected result, it is worth noting that careful consultation of good gene expression databases can provide clues for a wide range of future studies.

2.1.2 Development

Developmental studies based on cDNA arrays are relatively difficult to do and they are consequently few in number (Livesey 2002). Some of the most impressive applications of cDNA arrays have involved a marriage with complementary techniques designed to normalize samples and to eliminate redundant genes, such as subtractive hybridization or representational difference analysis (RDA), which are used to identify differential gene expression between tissues. RDA is a PCR-based subtractive hybridization technique that allows the isolation and cloning of differentially expressed transcripts between two complex cDNA populations, without prior knowledge of their functional or biochemical characteristics. In the classical cDNA–RDA method, two complex cDNA populations are restricted with *Dpn*II (restrictase that leaves 4-bp protruding 5' ends) and ligated to suitable linkers, the sequence of which is specific for the procedure. The amplicons, which represent the original RNA populations, are therefore called "representations" and provide the starting material. Next, the "tester" representation—from which isolation of specific messages is sought—is mixed with a large excess of the "driver" representation. After this subtractive hybridization, tester homoduplexes are enriched from the mixture by PCR (pre-PCR). Single strands are degraded using mung bean nuclease and a second PCR (final PCR) generates a complex population of amplicons that is named difference product. This protocol yields fragments visible as discrete bands on agarose gels that are cloned and sequenced for further analysis. As an example, stem cells from the CNS were analyzed before and after they were stimulated to differentiate with basic FGF (Geschwind et al. 2001). A preliminary list of some 6000 cDNAs was established by using an abbreviated RDA and these were subsequently represented on arrays. This approach provided a balance between subtraction efficiency and the need to conserve representation of all potentially relevant genes. In this case the resulting arrays represented relevant genes both in terms of the cell type and the potential changes associated with differentiation. Although the material is simpler than epithelial tissue, the strategy could be very powerful in tracking changes during development or regeneration of sensory epithelia. Now that greater representation of animal genomes is available on microarrays, one can argue that the results from subtraction or cell purification procedures can be correlated with whole genome screens without the need to make new arrays. Nevertheless, array experiments commonly reveal expression changes in many genes and it is essential to establish clear strategies for focusing on those that are most relevant to the processes being studied.

2.1.3 Degeneration and Regeneration

The effects of noise exposure on the chick basilar papilla, which is equivalent to the mammalian organ of Corti, and the rat cochlea have been analyzed by differential display, subtractive hybridization, and low-density Clontech Atlas™ arrays (Lomax et al. 2000, 2001; Cho et al. 2004). A group of control chicks not exposed to noise was compared with groups that were sacrificed either immediately after noise exposure or after a recovery period of 2 days. This is a relatively short period in terms of hair cell regeneration but it covers early events associated with noise damage, including apoptosis of hair cells. Rats were exposed to noise stimuli that would lead to either temporary or permanent threshold shifts. The results, extending those from an earlier study (Gong et al. 1996), revealed changes in numerous cDNAs after noise exposure. One of these was identified as a novel ubiquitin ligase, a small polypeptide that identifies selected proteins for degradation by proteasomes. The array data indicated upregulation of several early response genes such as c-*fos* after noise exposure that would normally lead to permanent threshold shifts (Lomax et al. 2001; Cho et al. 2004). The correlation between the different methods of expression analysis is often weak but can provide valuable, complementary data that helps to focus on the functionally relevant molecular events. The comparison between the effects of noise damage in birds and mammals is complex because there are many differences between the two species, including those related to the inherently different structural and functional properties of the sensory epithelia. However, because birds regenerate their hair cells but mammals do not, there is a reasonable chance that some applicable, practical insight into regenerative mechanisms might be obtained.

This idea has been addressed with a slightly different approach that removes the problem of species differences (Hawkins et al. 2003). Avian cochlear epithelia activate regenerative programs after damage whereas their vestibular epithelia constantly undergo regeneration. Thus a comparison of gene expression profiles between these tissues might reveal something of the nature of these programs. Hawkins et al. prepared two human cDNA arrays, the first representing 426 genes known to be expressed primarily in the inner ear or associated with hearing loss, and the second representing 1422 transcription factors. RNA was isolated from cochlear and vestibular epithelia taken from 10- to 21-day posthatch chicks and screened against the human cDNA sequences. The dissected epithelia were treated with thermolysin to help isolate epithelial sheets composed only of hair cells and supporting cells. A number of developmentally relevant genes were identified through this comparison, and some of these are certainly worth further study. In terms of transcription factors, the chick should provide highly relevant information for potential regenerative mechanisms in the mammalian ear.

2.2 Oligonucleotide Arrays

One disadvantage of cDNA arrays is that they always include an internal comparison to a reference sample, so hybridization results from different studies cannot be compared easily. This problem can be addressed by hybridizing test samples against a standard control. For example, "molecular portraits" have been produced from the RNA of 42 different breast tumors, each compared with a reference sample composed of RNA pooled from 11 different cell lines (Perou et al. 2000). Oligonucleotide arrays are technically more difficult to produce but they offer a simpler solution in terms of experimental design and a more powerful method of data analysis. As "whole-genome" arrays become more complete, the need to manufacture cDNA arrays from tissue-specific libraries may decrease.

High-density oligonucleotide arrays, also known as GeneChips, were originally developed by Affymetrix. GeneChips are made by simultaneous synthesis, base by base, of thousands of oligonucleotides on a glass support. This is achieved by using light-directed synthesis, which introduces photochemically removable protecting groups into the nucleotides and then deprotects them by selective illumination through a photolithographic mask. This process allows synthesis of approximately 10^7 copies of a single probe in a 24-μm square, although this is predicted to become 2 μm square in the future (Lockhart et al. 1996; Kozian and Kirschbaum 1999; Lipshutz et al. 1999), leaving space for representation of the entire genome.

Genes are represented as a combination of known genes and ESTs, which are partially sequenced regions of cDNAs representing unknown genes expressed in the tissue of origin. Until recently, GeneChips included genes from a broad range of different libraries and did not cover the entire genome. However, whole genome microarrays are quickly becoming available for most of the main model organisms, and custom-made microarrays can be made to order for most tissues and organs.

On Affymetrix GeneChips, each gene is represented by a grid of up to 40 features (Fig. 7.3). Ideally, 20 features contain a series of 25-mer oligonucleotides with a perfect match to nonoverlapping regions of the target sequence. Each "perfect match" feature is paired to a "mismatch" feature representing the same sequence but with a single base change at the 13th position. Thus each gene is represented by about 20 pairs of perfectly matched and mismatched oligonucleotides. Unlike the cDNA arrays, individual RNA samples are fluorescently labeled and hybridized to a single chip, so comparison between samples is achieved by comparing different chips, which offers a more flexible database for subsequent reference.

Analysis of GeneChips is quite different to that of cDNA arrays. Similar issues exist in terms of standardizing the fluorescent tag, hybridization, and sampling of the resultant fluorescence. However, the information content for each gene represented on the GeneChip is much greater. The difference in signal between each of the 20 feature pairs is averaged to generate an "average difference" value. This is combined with analysis of the variation between pairs to produce an estimate of gene expression level and a "presence/absence" call. The results can be used to determine whether or not a gene is expressed in a given tissue and to compare the expression levels between different samples. Comparisons between samples require that each chip be normalized to compensate for global differences in hybridization. The most common analytical tools are the MAS5.0 algorithm from Affymetrix (Affymetrix 2001) and the statistically more refined robust multiarray analysis (RMA; Irizarry et al. 2003). (See http://www.bioconductor.org/ for other useful software.) Additional probabilistic models can provide alternatives that perform well for single arrays and for genes expressed at lower levels (Milo et al. 2003).

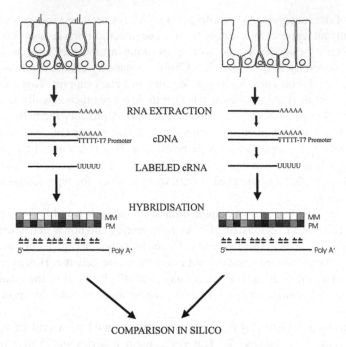

FIGURE 7.3. Representation of the same hypothetical experiment described in Fig. 7.2. RNA is extracted from the samples and reverse-transcribed into cDNA. This is done using oligodT primers including a T7 promoter extension. This cDNA is then used as a template for cRNA synthesis incorporating a label, typically biotin, which is subsequently detected with a fluorescent streptavidin conjugate. These samples are then hybridized in parallel to independent gene chips. In the chip, each gene is represented by a series of 16–20 probe pairs, short oligonucleotides designed from different regions of the transcript. Each pair includes a perfectly matching (PM) sequence and one including an internal mismatch (MM). The intensity of the signal is compared between the PM and the MM and averaged along the different probe pairs to provide an intensity value for a given gene. These values are later compared, using different algorithms, between the two experimental samples. Data from each hybridization can be used for many subsequent comparative studies.

2.2.1 Molecular Anatomy

The only published data for gene expression profiles at cellular resolution comes from a study of cristae ampullaris in young adult Brown Norway rats (Cristobal et al. 2005). More than 4000 hair cells, with a distinction between type I and type II hair cells, and 4000 supporting cells were isolated via laser capture microdissection from sectioned cristae. Total RNA was extracted and amplified before screening with CodeLink^{TM} rat whole genome Bioarrays (GE Healthcare, Amersham Biosciences); oligonucleotide arrays representing 29,842 unique UniGene identities. Genes encoding the hair cell markers myosin VIIa, the $\alpha 9$ subunit of the acetylcholine receptor (α9AChR), and a calmodulin were faithfully recorded in the hair cell sample. Interestingly, the α9AChR was found only in type II hair cells, as expected from the physiology. The cyclin-dependent kinase inhibitor p27^{kip1} was found in the supporting cell sample. The data from this study should provide a valuable resource. Another way of generating purified cell types is to isolate cells from transgenic mice carrying a fluorescent marker, such as green fluorescent protein (GFP), driven from a cell-specific promoter, and then to purify them with a cell sorter. This can yield high levels of cell purity, verified quantitatively by analysis of the sorted populations, and can provide sufficient material to avoid the need for RNA amplification. Mice carrying a *Atoh1/GFP* reporter have been used to isolate mammalian cochlear hair cells during the earliest stages of differentiation (Doetzlhofer et al. 2004) and a p27^{kip1}/GFP mouse has been used in a similar way to isolate supporting cells (Doetzlhofer et al. 2006).

2.2.2 Development

2.2.2.1 Temporal Profiles

Affymetrix GeneChips have been used to study developmental changes in tissue from the cochlea and utricle (Chen and Corey 2002a, b). The mouse Mu30k chip, representing 13,000 known genes and 21,000 ESTs, was used to screen two pooled samples of 10 cochleae dissected at postnatal days 2 and 32 (P2 and P32). At stage P2 the cochlea is still about 10 days from the onset of hearing function (Kros et al. 1998). Although the full complement of hair cells and neurons has been born (Ruben 1967), genes such as *prestin* (Belyantseva et al. 2000) and *TMC* (Kurima et al. 2002) are not normally expressed until about P6. Thus comparison of this sample with those from P32 should reveal expression of genes associated with adult cochlear function. Chen and Corey (2002b) provided a list of genes encoding ion channels, transporters, and calcium-binding proteins that were upregulated over this period. The tissue samples were complex, including most of the cellular content of the cochlea, and a maximum of only 33,000 hair cells, which would have accounted for a fraction of 1% of the total cell number. About 1.5 μg of total RNA was extracted from 10 ears. Despite these quantitative limitations, a number of hair cell specific genes were detected, including *myosins 1β* and *VI*. Other markers such as *myosin VIIa*, *Atoh1* and the calcium ATPase, *PCMA2*, were barely detectable or absent.

The study provides a valuable baseline for cochlear gene expression and the authors have served the community well by establishing a searchable database that can be downloaded from their laboratory website. Nevertheless, the complexity of the samples decreases the power of the comparison between developmental stages. One can conclude that certain genes are up- or downregulated but the lack of cellular resolution means that one only has an average output across a large number of different cell types. It is possible to extract discrete patterns of gene expression in complex tissue if one has some knowledge of functionally related gene clusters (Perou et al. 2000), but it is important to define the initial tissue preparations as tightly as possible. This can be done more easily in the macular epithelia from the vestibular system.

The mammalian utricular macula serves as a powerful model for development, degeneration, and regeneration. It can be dissected from the mouse inner ear as a sheet of supporting cells and hair cells by removing the surrounding connective tissue and nonsensory epithelia (Saffer et al. 1996). Taken from different developmental stages, it can be used to construct a temporal profile of mechanosensory epithelial development. Hair cells can be removed in organotypic cultures with the time-honored application of ototoxic drugs such as gentamicin, allowing the effects of such drugs to be explored in detail. It is possible to use this approach to obtain a sheet of supporting cells whose expression profiles can be subtracted from untreated epithelia to highlight those genes expressed specifically in hair cells. The problems of limited tissue remain, especially with the need to provide replica samples (Novak et al. 2002), but more recent methods for linear amplification of RNA can overcome this problem (Luzzi et al. 2005; Schindler et al. 2005).

Chen and Corey (2002a) screened preparations of mouse utricular maculae at E16, P2, and P6 and found some convincing temporal profiles for the expression of the inner ear specific genes α- and β-tectorin. The utricular maculae in this study were not removed from the surrounding connective tissue. However, the period studied covered that during which many hair cells are born and proceed to differentiate. Thus, with sufficient quantities of tissue, the preparation has the potential to reveal the genetic programs underlying the differentiation process. Unfortunately, the hair cells do not differentiate synchronously so the changes are less easy to detect. Nevertheless, the data indicated a potential role for the retinoblastoma gene $Rb1$ and subsequent studies revealed an extremely important function for this gene in controlling the proliferation of mature hair cells (Sage et al. 2005). This insight was not necessarily dependent on the application of microarrays (Mantela et al. 2005) but the experimental design did permit preliminary identification of genes that changed their expression during critical periods of development.

These attempts to use microarrays to study developmental processes were among the first to be published in any system. Given that mouse microarrays have been available for some time, it may seem surprising that more has not been published and that we have not already got good databases full of gene expression patterns through development (Livesey 2002). However, as experimental biologists explore this technology they have to take exceptional care with the experimental design and the data analysis is a complex and rapidly developing field. The experiment itself constitutes a very small investment of time compared to that spent on analysis and exploitation of the data.

A highly focused study on integrin expression in rat cochlear tissues was carried out across a developmental series with time points from E12, E14, E16, E18, and postnatal day 1 (Toyama et al. 2005). The dissected material was not closely characterized and the cellular composition would have changed substantially across this time period. Total RNA was analyzed with Affymetrix MG-U34 GeneChips, which did not represent all integrin subunits but revealed positive signals for $\alpha1$, αv, $\alpha7$, $\beta3$, and $\beta4$ integrins. Neither the arrays nor the analysis by RT-PCR revealed α3 and α6 subunits, both of which are known to be expressed in the cochlea, indicating the importance of defining the source tissue and selecting the appropriate arrays for the task. Nevertheless, this developmental profile should contain useful information concerning the remaining 24,000 mRNA transcripts and ESTs represented on the arrays.

2.2.2.2 Analysis of Mutant Mice

The application of microarrays to the analysis of transgenic or mutant mice is potentially powerful because gene expression can be compared between similar tissues or cells that differ only in terms of modifications to single genes. Nevertheless, the nature of the gene being studied has major implications for interpretation of the results. Complex tissues are still more difficult to interpret than single cells. Modifications to regulatory genes expressed in several tissues during early stages of development will cause complex indirect effects on gene expression profiles. The most precise expression patterns are likely to be obtained for genes expressed later and in relatively few cell types. One very successful example involved the POU-domain transcription factor *Pou4f3* (Hertzano et al. 2004), which is also known in the mouse as *Brn3.1* or *Brn3c* (Erkman et al. 1996). This gene is expressed during early stages of hair cell differentiation in auditory and vestibular epithelia and it is essential for their survival. Hertzano et al. used Affymetrix MG U74Av2 GeneChips to analyze total RNA from whole inner ears dissected from null and wild-type mice at embryonic days E16.5 and E18.5. They identified *gfi1* (growth factor-independent 1) as a target for *Pou4f3* and proposed that it enhances hair cell survival by increasing the activity of STAT3. This kind of information is crucial to our understanding of the signaling pathways involved in cell survival and thus protection from external factors such as excess noise or ototoxic agents. The array data must contain significant additional data relevant to otoprotection and possibly even regeneration. The nature of the study allowed potential targets to be identified for a single transcription factor. The construction of gene network models would require further analysis, for example, of additional transgenic mice, such as that for *gfi1*, or studies that involve modulation of the expression of *Pou4f3* with time in a tractable model system.

2.2.3 Degeneration and Regeneration

Aminoglycoside antibiotics are used extensively in many parts of the developing world but they have serious ototoxic properties. Protection from these side effects in the clinic is an important issue but the drugs are also used in vivo and in vitro for models of hair cell degeneration and regeneration. To explore early changes in gene expression in the cochlea, basal turns of neonatal (P5) rat organs of Corti, with the stria vascularis and spiral ganglion removed, were cultured for 4 and 8 hours after addition of 100 μM gentamicin (Nagy et al. 2004). Total RNA was analyzed with Affymetrix rat U34A GeneChips. After 4 hours with gentamicin, some 55 and 48 genes were down- and upregulated, respectively. A smaller number of changes occurred in both time points. The data implied that a number of genes involved in the electron transport chain were downregulated, reflecting a cellular response to attenuate the production of reactive oxygen species and to limit the oxidative stress. One gene related to apoptosis and of specific interest was the downregulation of the gene encoding the delta subunit of the F1F0 ATPase. In *Escherichia coli*, a functional deficiency in this gene is linked to aminoglycoside resistance. As with most array studies all of these conclusions must be confirmed by direct experiment.

3. Cell Lines

The application of microarrays to complex tissues during development, degeneration, or regeneration can highlight global changes that provide insights into relevant molecular mechanisms. However, complex tissues are not ideal for elucidating coherent signaling pathways within single cell types and for identifying the direct targets of specific transcription factors. An understanding of cell programming at this level is essential to the development of therapeutic strategies. One solution is to study simpler systems in suitable model organisms or to develop in vitro preparations such as organ or tissue-specific cell lines and stem cells.

In general terms, cell lines provide effective tools for identifying protein–protein associations (Kussel-Andermann et al. 2000; Velichkova et al. 2002), receptor signaling pathways (Hibino et al. 2002) and for studying the function of constructs such as the $\alpha9/\alpha10$ AChRs (Sgard et al. 2002). Ideally, it is best to use cells that reflect the background of the native tissue. A frog oocyte is a powerful tool for studying the molecular properties of a specific ion channel but in many cases it lacks associated molecules that plug that channel into the functional programs of the native cell. In the context of this chapter, we are interested in the actual signaling pathways used by sensory epithelial cells during development and potentially during regeneration. This places additional demands on the types of cell line that we can use. We also wish to model transient events because cellular responses during development and even regeneration occur against changing profiles of gene expression. To address these issues, a number of research groups have established conditionally immortal cell lines (Rivolta and Holley 2002b).

3.1 Conditionally Immortal Cell Lines

Most of the currently available inner ear cell lines are derived from the Immortomouse, a transgenic mouse that carries a conditionally expressed immortalizing gene (Jat et al. 1991; Noble et al. 1992). The transgene used in the Immortomouse is a temperature-sensitive variant of the large T, or tumor forming, protein from the SV40 virus. The protein product is stable at 33°C, below normal body temperature, but rapidly degrades at 37–39°C. It is encoded by the A gene under the control of the H2kb promoter, a major histocompatability complex class 1 promoter, which upregulates expression in the presence of γ-interferon. This means that the mouse is viable and that conditionally immortal cells can be derived from any tissue at any stage.

The fact that different cell lines possess the same genetic modification means that they can be compared more readily. The experimental design for derivation of appropriate cell lines revolves around the terminal mitoses for the desired cell type (Fig. 7.4). Immortalization halts developmental programs and induces cell proliferation, overriding the native cell programs. This appears to be less disruptive if it occurs before the cells have ceased proliferating and started to differentiate. Immortalization thus appears to be more coherent in embryonic cells than in adult cells that have differentiated and express genetic profiles not

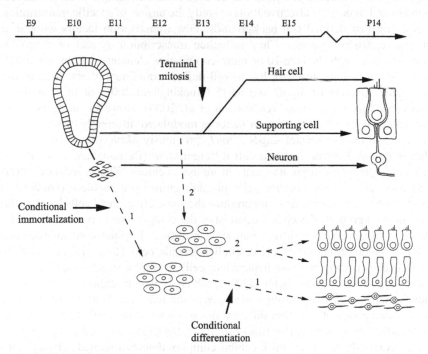

FIGURE 7.4. Conditionally immortal cell lines can be used to represent specific events in the differentiation of different cell types. This diagram represents otocyst development from E9 to E15 and to P14 when the auditory epithelium is functional. Birth is at about E19–20. Terminal mitosis for hair cells, supporting cells, and neurons in the cochlea is at E12–14 and it takes nearly 3 weeks for hair cells to differentiate fully. The timing and location of cells used to establish cell lines is critical, and two examples are illustrated with dotted lines. Delaminated cells taken at E10, if selected according to the appropriate markers, should represent neurons when conditionally differentiated (*1*). Those taken from the appropriate epithelium at E12 should be multipotent, with the ability to differentiate as hair cells and supporting cells (*2*). For developmental studies it is important to correlate the genetic profiles of cell lines and their temporal expression patterns under differentiating conditions with appropriate markers from studies in vivo.

normally compatible with cell division. Thus, these cell lines can offer reasonable models for cell differentiation. Published data is available for lines derived from the Immortomouse otocyst at E9.5 (Barald et al. 1997; Germiller et al. 2004), ventral otic epithelial cells and neuroblasts at E10.5 (Lawoko-Kerali et al. 2004b), the auditory epithelium at E13.5 (Rivolta et al. 1998), the utricular macula at P2 (Lawlor et al. 1999), and the organ of Corti 2 weeks after birth (Kalinec et al. 1999, 2003). Cells from the utricle have been conditionally immortalized by transfection in vitro (Zheng et al. 1998). These cells are now being used to study development, degeneration, and regeneration.

3.1.1 Development

Conditionally immortal cell lines can be used to model cellular events in development and to provide effective tools to study the targets of specific transcription factors against defined cell backgrounds. Few transcription factors work in a simple, consistent manner. They influence transcription as part of a protein complex that might include 10 or more cooperative elements (Morimoto 2002). In some cases a single factor has a well-defined, dominant effect, such as the myogenic function of *MyoD* and *Myf5* (Buckingham 2001) or the pancreatic endocrine function of *Pax4* (Dohrmann et al. 2000). However, in many cases the action of a given transcription factor is modulated differently between cells, tissues, and developmental stages. *Atoh1* (previously *Math1*) is a bHLH factor that is required for normal hair cell differentiation (Bermingham et al. 1999). It can also induce ectopic hair cells in mouse cochleae (Zheng and Gao 2000) and hair cells from supporting cells in adult guinea pig cochleae (Izumikawa et al. 2005). However, it does not produce the same effect in all cells because the genetic background of each cell establishes the competence to respond to *Atoh1*. *Atoh1* also governs the differentiation of millions of spinal cord interneurons (Helms and Johnson 1998) and cerebellar granule cells (Ben-Arie et al. 1997), so how does it know when to make a hair cell? A similar situation applies to the zinc finger transcription factor *GATA3*, which is widely expressed throughout development. It appears through all stages of the inner ear from the otic placode to the adult and in at least four different tissues (Karis et al. 2001, Lawoko-Kerali et al. 2002). How can the function of *Atoh1* or *GATA3* in any cell at a given time be analyzed? Null mutant mice exhibit complex defects summed across tissues through development. Conditional, targeted gene deletions can overcome these issues (Cohen-Salmon et al. 2002; Pirvola et al. 2002), but the tools are often not available to target the appropriate cells at the appropriate time.

Despite the limitations of in vitro culture and employment of an immortalizing gene, conditionally immortalized cell lines provide a useful way forward. They can be used to represent the behavior of specific cell types through defined time windows in development (Rivolta and Holley 2002b). To employ them in this way, several conditions must be met. Cells must be derived from a time and location as closely defined as possible and reliable molecular markers must be available for each cell type. Dissection of the embryonic inner ear is not easy but one of the greatest limitations is related to knowledge of molecular markers.

Critics are often quick to point out that a cell line expresses an unexpected combination of genes and is, therefore, not physiologically representative. However, we have surprisingly little information about expression patterns for individual genes during development. In many cases, data from in situ hybridization, gene expression studies, and antibody labeling do not correlate well. Further, the inner ear is a very complex structure and interpretation of data from null mutant animals can be extremely difficult. The solution is to define cells with a panel of as many markers as possible. Additional conditions are that immortalization must suspend cell differentiation in a stable manner and when the immortalizing gene is inactivated the cells should activate gene expression patterns similar to those observed during normal development. In this respect, cell lines derived from the Immortomouse are certainly not perfect, but they do behave remarkably well relative to their in vivo cousins (Rivolta and Holley 2002b). Further, their phenotype is easier to control and more closely related to the native cells than most stem cells.

3.1.1.1 Temporal Profiles of Gene Expression with Microarrays

Once characterized, a good cell line can be allied to microarray technologies to identify networks of genes, thus exploiting their most important feature, namely the ability to profile tens of thousands of genes simultaneously. Temporal profiles of gene expression derived from cells or tissues either through development or after a specific experimental manipulation can provide a powerful means of identifying clusters of functionally related genes (Rudie Hovland et al. 2001; Schulze et al. 2001). In one of the first and most impressive studies a human fibroblast cell line was starved of serum for 48 hours to induce quiescence (Iyer et al. 1999). At time zero, 10% fetal calf serum was added to the medium and 12 samples of RNA were prepared after periods ranging from 15 minutes to 24 hours. The samples were hybridized to cDNA arrays representing 8613 human genes. Discrete clusters of genes sharing common temporal expression profiles reflected the cellular responses as they reentered the cell cycle. In a different experiment HeLa cells were synchronized by arrest in S phase with a thymidine block or in mitosis with a thymidine/nocodazole block (Whitfield et al. 2002). cDNA arrays were then used to profile gene expression when the blocks were released. The temporal expression profiles for each gene were analyzed via Fourier transformation to reveal any inherent periodicity. A total of 874 genes expressed periodic variation and coexpressed groups included those related to processes such as DNA replication, chromosomal segregation, and cell adhesion. A similar approach based on oligonucleotide arrays representing 12,488 genes has been used to analyze circadian oscillations of gene expression in the liver and heart (Storch et al. 2002). These techniques could be applied productively to cell lines derived from inner ear epithelia.

3.1.1.2 Temporal Profiles from a Cochlear Cell Line

Affymetrix MU11k GeneChips were used to produce temporal, gene expression profiles during differentiation of a conditionally immortal, cochlear cell line

(Rivolta et al. 2002). The cell line UB/OC-1 was derived from Immortomouse organs of Corti to represent early stages of hair cell differentiation (Rivolta et al. 1998). It expresses the cytoskeletal proteins myosin VI and fimbrin and under differentiating conditions it upregulates *Brn3c*, *α9AChR*, and *myosin VIIa*. The expression profiles of these molecular markers, analyzed by RT-PCR, immunolabeling, and electrophysiology, suggest that UB/OC-1 represents key, dynamic features of early hair cell differentiation. There is no evidence for expression of markers for supporting cells, including OCP-2, a protein highly expressed by supporting cells in vivo (Yoho et al. 1997). Rivolta et al. analyzed the transcriptional profile of UB/OC-1 at daily intervals for 14 days under differentiating conditions. The arrays represented 11,000 genes, including about 6000 known genes and 5000 ESTs. The temporal component of the experimental design was punctuated by the fact that cells were released from the immortalizing gene and allowed to differentiate synchronously. Further, the cell line was clonal and all cells expressed *Brn3c* on differentiation (Rivolta et al. 1998). This means that there was a reasonable chance of clustering functionally related genes that shared similar temporal profiles and of modeling coherent programs within a single cell. The following sections illustrate some of the data that is available from the website. The entire data set obtained from this analysis has been deposited in the NCBI Gene Expression Omnibus Database (http://www.ncbi.nlm.nih.gov.geo/) under the series accession number GSE36 and sample numbers contained therein.

3.1.1.3 Proteases

The published data for UB/OC-1 reveals a remarkable upregulation of cathepsin D during differentiation. The specific probe set for cathepsin D, as well as two cathepsin D ESTs represented in the array, displayed a relative increase in transcript level of at least 30-fold (Fig. 7.5). Further experiments are needed to establish the possible involvement of cathepsin D in hair cell differentiation. An interaction worth exploring is the link between cathepsin D and the insulin-like growth factor (IGF) cascade. Cathepsin D is a regulatory protease that can modify the response to IGF-1 by hydrolyzing IGF binding proteins (IGFBPs; Conover 1995, Conover et al. 1995; Claussen et al. 1997). IGFBP2 is also notably upregulated during differentiation of UB/OC-1. The IGF-1 receptor is expressed in rat utricular sensory epithelia and may be downregulated after aminoglycoside treatment (Saffer et al. 1996). Lack of IGF-1 in null mutant mice affects postnatal survival, differentiation, and maturation of spiral ganglion cells, causing abnormal synaptogenesis and incomplete innervation (Camarero et al. 2001).

These results could be important because proteases are ideal therapeutic targets. They have a relatively small, active site against which to design inhibitory drugs. Between 1998 and 2000, the number of human proteases under investigation as therapeutic targets almost doubled to include about 15% of the total number of proteases described (Southan 2001). Cathepsins, for example, have been implicated in tumor invasion and progression (Berchem et al. 2002) and in

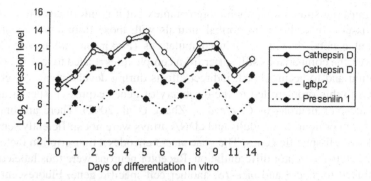

FIGURE 7.5. Temporal gene expression profiles for cathepsin D, insulin-like growth factor binding protein 2 (Igfbp2), and presenilin 1 in the cell line UB/OC-1 following analysis with Affymetrix oligonucleotide arrays (Rivolta et al. 2002). Two different probe sets for cathepsin D generated similar profiles that indicate more than a 30-fold increase in expression during the first 5–6 days (remember that each log2 unit represents a 2-fold change). These profiles correlated with those for Igfbp2 and presenilin 1. All genes were expressed at relatively high absolute levels against the background.

degenerative diseases (Tyynela et al. 2000). Cathepsin K is a classic example of a prototypic genomics-derived drug discovery target (Debouck and Metcalf 2000). To identify genes involved in bone resorption that could be responsible for osteo-porosis a large number of ESTs were analyzed from a human osteoclast cDNA library (Drake et al. 1996). Approximately 4% of the ESTs sequenced from this library encoded cathepsin K, which was independently proposed as the molecule underlying pycnodysostosis, an inherited syndrome combining osteopetrosis with bone fragility (Gelb et al. 1996). Inhibitors are being tested as a proposed therapy for bone resorption problems, such as the ones encountered in osteoporosis (Yamashita and Dodds 2000).

Cathepsin D may also be involved in the pathogenesis of Alzheimer's disease (AD) because a genetic polymorphism in this gene is associated with an increased risk of the disease (Papassotiropoulos et al. 1999). Moreover, it possesses β- and γ-secretase activities (Nixon 2000), necessary for the processing of amyloid precursor protein into Aβ peptides. Although BACE (Vassar et al. 1999) and presenilins (De Strooper et al. 1998; De Strooper et al. 1999; Herreman et al. 2000) have been identified as the respective main sources for the β- and γ-secretase activities, the probable important role of this lysosomal protease has justified efforts to design and explore specific inhibitors (Baldwin et al. 1993).

3.1.1.4 Data for Gene Networks

Gene networks can be constructed from a series of specific manipulations of single genes within a cell, tissue, or animal. Yeast cells have provided excellent models in this respect, as it is possible to analyze differences in gene expression profiles for a wide range of specific, known mutations (Jorgensen et al. 2002). In vertebrates the situation is much more complex but it is informative to compare cells, tissues, or embryos in normal animals with those from animals carrying inherited genetic defects or induced mutations (Parkinson and Brown 2002). However, the comparisons still suffer from complexities related to lack of cellular resolution and to accumulated, indirect effects during development. An elegant solution to this problem led to the discovery of genes expressed specifically in touch receptor neurons of *C. elegans* (Zhang et al. 2002). There are only six touch receptor neurons in adults and cDNA arrays were not sufficiently sensitive to pick up cell-specific genes. The gene *mec-3* encodes a transcription factor that regulates touch receptor differentiation. Receptor neurons were thus labeled with GFP linked to *mec-3* and *mec-18*, another cell-specific gene. Fluorescent cells were then selected from normal embryos and *mec-3* null embryos, cultured, and compared to identify genes specific for touch receptors. cDNA microarrays for *C. elegans* represent almost the entire genome, including 17,817 of the 18,967 known or predicted genes. The method led to the identification of 71 genes dependent on expression of *mec-3*. This approach could be adopted in the mammalian ear but it is also possible to use cell lines.

Loss-of-function and gain-of-function studies with specific genes can be conducted relatively easily in cell lines. The approach allows us to test hypotheses that emerge from the temporal profiles described in the preceding text. For example, the expression level for *GATA3* increased more than fivefold during differentiation of UB/OC-1 (Rivolta et al. 2002). The list of genes that cluster with *GATA3* provides candidates for functionally related genes, which can be tested by overexpressing or knocking down *GATA3* (Lawoko-Kerali et al. 2004a). This can be achieved with antisense oligonucleotides (Fig. 7.6) or antisense morpholinos (GeneTools Ltd.). The application of ribozymes (z) or of RNA interference (RNAi) may be more effective (Sandy et al. 2005). Replica samples for control and treated cells after given time periods can reveal direct and indirect targets of a cascade of transcription factors that are known to regulate cell differentiation. For example, we know that cochlear–vestibular ganglion cells depend upon expression of *neurogenin 1* (Ma et al. 1998), *GATA3* (Karis et al. 2001), *NeuroD* (Liu et al. 2000, Kim et al. 2001) and *Brn3a* (Huang et al. 2001). A series of loss of function studies in an appropriate cell line offers the chance to establish a true hierarchical network of downstream targets for these transcription factors. This kind of information will provide an extremely valuable, complementary dimension to our understanding of mammalian sensory cell differentiation. *Brn3c* has been discussed previously in the context of gene array analysis of functional null mice (Hertzano et al. 2004). A further piece of information relating to its function as a survival factor is the demonstration that *Brn3c* can regulate specific elements of the promoters for both brain-derived neurotrophic factor (BDNF) and neurotrophin-3 (NT-3) in cell lines (Clough et al. 2004). The results are based on the expression of promoter elements linked to a luciferase reporter in the cochlear cell lines UB/OC-1 and UB/OC-2. In the *dreidel* mutant mouse, in which *Brn3c* is nonfunctional, BDNF expression levels are also decreased.

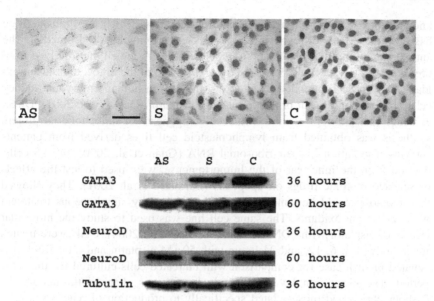

FIGURE 7.6. Knockdown of the transcription factor *GATA3* with antisense oligonu-cleotides in a cell line that represents delaminating cochlear neuroblasts. Immunolabeling shows the loss of *GATA3* after 36 hours with antisense (AS) as opposed to sense (S) or untreated controls (C). Immunoblots reveal that loss of *GATA3* is accompanied by loss of the neural transcription factor *NeuroD*. After 60 hours in culture without further treatment both factors are reexpressed. Scale bar = 100 μm. (From Lawoko-Kerali et al. 2004a.)

Retinoic acid and bone morphogenetic protein 4 (BMP4) are essential for normal inner ear development, including the formation of the semicircular canals. An interaction between them was discovered through experiments with the cell line 2B1, which was derived from the Immortomouse otocyst at E9.5 (Thompson et al. 2003). Activation of retinoic acid receptors α and γ led to an 80% decrease in the transcription of BMP4, mediated through two promoters distinct from that primarily used in bone tissue. BMP4 was also downregulated in developing otocysts treated with retinoic acid, and exogenous application rescued the inhibitory influence of retinoic acid on the formation of the semicircular canals. These interactions are tissue specific and the work depended on the use of cell lines that were derived from the inner ear at the appropriate time.

3.2 Degeneration

Cell lines have been used extensively in toxicity studies and they provide a first line of screening for ototoxic effects and the potential protective action of new drugs (Bertolaso et al. 2001; Devarajan et al. 2002; Jeong et al. 2005a, b). Numerous, commonly used drugs are ototoxic (see Forge and Van De Water, Chapter 6) and cell lines from the inner ear should provide insights into their function. The ionic platinum drug cisplatin is used to treat tumors despite its known ototoxic, nephrotoxic, and neurotoxic effects, which are thought to be based on the effects of reactive oxygen species on mitochondrial function. The antioxidant methionine was found to reduce cell death in cell lines derived from breast tumors (Reser et al. 1999) and overexpression of the mitochondrial antioxidant manganese superoxide dismutase (MnSOD) in human embryonic kidney cells provides significant protection against cisplatin (Davis et al. 2001). Direct evidence that aminoglycosides influence components of mitochondrial protein synthesis was obtained from lymphoblastoid cell lines derived from patients carrying a mutation in their ribosomal RNA (Guan et al. 2000). OC-k3 cells, derived from the inner ear of the Immortomouse, were used to test the effects of selected ototoxic drugs on apoptosis (Bertolaso et al. 2001). They showed that apoptosis could be almost completely blocked by simultaneous treatment with certain antioxidants. The same cell line was used to study the molecular effects of cisplatin with cDNA arrays (Previati et al. 2004). Cells were treated for periods of 3, 6, 12, and 24 hours with $50\,\mu$M cisplatin and total RNA was isolated in each case for comparison with untreated cells cultured for the same period. The expression levels of hundreds of genes changed throughout the 24-hour period and those related specifically to production of reactive oxygen species were identified. The cells were subsequently used to test the protective effects of several cytoprotective drugs, including butylated hydroxytoluene, dithiothreitol, and N-acetylcysteine, all of which reduced cell death caused by cisplatin. Temporal profiles of gene expression are extremely informative but dose-dependent changes can also be valuable, especially when trying to identify the specific effects of the given drug. There is increasing evidence that antioxidants and inhibitors of apoptotic pathways can protect both hair cells and their sensory nerves in vivo (Ylikoski et al. 2002) and cell lines should be of value in determining the molecular mechanisms involved. The differential sensitivity of different cell types means that cell lines derived from targeted inner ear epithelia should provide an important experimental asset.

3.3 Regeneration

Few studies with these lines have focused on regeneration directly although there is obvious potential to do so. However, Lawlor et al. (1999) showed that the cell line UB/UE-1, cloned from postnatal supporting cells from the utricular macula, has the potential to differentiate into neonatal hair cells and supporting cells. Characterization of the hair cell phenotype was based not only on RT-PCR and immunolabeling for hair cell markers, including Brn3c, myosin VIIa, and fimbrin, but also on electrophysiological recording of functional markers such as acetylcholine receptors and potassium channels. The origin of the cell line was tightly defined and the implication is that neonatal supporting cells in the mammalian utricle are multipotent, a property that correlates with asymmetric cell divisions during the early stages of differentiation (Rivolta and Holley 2002a). This cell line provides a potentially useful in vitro model for analyzing genes involved in cell cycle arrest and the early stages of hair cell differentiation. Weaknesses in the model include the lack of markers for supporting cells. In addition, the cells are more likely to reflect developmental rather than regenerative mechanisms.

However, modulation of signaling pathways that govern hair cell differentiation can certainly be explored. One such mechanism might involve the cell adhesion molecule E-cadherin, which is expressed by supporting cells but is downregulated in hair cells (Whitlon 1993; Whitlon et al. 1999; Hackett et al. 2002). Signaling through E-cadherin is complex and is modulated through homotypic interactions between cells, a route that governs epithelial structure and cell proliferation, and even the formation of homodimers on the same cell, which can activate growth factor receptors (Pece and Gutkind 2000). Constitutive expression of E-cadherin in UB/UE-1 inhibits progression of hair cell differentiation (Hackett et al. 2002). Elucidation of the associated signaling pathways, which could involve Wnt, is worth pursuing. Cadherins interact with cytoskeletal proteins in the hair bundles, including vezatin and myosin VIIa (Kussel-Andermann et al. 2000). This is particularly important because mutations in cadherin *Cdh23* and the protocadherin *Pcdh15* lead to disruption of the hair bundles and to hearing loss (Alagramam et al. 2001a, b, Bolz et al. 2001; Di Palma et al. 2001).

Cell lines from the Immortomouse have been used very successfully for cell transplantation into the retina (Lund et al. 2001) and the CNS (Virley et al. 1999). The cell line US/VOT-N33 represents neuroblasts delaminating from the ventral otocyst at E10.5 (Lawoko-Kerali et al. 2004b). It responds to FGF1 and FGF2 in vitro by forming bipolar cells resembling SGNs and associates with SGNs from primary organ culture in vitro (Fig. 7.7) (Nicholl et al. 2005). It differentiates into a neuronal phenotype in vitro and in vivo and expresses key transcriptional markers of native SGNs. This shows that conditionally immortal cell lines can retain their phenotype and differentiate in a similar way to their native equivalents, reinforcing their use in developmental studies of gene interactions in vitro.

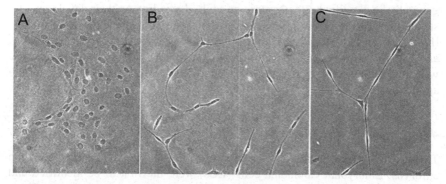

FIGURE 7.7. Differentiation of the cell line US/VOT-N33 (**A**) before and (**B, C**) after culture in vitro with $25\,\mathrm{ng\,ml^{-1}}$ FGF2 for 7 days. The cells form bipolar, neuronal morphologies similar to those of native SGNs. They associate with each other in vitro and with native SGNs in coculture (Nicholl et al. 2005).

3.4 Other Cell Lines

Other cell lines have been established but there is relatively little data regarding their application at this stage. Cells from the rat otocyst at E12 and from the rat organ of Corti at P5 have been immortalized by retroviral transfection with E6/E7 genes from human papillomavirus (Ozeki et al. 2003). The cells had mixed phenotypes and expressed numerous markers for hair cells, supporting cells, and neuronal tissue.

4. Stem Cells

Stem cells have been defined as "clonal precursors of both more stem cells of the same type, as well as a defined set of differentiated progeny" (Weissman et al. 2001, p. 388). The enormous potential of therapies employing stem cells has raised great hopes and expectations in almost every area of medicine. The possibilities range from cell replacement to the more ambitious and still futuristic idea of generating whole organs in vitro for transplantation. Stem cells can be classified into different types, depending on the source tissue, the time of derivation and the potential to produce different lineages. The primordial, master stem cell is the zygote. The fertilized egg is *totipotent,* that is, it has the potential to produce any cell type in the body, including extraembryonic tissue such as the trophoblast. *Pluripotent* stem cells have a slightly more restricted potential. They have the ability to generate cell types from all the three germ layers (endoderm, mesoderm, and ectoderm), including all the somatic lineages and germ cells. However, they do not normally produce extraembryonic lineages such as those from the placenta. Finally, *multipotent* stem cells have even more limited potential, producing cell types usually restricted to a single organ or germ layer.

Pluripotent stem cells have the widest range of potential applications. They can generally be classified as embryonic or adult, depending on their developmental stage of derivation. Some groups oppose the use of human embryonic stem cells (hESC) because of ethical and moral concerns. Our knowledge about the sources and potential of these cells is still very limited and further research is necessary with both stem cell types, embryonic and adult, to understand their possible applications.

4.1 Embryonic Pluripotent Cells

Three different cell types with pluripotency have been derived from mammalian embryos. Embryonal carcinoma cells (ECs) were the first to be identified and characterized during the 1960s and 1970s (reviewed in Andrews 2002). These cells are present in teratocarcinomas, which are gonadal tumors that can produce cell types from the three embryonic germ layers (Martin and Evans 1974). They are probably derived from primordial germ cells (PGCs), the embryonal precursors of the gametes (reviewed in Donovan and Gearhart 2001). Human EC cell lines were derived by isolating them from the tumors and growing them on mitotically inactive layers of fibroblasts (feeder layers). Although they had been considered for therapeutic applications, the presence of aneuploidy was a cause for concern (Andrews 1998, 2002). This limitation led Evans and Kaufman (1981) and Martin (1981) to isolate cells from normal mouse blastocysts, using culture conditions optimized for EC cells. These embryonic stem cells (ESCs) have a normal karyotype (Bradley et al. 1984) and are maintained in an undifferentiated state by the inclusion of leukemia inhibitory factor (LIF) in the culture media. Almost two decades later, Thomson et al. (1998) succeeded in establishing ESC lines from human blastocysts. A notable difference between human and mouse ESCs is that LIF is not sufficient to maintain human ESCs undifferentiated and it is necessary to grow them on feeder layers or feeder-conditioned medium. ESCs are derived from the inner cell mass (ICM) of the blastocyst, and they are roughly equivalent to these cells, although in vivo they do not persist for any great length of time. The apparent immortality and the maintenance of undifferentiated features of ESCs are a result of their establishment in vitro. One of the greatest challenges is to control their differentiation predictably to enable selection of specific cell types for therapeutic application (Lovell-Badge 2001). Finally, embryonic germ cells (EGCs) are derived from PGCs of the postimplantation embryo. They are maintained undifferentiated in the presence of feeder layers and have a similar potential to ESCs (Matsui et al. 1992; Resnick et al. 1992).

4.2 Fetal and Adult Stem Cells

Of the several major groups of fetal and adult stem cells, the hematopoietic, mesenchymal and neural stem cells are the best characterized. Given their higher numbers during development, fetal tissue is an ideal source for the initial isolation and setting up of cultures to expand cells in vitro.

4.2.1 Hematopoietic Stem Cells

The ability to produce multiple cell lineages is retained by certain cell types into adult life. Bone marrow hematopoietic stem cells have the capacity to reconstitute the blood cell progenies throughout the life of the individual. Long-term hematopoietic stem cells (LT-HSCs) are named according to their ability to give rise to the lymphoid and myeloerythroid lineages for life after transplantation into myelo-ablated, irradiated mice. LT-HSCs give rise to short-term hematopoietic stem cells (ST-HSCs), which can only give rise to blood lineages for 8–12 weeks when transplanted. ST-HSCs in turn are the source of multipotent progenitors, which progressively lose the potential for self-renewal. An intrinsic advantage of these kinds of stem cells is the possibility of using them for autologous transplants. They can be used to replenish the bone marrow of cancer patients that have undergone chemotherapy. By selecting HSCs based on their surface markers the possibility of contaminating the transplant with tumor cells and reinducing a spread can be virtually eliminated (Weissman 2000).

The potential applications of HSCs go beyond the fields of oncology and hematology. Lagasse et al. (2000) showed that HSCs can repair liver damage by giving raise to new hepatocytes. Furthermore, they used HSCs to treat fumary-lacetoacetate hydrolase (FAH)-deficient mice, which develop liver failure similar to human fatal hereditary tyrosinemia type I. Seven months after transplantation, 30%–50% of the liver mass in the treated animals was derived from donor cells. The mechanisms underlying this phenomenon are still unclear. Evidence suggests that the new hepatocytes may have been generated by cell fusion rather than by direct differentiation from HSCs (Vassilopoulos et al. 2003; Wang et al. 2003). However, more recent experiments appear to support the idea of differentiation without fusion (Jang et al. 2004). Regardless of the mechanism, these experiments show that stem cells can be successfully used to treat a genetic condition by replacing a critically targeted cell population.

4.2.2 Mesenchymal Stem Cells and Multipotent Adult Progenitor Cells

Mesenchymal stem cells (MSCs; also known as stromal cells) are a population located in the bone marrow that can grow as adherent cells in culture and can differentiate into osteoblasts, chondroblasts, and adipocytes in vitro and in vivo. They can be identified by a number of surface markers including CD29, CD44, CD90, and CD106, while they are generally negative for CD34, CD45, or CD14. Human MSCs can differentiate in vitro into mesodermal and neuroectodermal-derived tissues (Pittenger et al. 1999, Woodbury et al. 2002).

The derivation of multipotent adult progenitor cells (MAPCs) has created great expectations in the field. They were initially isolated from a subpopulation of highly plastic MSCs, and can also grow in vitro as adherent cells (Reyes et al. 2001, 2002; Jiang et al. 2002, 2003). Human MAPCs do not need leukemia inhibitory factor (LIF) for expansion, unlike their murine counterparts. They can proliferate for more than 100 population doublings without undergoing senescence. They lack most of the markers associated with MSCs or HSCs, such as CD34 or CD44, and express factors characteristic of ESCs, like *OCT4* and *REX1*. They have the potential to differentiate in vitro not only into mesenchymal progenies, but also into visceral mesodermal, neuroectodermal, and endodermal cell types (Jiang et al. 2002, 2003). When transplanted into early embryos, they contribute to most, if not all, of the somatic cell types. When grafted into an adult host they can differentiate into the hematopoietic lineages as well as contributing to the lung, gut, and liver epithelium.

These cells might prove fundamental in treating a broad range of diseases or conditions, regardless of the tissue involved. They could well have the potential to produce inner ear sensory cells if exposed to the right cues and introduced into the appropriate cellular environment. As with HSCs, a critical advantage is that they can be derived from the same patient, enabling autologous transplants that will avoid the complications of tissue rejection. Unfortunately, MAPCs appear to be notoriously difficult to culture and only a few labs worldwide have been able to maintain them.

4.2.3 Neural Stem Cells

The long-standing dogma that there were no cells in the adult CNS with proliferative capacity was shattered by the discovery of proliferating neuronal precursors (Reynolds and Weiss 1992; Lois and Alvarez-Buylla 1993). The main sources for derivation of adult neural stem cells (NSCs) are the subventricular zone (SVZ), the hippocampus (Gage 2000), and the olfactory bulb (Pagano et al. 2000). NSCs are normally grown as aggregates in suspension, known as neurospheres, although some labs have grown them as adherent monolayers. The multilineage potential of NSCs appears to stretch beyond the boundaries of neural tissue. Several reports have shown their ability to produce non-neural lineages such as blood (Bjornson et al. 1999; Shih et al. 2001; Vescovi et al. 2002) and even muscle (Galli et al. 2000). Morshead et al. (Morshead et al. 2002) have explained this non-neural plasticity of NSCs as artificially acquired by the culture in vitro. It is possible that the initial source of derivation would have a substantial effect on the different lineages and neural types produced, since it has been proposed that not all NSCs are equal, and some may be temporally and regionally restricted (Temple 2001). If this line of thinking proves correct, it would implicate the need to derive inner ear-specific NSCs to obtain fully functional, auditory sensory neurons. On the other hand, there is evidence that adult NSCs display a very broad repertoire for differentiation depending on their cellular environment (Clarke et al. 2000). Injected into the amniotic cavities of stage 4 chick embryos or in clonal culture derived from neurospheres, they can form a broad range of phenotypes including neural cells, muscle, mesonephric cells, and epithelial cells of liver and intestine. These results imply that stem cells in different adult tissues may be quite closely related and effective in "non-native" cellular environments.

4.3 Stem Cells from the Inner Ear

Evidence for the presence of stem cells in the mammalian inner ear has been difficult to obtain. The lack of regenerative capacity of mammalian sensory epithelia might suggest that they do not exist. However, until relatively recently this was believed to apply to other mammalian tissues, for example, the CNS and the retina. Stem cells tend to occupy defined, protective niches in which they can maintain their identity (Lovell-Badge 2001). There is no evidence for natural regeneration in the mammalian retina, either in the sensory neural epithelium or the retinal pigmented epithelium. Nevertheless, pigmented cells from the ciliary margin (PCM cells) of 2- to 3-month-old mice can form self-renewing colonies that can differentiate into retinal epithelial cells, including photoreceptors (Tropepe et al. 2000). These cells are distinct from the retinal progenitors that are produced during development and that have a limited lifespan and they are protected within a discrete anatomical niche.

Latent neural progenitors in the adult hippocampus can be stimulated by epidermal growth factor (EGF) and fibroblast growth factor 2 (FGF2) to repair ischemic damage to CA1 pyramidal neurons (Nakatomi et al. 2002). The response is subtle. If the growth factors are applied simultaneously with the ischemia they do not protect the CA1 neurons from subsequent degeneration and apoptotic death. However, they stimulate upregulation of cell-specific transcription factors within the first day. Proliferation of replacement neurons occurs within 4 days of treatment, preceding neuronal loss. By 28 days there are clear signs of both structural and functional recovery. This work suggests that stimulation with the appropriate growth factors at the appropriate time can activate an effective endogenous response.

Despite large numbers of experiments, including the application of many different factors to experimentally damaged epithelia, no such regenerative response has been observed in the mammalian cochlea. In contrast, the capacity for sensory repair in vestibular epithelia, even if limited, suggests the presence of sensory cell progenitors or even stem cells, for which there is now good experimental evidence (Li et al. 2003a). To place this work into a developmental context we will consider the origins of the sensory cells and the likely location of stem cell niche if one should exist in the inner ear.

4.3.1 Location of Sensory Progenitors

Over the past 15 years, supporting cells have been identified as the most likely source of new hair cells. In 1967, Ruben used tritiated thymidine to detect the last round of mitoses occurring in the mouse organ of Corti (Ruben 1967). He showed that hair cells and the surrounding supporting cells are born at around E14.5. The synchrony of their terminal mitoses suggested that hair cells and supporting cells probably share a common progenitor. This idea was supported by a study on the effects of retinoic acid (Kelley et al. 1993). Supernumerary hair cells and supporting cells were produced after embryonic cochlear explants were treated with exogenous retinoic acid. The additional cells appeared without signs of cell proliferation, implying that retinoic acid had changed the fate of a postmitotic cell population into one with the potential to produce hair cells and supporting cells. This population represents prosensory precursor cells. Laser ablation of hair cells in the developing mouse organ of Corti provided further evidence that new hair cells can be derived from supporting cells (Kelley et al. 1995). A few years later the lineage relationship between hair cells and supporting cells was demonstrated directly in the chicken basilar papilla (Fekete et al. 1998; Lang and Fekete 2001). Replication-defective retroviral vectors were used to label a few progenitor cells around the time of terminal mitoses and the vast majority of clones analyzed later contained both hair cells and supporting cells. Hair cells and their immediate supporting cells also share a clonal relationship with the neurons (Satoh and Fekete 2005). In mammals there is evidence for low-level regenerative capacity in the utricular macula (Forge et al. 1993; Warchol et al. 1993). Further, conditionally immortal supporting cell lines derived from the early postnatal mouse utricle have the potential to replace themselves and to produce cells with clear, neonatal hair cell phenotypes (Lawlor et al. 1999). White et al. (2006) have shown that postnatal supporting cells from the organ of Corti can proliferate in vitro and transdifferentiate into hair cells.

Pirvola et al. (2002) demonstrated with great elegance that FGF receptor 1 (FGFR1) is required for proliferation of the prosensory cell population in the developing mouse organ of Corti. It is not clear how the prosensory cells are specified or how FGF signaling is related to the expression of the cyclin-dependent kinase inhibitor p27. However, FGF signaling appears to regulate a population of "transit amplifying cells" in a manner that may shed light on the regulation of an endogenous stem cell population. The prosensory cells express the transcription factor *SOX2* (Kiernan et al. 2005). This gene has been associated with multipotency and with the proliferation and maintenance of stem cells from diverse origins. In the ear, however, it has been proposed as having an instructive role, helping to specify the prosensory field by acting upstream of *atoh1*.

4.3.2 Isolation of Stem Cells from the Inner Ear

Pluripotent stem cells have now been isolated from mammalian vestibular epithelia (Li et al. 2003a). They were isolated from the utricular macula of 3- to 4-month-old mice by their ability to form floating spheres. When disso-ciated and plated as adherent cultures, cells differentiated into hair cell and supporting cell phenotypes. They also expressed neuronal markers, and when grafted into chicken embryos, contributed to mesodermal, endodermal, and ectodermal derivatives. Similar approaches have provided no firm evidence for the existence of a stem cell population in the adult mammalian cochlea, but it is too early to rule out the possibility.

Malgrange et al. (2002) approached this issue by culturing cells from neonatal rat cochleae in the presence of EGF and FGF2. They used nestin as a marker for potential stem cells. Suspensions of nestin-positive cells formed "otospheres," reminiscent of the neurospheres formed from NSCs, and from these cells it was possible to differentiate a variety of cell phenotypes, including hair cells. In a similar study, but working with E13.5 *atoh1-GFP* transgenic mice, Doetzlhofer et al. (2004) were able to culture a population of progenitors that differentiate in vitro into cells displaying hair and supporting cell markers. These cells were only capable of producing *atoh1–GFP* cells for up to 3 weeks in vitro. The in vitro–generated hair cells needed EGF and the support of periotic mesenchyme for their survival. The number of *atoh1-GFP* cells dropped substantially when the cells were isolated from P2 cochleae. Oshima et al. (2007) compared the ability of different parts of the inner ear to generate floating spheres. The capacity of the cochlea to produce these spheres decreased substantially (100-fold) during the second and third postnatal week and it was completely abolished in older animals.

A population of nestin-positive cells is normally located in the most basal, supporting cell layer of the sensory epithelia and in the inner spiral sulcus, remaining in the inner spiral sulcus of the rat cochlea up to 2-weeks of age (Kojima et al. 2004). Using a GFP-nestin transgenic mouse, Lopez et al. (Lopez et al. 2004) have described a small population of Deiters' cells, located under-neath the outer hair cells, which remained positive for GFP-nestin as late as postnatal day 60. This work provides a preliminary indication that cochlear stem cells might exist in postnatal life. However, nestin alone cannot be considered an exclusive marker for stem cells.

Attempts to isolate populations from the adult cochlea have produced very limited results. A population of neural precursors has been isolated from adult guinea pig and human spiral ganglions, although with very limited proliferative capacity and restricted lineage potential (Rask-Andersen et al. 2005). Zhao (2001) attempted to derive stem cells from young adult guinea pigs. Cells from six to eight organs of Corti were cultured in a keratinocyte medium supplemented with EGF, bovine pituitary extract, and 10% fetal bovine serum. Epithelial clones were derived that appeared to have the potential to differentiate hair cells. Further experimental evidence is needed to identify the proliferative capacity and potency of these cells but the results should serve to encourage more studies of this kind.

It is not clear if the nestin-positive cells identified by Malgrange et al. (2002) form such a niche although they do lie in a region within which ectopic hair cells can be induced by transfection with *Atoh1* (Zheng and Gao 2000). Evidence for a true adult stem cell population should ideally come from a defined group of cells in adults with direct evidence that they can proliferate and are multipotent. Interestingly, PCM cells proliferate without exogenous growth factors but they do not normally produce a regenerative response. The suggestion is that, like the spinal cord (Schwab 2002), the cellular environment is inhibitory and that if the inhibition is removed then the endogenous stem cells might effect repair. However, this appears unlikely in the cochlea based on the loss of sphere-forming potential related to age as described by Oshima et al. (2007). The fact that cells do not proliferate once isolated from their environment implies that they have either disappeared or that their competence is lost, rather than an active inhibitory effect maintained by the surrounding cells.

4.4 Cell Transplantation

Give their immense capacity to proliferate and expand in vitro, embryonic stem cells are an ideal source to generate different cell types. Li et al. (2003b) demonstrated that it is possible to direct stepwise differentiation of murine ESCs into inner ear phenotypes. Initially, cells were allowed to aggregate into embryoid bodies in the presence of EGF and IGF and then the ear progenitors expanded by adding basic fibroblast growth factor (bFGF). A detailed experimental protocol can be found in Rivolta et al. (2006). These manipulations induce the coordinated expression of the hair cell transcription factors *Brn3c* and *atoh1* in a single cell. Transplantation into developing chicken otocysts was followed by further differentiation of hair cell characteristics. Given that progenitors are generated after the first stage of induction it is surprising that a vast majority of hair cell phenotypes was observed, with relatively few grafted cells that did not express hair cell markers. It is not yet clear if this is a peculiarity of the system or if other instructive signals are needed to support the differentiation of these progenitors into the remaining cell types, i.e., supporting cells and neurons.

Murine ESCs have been transplanted into mouse ears after treatment with stromal, cell-derived inducing activity (SDIA). This activity, obtained by growing the ES on PA6 feeder cells, promotes neural differentiation (Kawasaki et al. 2000). Cells survived for 4 weeks and expressed the neuronal marker βIII-tubulin, but no hair cell markers. Differentiation was not complete, as cells were still proliferating and expressing SSEA3, a marker of the undifferentiated state (Sakamoto et al. 2004). Neurons obtained this way have the ability to form connections with vestibular hair cells in vitro (Kim et al. 2005), and in a subsequent study, the ESCs-derived neurons showed an improvement of functionality in deafened guinea pigs as measured via auditory brainstem responses (ABRs; Okano et al. 2005). In an independent study, Hildenbrand et al. (2005) transplanted partially differentiated mouse ESCs into deafened guinea pigs cochleae. Although the cells survived in the cochlea, they failed to support functional recovery. Another group has described that untreated mouse ESCs transplanted into the vestibulocochlear nerve migrated centrally into the brain stem (Hu et al. 2004a; Regala et al. 2005). The survival of ESCs differentiated into TUJ1+ cells was improved by cotransplanting them with fetal DRGs (Hu et al. 2005a). ESC-derived neural progenitors expressing YGFP were transplanted into the cochleae of gerbils that have their SGNs destroyed with ouabain. After 2–3 months the transplanted cells had grown processes into the organ of Corti (Corrales et al. 2006).

NSCs derived from adult rat hippocampus have been transplanted into newborn rat cochleae in the hope that they might be incorporated into the sensory epithelia (Ito et al. 2001). The experimental evidence in this study is limited but there was some indication of survival and integration after 2–4 weeks. In a similar work the survival of adult NSCs was slightly improved by damaging the sensory epithelia with neomycin. NSCs that were transduced with the neurogenic gene *ngn2* showed better differentiation as neurons, but no cells were found that display hair cells markers (Hu et al. 2005b). Hakuba et al. (2005) transplanted fetal NSCs into gerbil cochleae that have been damaged by arterial occlusion. Nestin-positive cells that appeared to have differentiated into inner hair cells were evident in the transplanted animals. However, the grafted cells were not exogenously tagged, so it is impossible to exclude that nestin-positive cells were endogenously produced as a response to the transplant. In any case, an improvement of the hearing was recorded via ABRs.

Preliminary transplantation studies of naïve, untreated bone marrow MSCs into adult chinchilla cochleae show that although some cells have grafted, the proportion that present neuronal differentiation was low (about 0.4%; Naito et al. 2004). Differentiation of MSCs could be improved greatly by in vitro manipulations. In a more recent study, differentiation of glutamatergic sensory neurons was induced from MSCs after exposure in culture to sonic hedgehog and retinoic acid. It is unclear if the initial population isolated from the bone marrow included MAPCs, as no characterization of the surface markers was performed (Kondo et al. 2005). Jeon et al (Jeon et al. 2006) obtained neuroprogenitor-like cells by culturing MSCs in a media including neuralizing growth factors. When these cells were transfected with *atoh1* they adopted characteristics of hair cells.

Some attempts have been performed using allografic and xenografic implants of fetal dorsal root ganglion (DRG) into adult rat and guinea pig cochleae

(Olivius et al. 2003, 2004; Hu et al. 2004b). This tissue survived in the host for a few weeks, and cells were retained mainly in the scala tympani and along the auditory nerve fibers of the modiolus, but no evidence was provided on the formation of synaptic connections with the host hair cells. These are not stem cells or progenitors; hence they do not offer a expandable, renewable source. This type of experiment, however, could offer insights into the feasibility of integration and survival of donor tissue and to ascertain different surgical approaches.

Drawing information from other systems and the limited studies in the ear so far, a more successful approach might be achieved when stem cells, regardless of their origin, are exposed in vitro to specific signals that would trigger the initial programs of differentiation. Transplanting "naïve" stem cells, regardless of their ability to migrate and survive into different regions of the cochlea, may not produce the diversity of fully differentiated cells needed. It is likely that the necessary signals and cues to drive a particular lineage are no longer in place in the adult cochlea and the cells would need to be "jump-started" into a given lineage a priori. The pretransplantation "priming" of cells would be particularly important with embryonic stem cells and other pluripotent cell types, because highly undifferentiated cells could pose a tumorigenic risk.

Transplantation experiments depend largely on trial and error because there are so many unknown variables in the in vivo environment. This is particularly true in the inner ear, where pretreatment of the cells and host as well as the method of delivery could have a substantial influence on the outcome. These problems have complicated assessment of the clinical potential for cell transplantation in Parkinson's Disease (Winkler et al. 2005). Although some experiments (reviewed earlier) have been performed with different cell types in different host models, these are initial trials. A series of more systematic, quantitative studies is needed, where well-defined stem cell populations are compared in different animal models. The main targets for transplantation have been Parkinson's disease, Huntington's disease, epilepsy, and stroke (Bjorklund and Lindvall 2000). In these cases, clinical trials have been based primarily upon the use of primary fetal neural tissue, a rather ill-defined and controversial source. Successful experiments with retinal tissue have been discussed earlier. However, functional replacement of hair cells by transplantation is probably harder than replacement of brain cells, retinal cells, or pancreatic cells. This is because hair cells are highly structurally specialized and need to be placed with micron accuracy in order for their stereocilia bundles to be deflected by the mechanical input to the cochlea. Replacement of surrogate hair cells may be beneficial if they secrete the appropriate growth factors and thus help to retain the innervation. This kind of intervention would be most constructive in conjunction with cochlear implants. In the same context it may be easier to replace or regenerate SGNs.

4.4.1 Xenotransplantation

To transfer this technology to a clinical application, sources for stem cells will need to be scrutinized, not only in terms of tissue of origin, but also species. The use of animal tissue as donors for transplantation into humans, or xenotransplantation, is certainly a possibility. Pig cells, for instance, have been used to treat certain conditions such as diabetes (Groth et al. 1994) and Parkinson's disorder (Deacon et al. 1997). This approach, although attractive because of ready availability, is fraught with problems. Xenotransplants elicit a significant immune rejection. This is a formidable obstacle to overcome and requires substantial immunosuppression, even assuming that the inner ear could possess a modest level of immunoprivilege. Moreover, the possibility of pathogens crossing across species is a certain risk. Porcine endogenous retrovirus (PERV), for instance, has been shown to infect human cells (Patience et al. 1997) and more control experiments and closely monitored trials are required (Magre et al. 2003). Moreover, patients are becoming increasingly resistant to receive cell-based therapies from other species. Even in potentially life-threatening conditions such as diabetes type 1, more than 70% of the patients interviewed rejected the idea of pig islet xenotransplants (Deschamps et al. 2005).

The problems presented by xenotransplantation could be minimized by the use of human stem cells. With this aim, the development of protocols to induce differentiation of hESCs into auditory phenotypes as well as establishing human auditory stem cells from other tissue sources are important experimental approaches (Chen et al. 2005; 2007). hESC also appear to offer the peculiar advantage of possessing immunoprivileged properties, not eliciting an immune response (Li et al. 2004).

The therapeutic application is not the only reason to develop a human-based system. Basic differences in the biology of human stem cells are becoming more apparent when compared to other species. For instance, the surface antigens SSEA-3 and SSEA-4 are expressed by human but not mouse ES cells, while SSEA-1 is expressed by mouse but not hESCs (Henderson et al. 2002). More important is the dependence of undifferentiated mouse ES cells on leukemia inhibitory factor (LIF). hESCs do not require LIF but need to grow on feeder layers (Thomson et al. 1998). Comparison of the transcriptome of human and mouse ESCs by gene expression arrays (Sato et al. 2003; Ginis et al. 2004) as well as by massively parallel signature sequencing (MPSS; Wei et al. 2005) has shown substantial differences in the profile of transcripts and signaling cascades used. Human stem cells could therefore provide species-specific answers to fundamental biological questions.

4.4.2 Delivery of Stem Cells to the Inner Ear

The delivery of stem cells is likely to require improvement of current surgical techniques. A potential way of access could involve the round window, a route increasingly used for drug administration (Banerjee and Parnes 2004), or a cochleostomy near the round window as performed with cochlear implantation (Copeland and Pillsbury 2004). Experiments performed so far have delivered cells into the modiolus (Naito et al. 2004; Tamura et al. 2004) or into the perilymphatic space by either drilling a small hole into the scala tympani at the basal cochlear turn (Hu et al. 2004b), or into the lateral semicircular canal

(Iguchi et al. 2004). Sekiya et al. (2006), to avoid disturbing the delicate cochlear structures, transplanted ES cells in the internal auditory meatal portion of the auditory nerve. Cells delivered this way migrated into the cochlea, specifically through the Rosenthal's canal, but only when the auditory nerve had been damaged. Cells failed to migrate when the nerve was intact. These routes of delivery should be appropriate for neurons, but for the replacement of the sensory epithelium, cells would ideally have to be injected directly into the scala media. Iguchi et al. (2004) have experimented by drilling through the stria vascularis of the second turn. A considerable number of transplanted cells were located in scala media, but expectedly, a substantial elevation of the ABR thresholds was produced. Hildebrand et al. (2005) transplanted cells into scala media of guinea pig ears with a cannula inserted through the basilar membrane, via the scala tympani. This procedure appears to deliver cells in this compartment without producing a significant disruption of the auditory function.

5. Conclusions

Effective therapeutic approaches to repair or to regenerate sensory cells will depend upon greater knowledge of the molecules that govern cell proliferation, migration, and differentiation. Progress in inner ear research has accelerated in recent years as solutions have been found to deal with small, complex tissue samples. The combination of transgenic animals, organotypic cultures, cell lines, and genomic technologies will accelerate identification of key genes and their function within cells and tissues. Each experimental technique has its strengths and none is sufficient by itself. In the case of cell lines, their real power lies in their homogeneous nature and their application to the elucidation of coherent gene expression programs in conjunction with gene arrays and proteomics. Proteomics have not been discussed in this chapter because only a few studies concerning the inner ear have been published to date (Zheng et al. 2006). Because proteins reflect the function of the cell more directly than mRNA and many key posttranslational modifications, such as protein phosphorylation, can be taken into account, proteomics is an attractive approach.

Although the field of auditory stem cell research is still in its infancy; important advances are already taking place. The discovery of a population of pluripotent stem cells in the adult vestibular epithelia has opened the possibility of devising strategies to recruit these cells to repair injury. An equivalent cell type found in the adult cochlea would be a phenomenal therapeutic target, but all the attempts so far to prove if that population indeed exists have failed. Using alternative stem cell sources that can be coerced into inner ear cell types is thus a sensible complementary strategy and a few labs worldwide are working on finding ways to instruct these stem cells into the path of auditory fate. Improving the surgical techniques to facilitate their delivery represents another area of research that will need development. Finally, the identification and isolation of human auditory stem cells will take these technologies closer to a realistic clinical application.

The application of stem cells to the development of therapies for deafness is creating hopes and expectations. It has a potential that goes beyond other technologies. Gene therapy, for instance, aims to replace or correct a single defective gene. Unlike most metabolic disorders where the defect lies on an enzymatic gene, many cases of hereditary deafness are produced by mutations in genes encoding for cytoskeletal, structural, or channel proteins, whose lack of function leads to a direct or secondary degeneration of several cell types (Steel and Kros 2001; Kurima et al. 2002; Kudo et al. 2003; Rozengurt et al. 2003). Although exciting results including restoration of auditory function have been obtained by replacing the *atoh1* gene into acutely deafened guinea pigs (Izumikawa et al. 2005), this kind of approach alone may not work in many chronic conditions where the general cytoarchitecture of the inner ear has been disrupted, which is often the case with long-term hearing loss. A cell-based therapy must not only contribute to the regeneration of hair cells and neurons but also rebuild the entire cytological frame. Other important cell types, like those in the stria vascularis, could also be targeted in this kind of therapy.

References

Abe S, Katagiri T, Saito-Hisaminato A, Usami S, Inoue Y, Tsunoda T, Nakamura Y (2003) Identification of CRYM as a candidate responsible for nonsyndromic deafness, through cDNA microarray analysis of human cochlear and vestibular tissues. Am J Hum Genet 72:73–82.

Affymetrix (2001) Microarray Suite User Guide version 5,0. Santa Clara CA:Affymetrix, Inc.

Ahmad S, Chen S, Sun J, Lin X (2003) Connexins 26 and 30 are co-assembled to form gap junctions in the cochlea of mice. Biochem Biophys Res Commun 307:362–368.

Aigner A, Renneberg H, Bojunga J, Apel J, Nelson PS, Czubayko F (2002) Ribozyme-targeting of a secreted FGF-binding protein (FGF-BP) inhibits proliferation of prostate cancer cells in vitro and in vivo. Oncogene 21:5733–5742.

Alagramam KN, Murcia CL, Kwon HY, Pawlowski KS, Wright CG, Woychik RP (2001a) The mouse Ames *waltzer* hearing-loss mutant is caused by mutation of *Pcdh15*, a novel protocadherin gene. Nat Genet 27:99–102.

Alagramam KN, Yuan H, Kuehn MH, Murcia CL, Wayne S, Srisailpathy CR, Lowry RB, Knaus R, Van Laer L, Bernier FP, Schwartz S, Lee C, Morton CC, Mullins RF, Ramesh A, Van Camp G, Hageman GS, Woychik RP, Smith RJ (2001b) Mutations in the novel protocadherin *PCDH15* cause Usher syndrome type 1F. Hum Mol Genet 10:1709–1718.

Alsina B, Giraldez F, Varela-Nieto I (2003) Growth factors and early development of otic neurons: interactions between intrinsic and extrinsic signals. Curr Top Dev Biol 57:177–206.

Anagnostopoulos AV (2002) A compendium of mouse knockouts with inner ear defects. Trends Genet 18:499.

Andrews PW (1998) Teratocarcinomas and human embryology: pluripotent human EC cell lines. Review article. Apmis 106:158–167; discussion 167–158.

Andrews PW (2002) From teratocarcinomas to embryonic stem cells. Philos Trans R Soc Lond B Biol Sci 357:405–417.

Atar O, Avraham KB (2005) Therapeutics of hearing loss: expectations vs reality. Drug Discov Today 10:1323–1330.

Baldwin ET, Bhat TN, Gulnik S, Hosur MV, Sowder RC 2nd, Cachau RE, Collins J, Silva AM, Erickson JW (1993) Crystal structures of native and inhibited forms of human cathepsin D: implications for lysosomal targeting and drug design. Proc Natl Acad Sci U S A 90:6796–6800.

Banerjee A, Parnes LS (2004) The biology of intratympanic drug administration and pharmacodynamics of round window drug absorption. Otolaryngol Clin North Am 37:1035–1051.

Barald KF, Lindberg KH, Hardiman K, Kavka AI, Lewis JE, Victor JC, Gardner CA, Poniatowski A (1997) Immortalized cell lines from embryonic avian and murine otocysts: tools for molecular studies of the developing inner ear. Int J Dev Neurosci 15:523–540.

Beisel KW, Shiraki T, Morris KA, Pompeia C, Kachar B, Arakawa T, Bono H, Kawai J, Hayashizaki Y, Carninci P (2004) Identification of unique transcripts from a mouse full-length, subtracted inner ear cDNA library. Genomics 83:1012–1023.

Belyantseva IA, Adler HJ, Curi R, Frolenkov GI, Kachar B (2000) Expression and localization of prestin and the sugar transporter GLUT-5 during development of electromotility in cochlear outer hair cells. J Neurosci 20:RC116.

Ben-Arie N, Bellen HJ, Armstrong DL, McCall AE, Gordadze PR, Guo Q, Matzuk MM, Zoghbi HY (1997) *Math1* is essential for genesis of cerebellar granule neurons. Nature 390:169–172.

Berchem G, Glondu M, Gleizes M, Brouillet JP, Vignon F, Garcia M, Liaudet-Coopman E (2002) Cathepsin-D affects multiple tumor progression steps in vivo: proliferation, angiogenesis and apoptosis. Oncogene 21:5951–5955.

Bermingham NA, Hassan BA, Price SD, Vollrath MA, Ben-Arie N, Eatock RA, Bellen HJ, Lysakowski A, Zoghbi HY (1999) *Math1*: an essential gene for the generation of inner ear hair cells. Science 284:1837–1841.

Bertolaso L, Martini A, Bindini D, Lanzoni I, Parmeggiani A, Vitali C, Kalinec G, Kalinec F, Capitani S, Previati M (2001) Apoptosis in the OC-k3 immortalized cell line treated with different agents. Audiology 40:327–335.

Bertrand N, Castro DS, Guillemot F (2002) Proneural genes and the specification of neural cell types. Nat Rev Neurosci 3:517–530.

Bjorklund A, Lindvall O (2000) Cell replacement therapies for central nervous system disorders. Nat Neurosci 3:537–544.

Bjornson CR, Rietze RL, Reynolds BA, Magli MC, Vescovi AL (1999) Turning brain into blood: a hematopoietic fate adopted by adult neural stem cells in vivo. Science 283:534–537.

Blackshaw S, Livesey R (2002) Applying genomics technologies to neural development. Curr Opin Neurobiol 12:110–114.

Bolz H, von Brederlow B, Ramirez A, Bryda EC, Kutsche K, Nothwang HG, Seeliger M, del CSCM, Vila MC, Molina OP, Gal A, Kubisch C (2001) Mutation of CDH23, encoding a new member of the cadherin gene family, causes Usher syndrome type 1D. Nat Genet 27:108–112.

Bradley A, Evans M, Kaufman MH, Robertson E (1984) Formation of germ-line chimaeras from embryo-derived teratocarcinoma cell lines. Nature 309:255–256.

Buckingham M (2001) Skeletal muscle formation in vertebrates. Curr Opin Genet Dev 11:440–448.

Camarero G, Avendano C, Fernandez-Moreno C, Villar A, Contreras J, de Pablo F, Pichel JG, Varela-Nieto I (2001) Delayed inner ear maturation and neuronal loss in postnatal Igf-1–deficient mice. J Neurosci 21:7630–7641.

Chen P, Zindy F, Abdala C, Liu F, Li X, Roussel MF, Segil N (2003) Progressive hearing loss in mice lacking the cyclin-dependent kinase inhibitor Ink4d. Nat Cell Biol 5:422–426.

Chen W, Moore H, Andrews PW, Rivolta MN (2005) Isolation and characterization of human auditory stem cells and multipotent progenitors. In: 3rd Annual Meeting of the International Society for Stem Cell Research. San Francisco, CA.

Chen W, Cacciabue-Rivolta DI, Moore H, Rivolta MN (2007) The human fetal cochlea can be a source for auditory progenitors/stem cells isolation. Hearing Res 233:23-29.

Chen ZY, Corey DP (2002a) Understanding inner ear development with gene expression profiling. J Neurobiol 53:276–285.

Chen ZY, Corey DP (2002b) An inner ear gene expression database. J Assoc Res Otolaryngol 3:140–148.

Cho Y, Gong TW, Stover T, Lomax MI, Altschuler RA (2002) Gene expression profiles of the rat cochlea, cochlear nucleus, and inferior colliculus. J Assoc Res Otolaryngol 3:54–67.

Cho Y, Gong TW, Kanicki A, Altschuler RA, Lomax MI (2004) Noise overstimulation induces immediate early genes in the rat cochlea. Brain Res Mol Brain Res 130:134–148.

Clarke DL, Johansson CB, Wilbertz J, Veress B, Nilsson E, Karlstrom H, Lendahl U, Frisen J (2000) Generalized potential of adult neural stem cells. Science 288:1660–1663.

Claussen M, Kubler B, Wendland M, Neifer K, Schmidt B, Zapf J, Braulke T (1997) Proteolysis of insulin-like growth factors (IGF) and IGF binding proteins by cathepsin D. Endocrinology 138:3797–3803.

Clough RL, Sud R, Davis-Silberman N, Hertzano R, Avraham KB, Holley M, Dawson SJ (2004) Brn-3c (POU4F3) regulates BDNF and NT-3 promoter activity. Biochem Biophys Res Commun 324:372–381.

Cohen P (2002) Protein kinases—the major drug targets of the twenty-first century? Nat Rev Drug Discov 1:309–315.

Cohen-Salmon M, Ott T, Michel V, Hardelin JP, Perfettini I, Eybalin M, Wu T, Marcus DC, Wangemann P, Willecke K, Petit C (2002) Targeted ablation of connexin26 in the inner ear epithelial gap junction network causes hearing impairment and cell death. Curr Biol 12:1106–1111.

Conover CA (1995) Insulin-like growth factor binding protein proteolysis in bone cell models. Prog Growth Factor Res 6:301–309.

Conover CA, Perry JE, Tindall DJ (1995) Endogenous cathepsin D-mediated hydrolysis of insulin-like growth factor-binding proteins in cultured human prostatic carcinoma cells. J Clin Endocrinol Metab 80:987–993.

Copeland BJ, Pillsbury HC, 3rd (2004) Cochlear implantation for the treatment of deafness. Annu Rev Med 55:157–167.

Corrales CE, Pan L, Li H, Liberman MC, Heller S, Edge AS (2006) Engraftment and differentiation of embryonic stem cell-derived neural progenitor cells in the cochlear nerve trunk: growth of processes into the organ of Corti. J Neurobiol 66:1489–1500.

Corwin JT, Oberholtzer JC (1997) Fish n' chicks: model recipes for hair-cell regeneration? Neuron 19:951–954.

Cristobal R, Wackym PA, Cioffi JA, Erbe CB, Roche JP, Popper P (2005) Assessment of differential gene expression in vestibular epithelial cell types using microarray analysis. Brain Res Mol Brain Res 133:19–36.

Dahl SG, Edvardsen O, Kristiansen K, Sylte I (2001) Bioinformatics and receptor mechanisms of psychotropic drugs. Biotechnol Annu Rev 7:165–177.

Davis A (1995) Hearing in Adults. London: Whurr.

Davis A (1998) Epidemiology of Hearing Impairment. In Ludman H, Wright A (eds) Diseases of the Ear, 6th ed. London: Arnold, pp 129–137.

Davis CA, Nick HS, Agarwal A (2001) Manganese superoxide dismutase attenuates cisplatin-induced renal injury: importance of superoxide. J Am Soc Nephrol 12:2683–2690.

De Strooper B, Saftig P, Craessaerts K, Vanderstichele H, Guhde G, Annaert W, Von Figura K, Van Leuven F (1998) Deficiency of presenilin-1 inhibits the normal cleavage of amyloid precursor protein. Nature 391:387–390.

De Strooper B, Annaert W, Cupers P, Saftig P, Craessaerts K, Mumm JS, Schroeter EH, Schrijvers V, Wolfe MS, Ray WJ, Goate A, Kopan R (1999) A presenilin-1–dependent gamma-secretase-like protease mediates release of Notch intracellular domain. Nature 398:518–522.

Deacon T, Schumacher J, Dinsmore J, Thomas C, Palmer P, Kott S, Edge A, Penney D, Kassissieh S, Dempsey P, Isacson O (1997) Histological evidence of fetal pig neural cell survival after transplantation into a patient with Parkinson's disease. Nat Med 3:350–353.

Debouck C, Metcalf B (2000) The impact of genomics on drug discovery. Annu Rev Pharmacol Toxicol 40:193–207.

Deschamps JY, Roux FA, Gouin E, Sai P (2005) Reluctance of French patients with type 1 diabetes to undergo pig pancreatic islet xenotransplantation. Xenotransplantation 12:175–180.

Devarajan P, Savoca M, Castaneda MP, Park MS, Esteban-Cruciani N, Kalinec G, Kalinec F (2002) Cisplatin-induced apoptosis in auditory cells: role of death receptor and mitochondrial pathways. Hear Res 174:45–54.

Di Palma F, Holme RH, Bryda EC, Belyantseva IA, Pellegrino R, Kachar B, Steel KP, Noben-Trauth K (2001) Mutations in *Cdh23*, encoding a new type of cadherin, cause stereocilia disorganization in *waltzer*, the mouse model for Usher syndrome type 1D. Nat Genet 27:103–107.

Ding D, Stracher A, Salvi RJ (2002) Leupeptin protects cochlear and vestibular hair cells from gentamicin ototoxicity. Hear Res 164:115–126.

Doetzlhofer A, White PM, Johnson JE, Segil N, Groves AK (2004) In vitro growth and differentiation of mammalian sensory hair cell progenitors: a requirement for EGF and periotic mesenchyme. Dev Biol 272:432–447.

Doetzlhofer A, White P, Lee YS, Groves A, Segil N (2006) Prospective identification and purification of hair cell and supporting cell progenitors from the embryonic cochlea. Brain Res 1091:282–288.

Dohrmann C, Gruss P, Lemaire L (2000) Pax genes and the differentiation of hormone-producing endocrine cells in the pancreas. Mech Dev 92:47–54.

Donovan PJ, Gearhart J (2001) The end of the beginning for pluripotent stem cells. Nature 414:92–97.

Drake FH, Dodds RA, James IE, Connor JR, Debouck C, Richardson S, Lee-Rykaczewski E, Coleman L, Rieman D, Barthlow R, Hastings G, Gowen M (1996) Cathepsin K, but not cathepsins B, L, or S, is abundantly expressed in human osteoclasts. J Biol Chem 271:12511–12516.

Erkman L, McEvilly RJ, Luo L, Ryan AK, Hooshmand F, O'Connell SM, Keithley EM, Rapaport DH, Ryan AF, Rosenfeld MG (1996) Role of transcription factors *Brn-3.1* and *Brn-3.2* in auditory and visual system development. Nature 381:603–606.

Evans MJ, Kaufman MH (1981) Establishment in culture of pluripotential cells from mouse embryos. Nature 292:154–156.

Fekete DM (1999) Development of the vertebrate ear: insights from knockouts and mutants. Trends Neurosci 22:263–269.

Fekete DM, Wu DK (2002) Revisiting cell fate specification in the inner ear. Curr Opin Neurobiol 12:35–42.

Fekete DM, Muthukumar S, Karagogeos D (1998) Hair cells and supporting cells share a common progenitor in the avian inner ear. J Neurosci 18:7811–7821.

Forge A, Li L, Corwin JT, Nevill G (1993) Ultrastructural evidence for hair cell regeneration in the mammalian inner ear. Science 259:1616–1619.

Forge A, Li L, Nevill G (1998) Hair cell recovery in the vestibular sensory epithelia of mature guinea pigs. J Comp Neurol 397:69–88.

Fritzsch B (2003) Development of inner ear afferent connections: forming primary neurons and connecting them to the developing sensory epithelia. Brain Res Bull 60:423–433.

Gage FH (2000) Mammalian neural stem cells. Science 287:1433–1438.

Galli R, Borello U, Gritti A, Minasi MG, Bjornson C, Coletta M, Mora M, De Angelis MG, Fiocco R, Cossu G, Vescovi AL (2000) Skeletal myogenic potential of human and mouse neural stem cells. Nat Neurosci 3:986–991.

Gelb BD, Shi GP, Chapman HA, Desnick RJ (1996) Pycnodysostosis, a lysosomal disease caused by cathepsin K deficiency. Science 273:1236–1238.

Germiller JA, Smiley EC, Ellis AD, Hoff JS, Deshmukh I, Allen SJ, Barald KF (2004) Molecular characterization of conditionally immortalized cell lines derived from mouse early embryonic inner ear. Dev Dyn 231:815–827.

Geschwind DH, Ou J, Easterday MC, Dougherty JD, Jackson RL, Chen Z, Antoine H, Terskikh A, Weissman IL, Nelson SF, Kornblum HI (2001) A genetic analysis of neural progenitor differentiation. Neuron 29:325–339.

Ginis I, Luo Y, Miura T, Thies S, Brandenberger R, Gerecht-Nir S, Amit M, Hoke A, Carpenter MK, Itskovitz-Eldor J, Rao MS (2004) Differences between human and mouse embryonic stem cells. Dev Biol 269:360–380.

Gong TW, Hegeman AD, Shin JJ, Adler HJ, Raphael Y, Lomax MI (1996) Identification of genes expressed after noise exposure in the chick basilar papilla. Hear Res 96:20–32.

Groth CG, Korsgren O, Tibell A, Tollemar J, Moller E, Bolinder J, Ostman J, Reinholt FP, Hellerstrom C, Andersson A (1994) Transplantation of porcine fetal pancreas to diabetic patients. Lancet 344:1402–1404.

Guan MX, Fischel-Ghodsian N, Attardi G (2000) A biochemical basis for the inherited susceptibility to aminoglycoside ototoxicity. Hum Mol Genet 9:1787–1793.

Hackett L, Davies D, Helyer R, Kennedy H, Kros C, Lawlor P, Rivolta MN, Holley M (2002) E-cadherin and the differentiation of mammalian vestibular hair cells. Exp Cell Res 278:19–30.

Hakuba N, Hata R, Morizane I, Feng G, Shimizu Y, Fujita K, Yoshida T, Sakanaka M, Gyo K (2005) Neural stem cells suppress the hearing threshold shift caused by cochlear ischemia. NeuroReport 16:1545–1549.

Hawkins RD, Bashiardes S, Helms CA, Hu L, Saccone NL, Warchol ME, Lovett M (2003) Gene expression differences in quiescent versus regenerating hair cells of avian sensory epithelia: implications for human hearing and balance disorders. Hum Mol Genet 12:1261–1272.

Heller S (2002) Molecular screens for inner ear genes. J Neurobiol 53:265–275.

Helms AW, Johnson JE (1998) Progenitors of dorsal commissural interneurons are defined by *MATH1* expression. Development 125:919–928.

Henderson JK, Draper JS, Baillie HS, Fishel S, Thomson JA, Moore H, Andrews PW (2002) Preimplantation human embryos and embryonic stem cells show comparable expression of stage-specific embryonic antigens. Stem Cells 20:329–337.

Herreman A, Serneels L, Annaert W, Collen D, Schoonjans L, De Strooper B (2000) Total inactivation of gamma-secretase activity in presenilin-deficient embryonic stem cells. Nat Cell Biol 2:461–462.

Hertzano R, Montcouquiol M, Rashi-Elkeles S, Elkon R, Yucel R, Frankel WN, Rechavi G, Moroy T, Friedman TB, Kelley MW, Avraham KB (2004) Transcription profiling of inner ears from *Pou4f3(ddl/ddl)* identifies Gfi1 as a target of the *Pou4f3* deafness gene. Hum Mol Genet 13:2143–2153.

Hibino H, Pironkova R, Onwumere O, Vologodskaia M, Hudspeth AJ, Lesage F (2002) RIM binding proteins (RBPs) couple Rab3–interacting molecules (RIMs) to voltage-gated Ca(2+) channels. Neuron 34:411–423.

Hildebrand MS, de Silva MG, Klockars T, Rose E, Price M, Smith RJ, McGuirt WT, Christopoulos H, Petit C, Dahl HH (2004) Characterisation of DRASIC in the mouse inner ear. Hear Res 190:149–160.

Hildebrand MS, Dahl HH, Hardman J, Coleman B, Shepherd RK, de Silva MG (2005) Survival of partially differentiated mouse embryonic stem cells in the scala media of the guinea pig cochlea. J Assoc Res Otolaryngol 6:341–354.

Hirose K, Liberman MC (2003) Lateral wall histopathology and endocochlear potential in the noise-damaged mouse cochlea. J Assoc Res Otolaryngol 4:339–352.

Holley MC (2005) Keynote review: The auditory system, hearing loss and potential targets for drug development. Drug Discov Today 10:1269–1282.

Hu Z, Ulfendahl M, Olivius NP (2004a) Central migration of neuronal tissue and embryonic stem cells following transplantation along the adult auditory nerve. Brain Res 1026:68–73.

Hu Z, Ulfendahl M, Olivius NP (2004b) Survival of neuronal tissue following xenograft implantation into the adult rat inner ear. Exp Neurol 185:7–14.

Hu Z, Andang M, Ni D, Ulfendahl M (2005a) Neural cograft stimulates the survival and differentiation of embryonic stem cells in the adult mammalian auditory system. Brain Res 1051:137–144.

Hu Z, Wei D, Johansson CB, Holmstrom N, Duan M, Frisen J, Ulfendahl M (2005b) Survival and neural differentiation of adult neural stem cells transplanted into the mature inner ear. Exp Cell Res 302:40–47.

Huang EJ, Liu W, Fritzsch B, Bianchi LM, Reichardt LF, Xiang M (2001) Brn3a is a transcriptional regulator of soma size, target field innervation and axon pathfinding of inner ear sensory neurons. Development 128:2421–2432.

Iguchi F, Nakagawa T, Tateya I, Endo T, Kim TS, Dong Y, Kita T, Kojima K, Naito Y, Omori K, Ito J (2004) Surgical techniques for cell transplantation into the mouse cochlea. Acta Otolaryngol Suppl:43–47.

Irizarry RA, Bolstad BM, Collin F, Cope LM, Hobbs B, Speed TP (2003) Summaries of Affymetrix GeneChip probe level data. Nucleic Acids Res 31:e15.

Ito J, Kojima K, Kawaguchi S (2001) Survival of neural stem cells in the cochlea. Acta Otolaryngol 121:140–142.

Iyer VR, Eisen MB, Ross DT, Schuler G, Moore T, Lee JC, Trent JM, Staudt LM, Hudson J, Jr., Boguski MS, Lashkari D, Shalon D, Botstein D, Brown PO (1999)

The transcriptional program in the response of human fibroblasts to serum. Science 283:83–87.

Izumikawa M, Minoda R, Kawamoto K, Abrashkin KA, Swiderski DL, Dolan DF, Brough DE, Raphael Y (2005) Auditory hair cell replacement and hearing improvement by *Atoh1* gene therapy in deaf mammals. Nat Med 11:271–276.

Jang YY, Collector MI, Baylin SB, Diehl AM, Sharkis SJ (2004) Hematopoietic stem cells convert into liver cells within days without fusion. Nat Cell Biol 6:532–539.

Jat PS, Noble MD, Ataliotis P, Tanaka Y, Yannoutsos N, Larsen L, Kioussis D (1991) Direct derivation of conditionally immortal cell lines from an H-2Kb-tsA58 transgenic mouse. Proc Natl Acad Sci U S A 88:5096–5100.

Jeon SJ, Oshima K, Heller S, Edge AS (2006) Bone marrow mesenchymal stem cells are progenitors in vitro for inner ear hair cells. Mol Cell Neurosci.

Jeong HJ, Hong SH, Park RK, Shin T, An NH, Kim HM (2005a) Hypoxia-induced IL-6 production is associated with activation of MAP kinase, HIF-1, and NF-kappaB on HEI-OC1 cells. Hear Res.

Jeong HJ, Kim JB, Hong SH, An NH, Kim MS, Park BR, Park RK, Kim HM (2005b) Vascular endothelial growth factor is regulated by hypoxic stress via MAPK and HIF-1alpha in the inner ear. J Neuroimmunol 163:84–91.

Jiang Y, Jahagirdar BN, Reinhardt RL, Schwartz RE, Keene CD, Ortiz-Gonzalez XR, Reyes M, Lenvik T, Lund T, Blackstad M, Du J, Aldrich S, Lisberg A, Low WC, Largaespada DA, Verfaillie CM (2002) Pluripotency of mesenchymal stem cells derived from adult marrow. Nature 418:41–49.

Jiang Y, Henderson D, Blackstad M, Chen A, Miller RF, Verfaillie CM (2003) Neuroectodermal differentiation from mouse multipotent adult progenitor cells. Proc Natl Acad Sci USA 100 Suppl 1:11854–11860.

Jones JE, Corwin JT (1996) Regeneration of sensory cells after laser ablation in the lateral line system: hair cell lineage and macrophage behavior revealed by time-lapse video microscopy. J Neurosci 16:649–662.

Jorgensen P, Nishikawa JL, Breitkreutz BJ, Tyers M (2002) Systematic identification of pathways that couple cell growth and division in yeast. Science 297:395–400.

Kalinec F, Kalinec G, Boukhvalova M, Kachar B (1999) Establishment and characterization of conditionally immortalized organ of corti cell lines. Cell Biol Int 23:175–184.

Kalinec GM, Webster P, Lim DJ, Kalinec F (2003) A cochlear cell line as an in vitro system for drug ototoxicity screening. Audiol Neurootol 8:177–189.

Karis A, Pata I, van Doorninck JH, Grosveld F, de Zeeuw CI, de Caprona D, Fritzsch B (2001) Transcription factor GATA-3 alters pathway selection of olivocochlear neurons and affects morphogenesis of the ear. J Comp Neurol 429:615–630.

Kawasaki H, Mizuseki K, Nishikawa S, Kaneko S, Kuwana Y, Nakanishi S, Nishikawa SI, Sasai Y (2000) Induction of midbrain dopaminergic neurons from ES cells by stromal cell-derived inducing activity. Neuron 28:31–40.

Kelley MW, Xu XM, Wagner MA, Warchol ME, Corwin JT (1993) The developing organ of Corti contains retinoic acid and forms supernumerary hair cells in response to exogenous retinoic acid in culture. Development 119:1041–1053.

Kelley MW, Talreja DR, Corwin JT (1995) Replacement of hair cells after laser microbeam irradiation in cultured organs of corti from embryonic and neonatal mice. J Neurosci 15:3013–3026.

Kiernan AE, Pelling AL, Leung KK, Tang AS, Bell DM, Tease C, Lovell-Badge R, Steel KP, Cheah KS (2005) *Sox2* is required for sensory organ development in the mammalian inner ear. Nature 434:1031–1035.

Kim WY, Fritzsch B, Serls A, Bakel LA, Huang EJ, Reichardt LF, Barth DS, Lee JE (2001) *NeuroD*-null mice are deaf due to a severe loss of the inner ear sensory neurons during development. Development 128:417–426.

Kim TS, Nakagawa T, Kita T, Higashi T, Takebayashi S, Matsumoto M, Kojima K, Sakamoto T, Ito J (2005) Neural connections between embryonic stem cell-derived neurons and vestibular hair cells in vitro. Brain Res.

Kojima K, Takebayashi S, Nakagawa T, Iwai K, Ito J (2004) Nestin expression in the developing rat cochlea sensory epithelia. Acta Otolaryngol Suppl:14–17.

Kondo T, Johnson SA, Yoder MC, Romand R, Hashino E (2005) Sonic hedgehog and retinoic acid synergistically promote sensory fate specification from bone marrow-derived pluripotent stem cells. Proc Natl Acad Sci USA 102:4789–4794.

Kozian DH, Kirschbaum BJ (1999) Comparative gene-expression analysis. Trends Biotechnol 17:73–78.

Kros CJ, Ruppersberg JP, Rusch A (1998) Expression of a potassium current in inner hair cells during development of hearing in mice. Nature 394:281–284.

Kudo T, Kure S, Ikeda K, Xia AP, Katori Y, Suzuki M, Kojima K, Ichinohe A, Suzuki Y, Aoki Y, Kobayashi T, Matsubara Y (2003) Transgenic expression of a dominant-negative connexin26 causes degeneration of the organ of Corti and non-syndromic deafness. Hum Mol Genet 12:995–1004.

Kuntz AL, Oesterle EC (1998) Transforming growth factor alpha with insulin stimulates cell proliferation in vivo in adult rat vestibular sensory epithelium. J Comp Neurol 399:413–423.

Kurima K, Peters LM, Yang Y, Riazuddin S, Ahmed ZM, Naz S, Arnaud D, Drury S, Mo J, Makishima T, Ghosh M, Menon PS, Deshmukh D, Oddoux C, Ostrer H, Khan S, Deininger PL, Hampton LL, Sullivan SL, Battey JF, Jr., Keats BJ, Wilcox ER, Friedman TB, Griffith AJ (2002) Dominant and recessive deafness caused by mutations of a novel gene, *TMC1*, required for cochlear hair-cell function. Nat Genet 30:277–284.

Kussel-Andermann P, El-Amraoui A, Safieddine S, Nouaille S, Perfettini I, Lecuit M, Cossart P, Wolfrum U, Petit C (2000) Vezatin, a novel transmembrane protein, bridges myosin VIIA to the cadherin-catenins complex. Embo J 19:6020–6029.

Lagasse E, Connors H, Al-Dhalimy M, Reitsma M, Dohse M, Osborne L, Wang X, Finegold M, Weissman IL, Grompe M (2000) Purified hematopoietic stem cells can differentiate into hepatocytes in vivo. Nat Med 6:1229–1234.

Lang H, Fekete DM (2001) Lineage analysis in the chicken inner ear shows differences in clonal dispersion for epithelial, neuronal, and mesenchymal cells. Dev Biol 234:120–137.

Lawlor P, Marcotti W, Rivolta MN, Kros CJ, Holley MC (1999) Differentiation of mammalian vestibular hair cells from conditionally immortal, postnatal supporting cells. J Neurosci 19:9445–9458.

Lawoko-Kerali G, Rivolta MN, Holley M (2002) Expression of the transcription factors *GATA3* and *Pax2* during development of the mammalian inner ear. J Comp Neurol 442:378–391.

Lawoko-Kerali G, Rivolta MN, Lawlor P, Cacciabue-Rivolta DI, Langton-Hewer C, van Doorninck JH, Holley MC (2004a) *GATA3* and *NeuroD* distinguish auditory and vestibular neurons during development of the mammalian inner ear. Mech Dev 121:287–299.

Lawoko-Kerali G, Milo M, Davies D, Halsall A, Helyer R, Johnson CM, Rivolta MN, Tones MA, Holley MC (2004b) Ventral otic cell lines as developmental models of auditory epithelial and neural precursors. Dev Dyn 231:801–814.

Levine M, Davidson EH (2005) Gene regulatory networks for development. Proc Natl Acad Sci U S A 102:4936–4942.

Li H, Liu H, Heller S (2003a) Pluripotent stem cells from the adult mouse inner ear. Nat Med 9:1293–1299.

Li H, Roblin G, Liu H, Heller S (2003b) Generation of hair cells by stepwise differentiation of embryonic stem cells. Proc Natl Acad Sci USA 100:13495–13500.

Li L, Baroja ML, Majumdar A, Chadwick K, Rouleau A, Gallacher L, Ferber I, Lebkowski J, Martin T, Madrenas J, Bhatia M (2004) Human embryonic stem cells possess immune-privileged properties. Stem Cells 22:448–456.

Lin J, Ozeki M, Javel E, Zhao Z, Pan W, Schlentz E, Levine S (2003) Identification of gene expression profiles in rat ears with cDNA microarrays. Hear Res 175:2–13.

Lipshutz RJ, Fodor SP, Gingeras TR, Lockhart DJ (1999) High density synthetic oligonucleotide arrays. Nat Genet 21:20–24.

Liu M, Pereira FA, Price SD, Chu MJ, Shope C, Himes D, Eatock RA, Brownell WE, Lysakowski A, Tsai MJ (2000) Essential role of BETA2/NeuroD1 in development of the vestibular and auditory systems. Genes Dev 14:2839–2854.

Liu X, Mohamed JA, Ruan R (2004) Analysis of differential gene expression in the cochlea and kidney of mouse by cDNA microarrays. Hear Res 197:35–43.

Livesey R (2002) Have microarrays failed to deliver for developmental biology? Genome Biol 3:comment2009.

Lockhart DJ, Dong H, Byrne MC, Follettie MT, Gallo MV, Chee MS, Mittmann M, Wang C, Kobayashi M, Horton H, Brown EL (1996) Expression monitoring by hybridization to high-density oligonucleotide arrays. Nat Biotechnol 14:1675–1680.

Lois C, Alvarez-Buylla A (1993) Proliferating subventricular zone cells in the adult mammalian forebrain can differentiate into neurons and glia. Proc Natl Acad Sci USA 90:2074–2077.

Lomax MI, Huang L, Cho Y, Gong TL, Altschuler RA (2000) Differential display and gene arrays to examine auditory plasticity. Hear Res 147:293–302.

Lomax MI, Gong TW, Cho Y, Huang L, Oh SH, Adler HJ, Raphael Y, Altschuler RA (2001) Differential gene expression following noise trauma in birds and mammals. Noise Health 3:19–35.

Lopez IA, Zhao PM, Yamaguchi M, de Vellis J, Espinosa-Jeffrey A (2004) Stem/progenitor cells in the postnatal inner ear of the GFP-nestin transgenic mouse. Int J Dev Neurosci 22:205–213.

Lovell-Badge R (2001) The future for stem cell research. Nature 414:88–91.

Lund RD, Adamson P, Sauve Y, Keegan DJ, Girman SV, Wang S, Winton H, Kanuga N, Kwan AS, Beauchene L, Zerbib A, Hetherington L, Couraud PO, Coffey P, Greenwood J (2001) Subretinal transplantation of genetically modified human cell lines attenuates loss of visual function in dystrophic rats. Proc Natl Acad Sci USA 98:9942–9947.

Luzzi VI, Holtschlag V, Watson MA (2005) Gene expression profiling of primary tumor cell populations using laser capture microdissection, RNA transcript amplification, and GeneChip microarrays. Methods Mol Biol 293:187–207.

Ma Q, Chen Z, del Barco Barrantes I, de la Pompa JL, Anderson DJ (1998) neurogenin1 is essential for the determination of neuronal precursors for proximal cranial sensory ganglia. Neuron 20:469–482.

Magre S, Takeuchi Y, Bartosch B (2003) Xenotransplantation and pig endogenous retroviruses. Rev Med Virol 13:311–329.

Malgrange B, Belachew S, Thiry M, Nguyen L, Rogister B, Alvarez ML, Rigo JM, Van De Water TR, Moonen G, Lefebvre PP (2002) Proliferative generation of mammalian auditory hair cells in culture. Mech Dev 112:79–88.

Mantela J, Jiang Z, Ylikoski J, Fritzsch B, Zacksenhaus E, Pirvola U (2005) The retinoblastoma gene pathway regulates the postmitotic state of hair cells of the mouse inner ear. Development 132:2377–2388.

Martin GR (1981) Isolation of a pluripotent cell line from early mouse embryos cultured in medium conditioned by teratocarcinoma stem cells. Proc Natl Acad Sci USA 78:7634–7638.

Martin GR, Evans MJ (1974) The morphology and growth of a pluripotent teratocarcinoma cell line and its derivatives in tissue culture. Cell 2:163–172.

Matsui Y, Zsebo K, Hogan BL (1992) Derivation of pluripotential embryonic stem cells from murine primordial germ cells in culture. Cell 70:841–847.

Milo M, Fazeli A, Niranjan M, Lawrence ND (2003) A probabilistic model for the extraction of expression levels from oligonucleotide arrays. Biochem Soc Trans 31:1510–1512.

Morimoto RI (2002) Dynamic remodeling of transcription complexes by molecular chaperones. Cell 110:281–284.

Morris KA, Snir E, Pompeia C, Koroleva IV, Kachar B, Hayashizaki Y, Carninci P, Soares MB, Beisel KW (2005) Differential expression of genes within the cochlea as defined by a custom mouse inner ear microarray. J Assoc Res Otolaryngol 6:75–89.

Morshead CM, Benveniste P, Iscove NN, van der Kooy D (2002) Hematopoietic competence is a rare property of neural stem cells that may depend on genetic and epigenetic alterations. Nat Med 8:268–273.

Morton CC (2002) Genetics, genomics and gene discovery in the auditory system. Hum Mol Genet 11:1229–1240.

Morton NE (1991) Genetic epidemiology of hearing impairment. Ann NY Acad Sci 630:16–31.

Nagy I, Bodmer M, Brors D, Bodmer D (2004) Early gene expression in the organ of Corti exposed to gentamicin. Hear Res 195:1–8.

Naito Y, Nakamura T, Nakagawa T, Iguchi F, Endo T, Fujino K, Kim TS, Hiratsuka Y, Tamura T, Kanemaru S, Shimizu Y, Ito J (2004) Transplantation of bone marrow stromal cells into the cochlea of chinchillas. NeuroReport 15:1–4.

Nakatomi H, Kuriu T, Okabe S, Yamamoto S, Hatano O, Kawahara N, Tamura A, Kirino T, Nakafuku M (2002) Regeneration of hippocampal pyramidal neurons after ischemic brain injury by recruitment of endogenous neural progenitors. Cell 110:429–441.

Neubig RR, Siderovski DP (2002) Regulators of G-protein signalling as new central nervous system drug targets. Nat Rev Drug Discov 1:187–197.

Nicholl AJ, Kneebone A, Davies D, Cacciabue-Rivolta DI, Rivolta MN, Coffey P, Holley MC (2005) Differentiation of an auditory neuronal cell line suitable for cell transplantation. Eur J Neurosci 22:343–353.

Nixon RA (2000) A "protease activation cascade" in the pathogenesis of Alzheimer's disease. Ann NY Acad Sci 924:117–131.

Noble M, Groves AK, Ataliotis P, Jat PS (1992) From chance to choice in the generation of neural cell lines. Brain Pathol 2:39–46.

Novak JP, Sladek R, Hudson TJ (2002) Characterization of variability in large-scale gene expression data: implications for study design. Genomics 79:104–113.

Okano T, Nakagawa T, Endo T, Kim TS, Kita T, Tamura T, Matsumoto M, Ohno T, Sakamoto T, Iguchi F, Ito J (2005) Engraftment of embryonic stem cell-derived neurons into the cochlear modiolus. NeuroReport 16:1919–1922.

Olivius P, Alexandrov L, Miller J, Ulfendahl M, Bagger-Sjoback D, Kozlova EN (2003) Allografted fetal dorsal root ganglion neuronal survival in the guinea pig cochlea. Brain Res 979:1–6.

Olivius P, Alexandrov L, Miller JM, Ulfendahl M, Bagger-Sjoback D, Kozlova EN (2004) A model for implanting neuronal tissue into the cochlea. Brain Res Brain Res Protoc 12:152–156.

Oshima K, Grimm CM, Corrales CE, Senn P, Martinez Monedero R, Geleoc GS, Edge A, Holt JR, Heller S (2007) Differential distribution of stem cells in the auditory and vestibular organs of the inner ear. J Assoc Res Otolaryngol 8:18–31.

Ozeki M, Duan L, Hamajima Y, Obritch W, Edson-Herzovi D, Lin J (2003) Establishment and characterization of rat progenitor hair cell lines. Hear Res 179:43–52.

Pagano SF, Impagnatiello F, Girelli M, Cova L, Grioni E, Onofri M, Cavallaro M, Etteri S, Vitello F, Giombini S, Solero CL, Parati EA (2000) Isolation and characterization of neural stem cells from the adult human olfactory bulb. Stem Cells 18:295–300.

Papassotiropoulos A, Bagli M, Feder O, Jessen F, Maier W, Rao ML, Ludwig M, Schwab SG, Heun R (1999) Genetic polymorphism of cathepsin D is strongly associated with the risk for developing sporadic Alzheimer's disease. Neurosci Lett 262:171–174.

Parkinson N, Brown SD (2002) Focusing on the genetics of hearing: you ain't heard nothin' yet. Genome Biol 3:Comment 2006.

Patience C, Takeuchi Y, Weiss RA (1997) Infection of human cells by an endogenous retrovirus of pigs. Nat Med 3:282–286.

Pece S, Gutkind JS (2000) Signaling from E-cadherins to the MAPK pathway by the recruitment and activation of epidermal growth factor receptors upon cell-cell contact formation. J Biol Chem 275:41227–41233.

Perou CM, Sorlie T, Eisen MB, van de Rijn M, Jeffrey SS, Rees CA, Pollack JR, Ross DT, Johnsen H, Akslen LA, Fluge O, Pergamenschikov A, Williams C, Zhu SX, Lonning PE, Borresen-Dale AL, Brown PO, Botstein D (2000) Molecular portraits of human breast tumours. Nature 406:747–752.

Pirvola U, Ylikoski J, Trokovic R, Hebert JM, McConnell SK, Partanen J (2002) FGFR1 is required for the development of the auditory sensory epithelium. Neuron 35:671–680.

Pittenger MF, Mackay AM, Beck SC, Jaiswal RK, Douglas R, Mosca JD, Moorman MA, Simonetti DW, Craig S, Marshak DR (1999) Multilineage potential of adult human mesenchymal stem cells. Science 284:143–147.

Pompeia C, Hurle B, Belyantseva IA, Noben-Trauth K, Beisel K, Gao J, Buchoff P, Wistow G, Kachar B (2004) Gene expression profile of the mouse organ of Corti at the onset of hearing. Genomics 83:1000–1011.

Powles N, Babbs C, Ficker M, Schimmang T, Maconochie M (2004) Identification and analysis of genes from the mouse otic vesicle and their association with developmental subprocesses through in situ hybridization. Dev Biol 268:24–38.

Previati M, Lanzoni I, Corbacella E, Magosso S, Giuffre S, Francioso F, Arcelli D, Volinia S, Barbieri A, Hatzopoulos S, Capitani S, Martini A (2004) RNA expression induced by cisplatin in an organ of Corti-derived immortalized cell line. Hear Res 196:8–18.

Rask-Andersen H, Bostrom M, Gerdin B, Kinnefors A, Nyberg G, Engstrand T, Miller JM, Lindholm D (2005) Regeneration of human auditory nerve. In vitro/in video demonstration of neural progenitor cells in adult human and guinea pig spiral ganglion. Hear Res 203:180–191.

Regala C, Duan M, Zou J, Salminen M, Olivius P (2005) Xenografted fetal dorsal root ganglion, embryonic stem cell and adult neural stem cell survival following implantation into the adult vestibulocochlear nerve. Exp Neurol 193:326–333.

Reser D, Rho M, Dewan D, Herbst L, Li G, Stupak H, Zur K, Romaine J, Frenz D, Goldbloom L, Kopke R, Arezzo J, Van De Water T (1999) L- and D-methionine provide equivalent long term protection against CDDP-induced ototoxicity in vivo, with partial in vitro and in vivo retention of antineoplastic activity. Neurotoxicology 20:731–748.

Resnick JL, Bixler LS, Cheng L, Donovan PJ (1992) Long-term proliferation of mouse primordial germ cells in culture. Nature 359:550–551.

Reyes M, Lund T, Lenvik T, Aguiar D, Koodie L, Verfaillie CM (2001) Purification and ex vivo expansion of postnatal human marrow mesodermal progenitor cells. Blood 98:2615–2625.

Reyes M, Dudek A, Jahagirdar B, Koodie L, Marker PH, Verfaillie CM (2002) Origin of endothelial progenitors in human postnatal bone marrow. J Clin Invest 109:337–346.

Reynolds BA, Weiss S (1992) Generation of neurons and astrocytes from isolated cells of the adult mammalian central nervous system. Science 255:1707–1710.

Rivolta MN, Holley MC (2002a) Asymmetric segregation of mitochondria and mortalin correlates with the multi-lineage potential of inner ear sensory cell progenitors in vitro. Brain Res Dev Brain Res 133:49–56.

Rivolta MN, Holley MC (2002b) Cell lines in inner ear research. J Neurobiol 53:306–318.

Rivolta MN, Grix N, Lawlor P, Ashmore JF, Jagger DJ, Holley MC (1998) Auditory hair cell precursors immortalized from the mammalian inner ear. Proc Biol Sci 265:1595–1603.

Rivolta MN, Halsall A, Johnson CM, Tones MA, Holley MC (2002) Transcript profiling of functionally related groups of genes during conditional differentiation of a mammalian cochlear hair cell line. Genome Res 12:1091–1099.

Rivolta MN, Li H, Heller S (2006) Generation of inner ear cell types from embryonic stem cells. In Turksen K (ed) Embryonic Stem Cells: Methods and Protocols, 2nd ed. Totowa, NJ: Humana Press.

Rozengurt N, Lopez I, Chiu CS, Kofuji P, Lester HA, Neusch C (2003) Time course of inner ear degeneration and deafness in mice lacking the Kir4.1 potassium channel subunit. Hear Res 177:71–80.

Ruan Y, Le Ber P, Ng HH, Liu ET (2004) Interrogating the transcriptome. Trends Biotechnol 22:23–30.

Ruben RJ (1967) Development of the inner ear of the mouse: a radioautographic study of terminal mitoses. Acta Otolaryngol:Suppl 220:221–244.

Rudie Hovland A, Nahreini P, Andreatta CP, Edwards-Prasad J, Prasad KN (2001) Identifying genes involved in regulating differentiation of neuroblastoma cells. J Neurosci Res 64:302–310.

Saffer LD, Gu R, Corwin JT (1996) An RT-PCR analysis of mRNA for growth factor receptors in damaged and control sensory epithelia of rat utricles. Hear Res 94:14–23.

Sage C, Huang M, Karimi K, Gutierrez G, Vollrath MA, Zhang DS, Garcia-Anoveros J, Hinds PW, Corwin JT, Corey DP, Chen ZY (2005) Proliferation of functional hair cells in vivo in the absence of the retinoblastoma protein. Science 307:1114–1118.

Sakamoto T, Nakagawa T, Endo T, Kim TS, Iguchi F, Naito Y, Sasai Y, Ito J (2004) Fates of mouse embryonic stem cells transplanted into the inner ears of adult mice and embryonic chickens. Acta Otolaryngol Suppl:48–52.

Sandy P, Ventura A, Jacks T (2005) Mammalian RNAi: a practical guide. Biotechniques 39:215–224.

Sato N, Sanjuan IM, Heke M, Uchida M, Naef F, Brivanlou AH (2003) Molecular signature of human embryonic stem cells and its comparison with the mouse. Dev Biol 260:404–413.

Satoh T, Fekete DM (2005) Clonal analysis of the relationships between mechanosensory cells and the neurons that innervate them in the chicken ear. Development 132:1687–1697.

Schindler H, Wiese A, Auer J, Burtscher H (2005) cRNA target preparation for microarrays: Comparison of gene expression profiles generated with different amplification procedures. Anal Biochem 344:92–101.

Schlitt T, Brazma A (2005) Modelling gene networks at different organisational levels. FEBS Lett 579:1859–1866.

Schulze A, Lehmann K, Jefferies HB, McMahon M, Downward J (2001) Analysis of the transcriptional program induced by Raf in epithelial cells. Genes Dev 15:981–994.

Schwab ME (2002) Repairing the injured spinal cord. Science 295:1029–1031.

Sekiya T, Kojima K, Matsumoto M, Kim TS, Tamura T, Ito J (2006) Cell transplantation to the auditory nerve and cochlear duct. Exp Neurol 198:12–24.

Sgard F, Charpantier E, Bertrand S, Walker N, Caput D, Graham D, Bertrand D, Besnard F (2002) A novel human nicotinic receptor subunit, alpha10, that confers functionality to the alpha9–subunit. Mol Pharmacol 61:150–159.

Shih CC, Weng Y, Mamelak A, LeBon T, Hu MC, Forman SJ (2001) Identification of a candidate human neurohematopoietic stem-cell population. Blood 98:2412–2422.

Sobkowicz HM, Loftus JM, Slapnick SM (1993) Tissue culture of the organ of Corti. Acta Otolaryngol Suppl 502:3–36.

Southan C (2001) A genomic perspective on human proteases. FEBS Lett 498:214–218.

Steel KP, Kros CJ (2001) A genetic approach to understanding auditory function. Nat Genet 27:143–149.

Stone JS, Rubel EW (1999) Delta1 expression during avian hair cell regeneration. Development 126:961–973.

Stone JS, Rubel EW (2000) Temporal, spatial, and morphologic features of hair cell regeneration in the avian basilar papilla. J Comp Neurol 417:1–16.

Storch KF, Lipan O, Leykin I, Viswanathan N, Davis FC, Wong WH, Weitz CJ (2002) Extensive and divergent circadian gene expression in liver and heart. Nature 417:78–83.

Tamura T, Nakagawa T, Iguchi F, Tateya I, Endo T, Kim TS, Dong Y, Kita T, Kojima K, Naito Y, Omori K, Ito J (2004) Transplantation of neural stem cells into the modiolus of mouse cochleae injured by cisplatin. Acta Otolaryngol Suppl:65–68.

Temple S (2001) The development of neural stem cells. Nature 414:112–117.

Thomson JA, Itskovitz-Eldor J, Shapiro SS, Waknitz MA, Swiergiel JJ, Marshall VS, Jones JM (1998) Embryonic stem cell lines derived from human blastocysts. Science 282:1145–1147.

Thompson DL, Gerlach-Bank LM, Barald KF, Koenig RJ (2003) Retinoic acid repression of bone morphogenetic protein 4 in inner ear development. Mol Cell Biol 23:2277–2286.

Toyama K, Ozeki M, Hamajima Y, Lin J (2005) Expression of the integrin genes in the developing cochlea of rats. Hear Res 201:21–26.

Tropepe V, Coles BL, Chiasson BJ, Horsford DJ, Elia AJ, McInnes RR, van der Kooy D (2000) Retinal stem cells in the adult mammalian eye. Science 287:2032–2036.

Tyynela J, Sohar I, Sleat DE, Gin RM, Donnelly RJ, Baumann M, Haltia M, Lobel P (2000) A mutation in the ovine cathepsin D gene causes a congenital lysosomal storage disease with profound neurodegeneration. Embo J 19:2786–2792.

Van Camp G, Smith RJH. Hereditary Hearing Loss Homepage. URL: http://webh01.ua.ac.be/hhh/ (2007).

Vandenberg JI, Lummis SC (2000) Ion channels—a plethora of pharmaceutical targets. Trends Pharmacol Sci 21:409–410.

Vassar R, Bennett BD, Babu-Khan S, Kahn S, Mendiaz EA, Denis P, Teplow DB, Ross S, Amarante P, Loeloff R, Luo Y, Fisher S, Fuller J, Edenson S, Lile J, Jarosinski MA, Biere AL, Curran E, Burgess T, Louis JC, Collins F, Treanor J, Rogers G, Citron M (1999) Beta-secretase cleavage of Alzheimer's amyloid precursor protein by the trans-membrane aspartic protease BACE. Science 286:735–741.

Vassilopoulos G, Wang PR, Russell DW (2003) Transplanted bone marrow regenerates liver by cell fusion. Nature 422:901–904.

Velichkova M, Guttman J, Warren C, Eng L, Kline K, Vogl AW, Hasson T (2002) A human homologue of Drosophila kelch associates with myosin-VIIa in specialized adhesion junctions. Cell Motil Cytoskel 51:147–164.

Vescovi AL, Rietze R, Magli MC, Bjornson C (2002) Hematopoietic potential of neural stem cells. Nat Med 8:535; author reply 536–537.

Virley D, Ridley RM, Sinden JD, Kershaw TR, Harland S, Rashid T, French S, Sowinski P, Gray JA, Lantos PL, Hodges H (1999) Primary CA1 and conditionally immortal MHP36 cell grafts restore conditional discrimination learning and recall in marmosets after excitotoxic lesions of the hippocampal CA1 field. Brain 122 (Pt 12):2321–2335.

Wallis D, Hamblen M, Zhou Y, Venken KJ, Schumacher A, Grimes HL, Zoghbi HY, Orkin SH, Bellen HJ (2003) The zinc finger transcription factor Gfi1, implicated in lymphomagenesis, is required for inner ear hair cell differentiation and survival. Development 130:221–232.

Wang X, Willenbring H, Akkari Y, Torimaru Y, Foster M, Al-Dhalimy M, Lagasse E, Finegold M, Olson S, Grompe M (2003) Cell fusion is the principal source of bone-marrow-derived hepatocytes. Nature 422:897–901.

Warchol ME, Lambert PR, Goldstein BJ, Forge A, Corwin JT (1993) Regenerative proliferation in inner ear sensory epithelia from adult guinea pigs and humans. Science 259:1619–1622.

Wei CL, Miura T, Robson P, Lim SK, Xu XQ, Lee MY, Gupta S, Stanton L, Luo Y, Schmitt J, Thies S, Wang W, Khrebtukova I, Zhou D, Liu ET, Ruan YJ, Rao M, Lim B (2005) Transcriptome profiling of human and murine ESCs identifies divergent paths required to maintain the stem cell state. Stem Cells 23:166–185.

Weissman IL (2000) Translating stem and progenitor cell biology to the clinic: barriers and opportunities. Science 287:1442–1446.

Weissman IL, Anderson DJ, Gage F (2001) Stem and progenitor cells: origins, phenotypes, lineage commitments, and transdifferentiations. Annu Rev Cell Dev Biol 17:387–403.

White PM, Doetzlhofer A, Lee YS, Groves AK, Segil N (2006) Mammalian cochlear supporting cells can divide and trans-differentiate into hair cells. Nature 441:984–987.

Whitfield ML, Sherlock G, Saldanha AJ, Murray JI, Ball CA, Alexander KE, Matese JC, Perou CM, Hurt MM, Brown PO, Botstein D (2002) Identification of genes periodically expressed in the human cell cycle and their expression in tumors. Mol Biol Cell 13:1977–2000.

Whitlon DS (1993) E-cadherin in the mature and developing organ of Corti of the mouse. J Neurocytol 22:1030–1038.

Whitlon DS, Zhang X, Pecelunas K, Greiner MA (1999) A temporospatial map of adhesive molecules in the organ of Corti of the mouse cochlea. J Neurocytol 28:955–968.

Winkler C, Kirik D, Bjorklund A (2005) Cell transplantation in Parkinson's disease: how can we make it work? Trends Neurosci 28:86–92.

Woodbury D, Reynolds K, Black IB (2002) Adult bone marrow stromal stem cells express germline, ectodermal, endodermal, and mesodermal genes prior to neurogenesis. J Neurosci Res 69:908–917.

Yamashita DS, Dodds RA (2000) Cathepsin K and the design of inhibitors of cathepsin K. Curr Pharm Des 6:1–24.

Ylikoski J, Xing-Qun L, Virkkala J, Pirvola U (2002) Blockade of c-Jun N-terminal kinase pathway attenuates gentamicin-induced cochlear and vestibular hair cell death. Hear Res 163:71–81.

Yoho ER, Thomopoulos GN, Thalmann I, Thalmann R, Schulte BA (1997) Localization of organ of Corti protein II in the adult and developing gerbil cochlea. Hear Res 104:47–56.

Zhang Y, Ma C, Delohery T, Nasipak B, Foat BC, Bounoutas A, Bussemaker HJ, Kim SK, Chalfie M (2002) Identification of genes expressed in *C. elegans* touch receptor neurons. Nature 418:331–335.

Zhao HB (2001) Long-term natural culture of cochlear sensory epithelia of guinea pigs. Neurosci Lett 315:73–76.

Zheng JL, Gao WQ (2000) Overexpression of *Math1* induces robust production of extra hair cells in postnatal rat inner ears. Nat Neurosci 3:580–586.

Zheng JL, Lewis AK, Gao WQ (1998) Establishment of conditionally immortalized rat utricular epithelial cell lines using a retrovirus-mediated gene transfer technique. Hear Res 117:13–23.

Zheng QY, Rozanas CR, Thalmann I, Chance MR, Alagramam KN (2006) Inner ear proteomics of mouse models for deafness, a discovery strategy. Brain Res 1091:113–121.

Index

309

For more information about the series, please visit www.springer-ny.com/shar.